Breast Cancer Chemosensitivity

ADVANCES IN EXPERIMENTAL MEDICINE AND BIOLOGY

Recent Volumes in this Series

Volume 600
SEMAPHORINS: RECEPTOR AND INTRACELLULAR SIGNALING MECHANISMS
Edited by R. Jeroen Pasterkamp

Volume 601
IMMUNE MEDIATED DISEASES: FROM THEORY TO THERAPY
Edited by Michael R. Shurin

Volume 602
OSTEOIMMUNOLOGY: INTERACTIONS OF THE IMMUNE AND SKELETAL SYSTEMS
Edited by Yongwon Choi

Volume 603
THE GENUS YERSINIA: FROM GENOMICS TO FUNCTION
Edited by Robert D. Perry and Jacqueline D. Fetherson

Volume 604
ADVANCES IN MOLECULAR ONCOLOGY
Edited by Fabrizio d'Adda di Gagagna, Susanna Chiocca, Fraser McBlane and Ugo Cavallaro

Volume 605
INTEGRATION IN RESPIRATORY CONTROL: FROM GENES TO SYSTEMS
Edited by Marc Poulin and Richard Wilson

Volume 606
BIOACTIVE COMPONENTS OF MILK
Edited by Zsuzsanna Bősze

Volume 607
EUKARYOTIC MEMBRANES AND CYTOSKELETON: ORIGINS AND EVOLUTION
Edited by Gáspár Jékely

Volume 608
BREAST CANCER CHEMOSENSITIVITY
Edited by Dihua Yu and Mien-Chie Hung

A Continuation Order Plan is available for this series. A continuation order will bring delivery of each new volume immediately upon publication. Volumes are billed only upon actual shipment. For further information please contact the publisher.

Breast Cancer Chemosensitivity

Edited by

Dihua Yu

Department of Surgical Oncology, The University of Texas M.D. Anderson Cancer Center, Houston, Texas, U.S.A.

Mien-Chie Hung

Department of Molecular and Cellular Oncology, The University of Texas M.D. Anderson Cancer Center, Houston, Texas, U.S.A.

Springer Science+Business Media, LLC

Landes Bioscience

Springer Science+Business Media, LLC
Landes Bioscience

Printed in the U.S.A.

Springer Science+Business Media, LLC, 233 Spring Street, New York, New York 10013, U.S.A.
http://www.springer.com

Please address all inquiries to the Publishers:
Landes Bioscience, 1002 West Avenue, 2nd Floor, Austin, Texas 78701, U.S.A.
Phone: 512/ 637 6050; FAX: 512/ 637 6079
http://www.landesbioscience.com

Breast Cancer Chemosensitivity edited by Dihua Yu and Mien-Chie Hung, Landes Bioscience / Springer Science+Business Media, LLC dual imprint / Springer series: Advances in Experimental Medicine and Biology

ISBN 978-1-4419-2542-8 e-ISBN 978-0-387-74039-3

While the authors, editors and publisher believe that drug selection and dosage and the specifications and usage of equipment and devices, as set forth in this book, are in accord with current recommendations and practice at the time of publication, they make no warranty, expressed or implied, with respect to material described in this book. In view of the ongoing research, equipment development, changes in governmental regulations and the rapid accumulation of information relating to the biomedical sciences, the reader is urged to carefully review and evaluate the information provided herein.

A C.I.P. catalog is available from the Library of Congress.

About the Editor...

DIHUA YU, M.D., Ph.D. is the Nylene Eckles Distinguished Professor in Breast Cancer Research at the The University of Texas, M.D. Anderson Cancer Center. She is the Director of the Cancer Biology Program, The University of Texas Graduate School of Biomedical Sciences in Houston. Dr. Yu's research focuses on molecular mechanisms of breast cancer initiation, progression, metastasis, and therapeutic resistance. Her research interest covers broad areas including receptor tyrosine kinases and downstream signal transduction, cell cycle deregulation, apoptosis, cell polarity, angiogenesis, and cancer metastasis. She uses 2-Dimensional and 3-Dimensional cell culture models, SCID and nude mouse models, transgenic and knockout mouse models, as well as human tumor samples as research models. She has over 100 publications, has trained many Ph.D. students in her laboratory, and has received several awards on research and education excellence at the The University of Texas, M.D. Anderson Cancer Center. She serves on several editorial boards, including *Cancer Research.*

About the Editor...

DR. MIEN-CHIE HUNG is Professor and Chair for the Department of Molecular and Cellular Oncology at The University of Texas M.D. Anderson Cancer Center, Houston, Texas. He received his Ph.D. from Brandeis University in Massachusetts. Currently, he also serves as the Director of the Breast Cancer Basic Research Program and is the Ruth Legett Jones Distinguished Chair. Dr. Hung's laboratory has focused on signaling transduction pathways of tyrosine kinase growth factor receptors such as EGFR and HER-2/neu; molecular mechanisms of oncogenes, including transformation and tumorigenesis; and molecular mechanisms of tumor suppressor gene-mediated anti-tumor activities. His group made a critical breakthrough in showing that the transmembrane tyrosine kinase receptor EGFR can bind to a specific DNA sequence in the nucleus and that it functions as a transcription factor that can activate genes required for cell proliferation. The other main research in Dr. Hung's laboratory is in the area of cancer gene therapy that includes development of preclinical gene therapy animal models, including breast, ovarian and pancreatic cancers; identification of therapeutic genes suitable for cancer gene therapy; and development of gene delivery systems for cancer gene therapy. In addition to numerous publications and research endeavors, Dr. Hung became an Academician of the Academia Sinica in Taiwan in July 2002. Dr. Hung serves as a founding Editorial Member on *Cancer Cell* as well as an Associate Editor on *Cancer Research, Clinical Cancer Research, Molecular Cancer Research* and *Molecular Carcinogenesis.*

PREFACE

Breast cancer is the most common cancer and the second leading cause of cancer death in American women. Despite advances in early detection and the improved understanding of the molecular basis of breast cancer biology, about 30% of patients with early-stage breast cancer have recurrent disease and need effective systemic treatment. Cytotoxic chemotherapies, hormonal therapies, and immunotherapeutic agents are used in the adjuvant, neoadjuvant, and metastatic setting. Systemic agents are generally active at the beginning of therapy in the majority of breast cancers. However, progression occurs after a variable period of time when resistance to therapy develops. In this book, a group of world leading experts review critical aspects of resistance to systemic therapy in breast cancer patients. The book begins with a clinical overview of the problem. The following chapters focus on the latest findings of molecular mechanisms of drug resistance. These include in-depth discussions on multidrug resistance by P-glycoprotein and the multidrug resistance protein family, resistance to therapeutic agent-induced apoptosis, cell cycle deregulation, deregulation of DNA repair, loss of tumor suppressor genes, integrin-mediated adhesion, insulin-like growth factors, epidermal growth factor, and ErbB2 in modulating breast cancer response to systemic therapy, especially certain chemotherapeutic agents. Mechanisms of resistance to hormonal therapy are also discussed. Finally, an example of using novel approaches for chemosensitization of breast cancer cells is described that gives readers an idea about the future direction in breast cancer treatment. This book is not an encyclopedia of systemic therapy resistance, but a handy reference to some of the most important aspects of systemic therapy resistance which allows those who are interested in breast cancer therapy to get a jump-start on critical issues in breast cancer therapeutic resistance.

Dihua Yu and Mien-Chie Hung

PARTICIPANTS

Tushar Baran Deb
Lombardi Comprehensive Cancer Center
Georgetown University Medical Center
Washington, DC
U.S.A.

Robert B. Dickson
Georgetown University Medical Center
Washington, DC
U.S.A.

Wafik S. El-Deiry
University of Pennsylvania
Medical Center
Philadelphia, Pennsylvania
U.S.A.

Suzanne A.W. Fuqua
Departments of Medicine, Breast Center
Molecular and Cellular Biology
Baylor College of Medicine
Houston, Texas
U.S.A.

Ana Maria Gonzalez-Angulo
Department of Breast Medical Oncology
The University of Texas M.D. Anderson
Cancer Center
Houston, Texas
U.S.A.

Matthew H. Herynk
Breast Center
Baylor College of Medicine
and the Methodist Hospital
Houston, Texas
U.S.A.

Gabriel N. Hortobagyi
Department of Breast Medical Oncology
The University of Texas M.D. Anderson
Cancer Center
Houston, Texas
U.S.A.

Mien-Chie Hung
Department of Molecular and Cellular
Oncology
The University of Texas M.D. Anderson
Cancer Center
Houston, Texas
U.S.A.

Khandan Keyomarsi
Departments of Surgical Oncology
and Experimental Radiation Oncology
The University of Texas M.D. Anderson
Cancer Center
Houston, Texas
U.S.A.

M. Tien Kuo
Department of Molecular Pathology
The University of Texas M.D. Anderson
Cancer Center
Houston, Texas
U.S.A.

Laura Lambert
Departments of Surgical Oncology
and Experimental Radiation Oncology
The University of Texas M.D. Anderson
Cancer Center
Houston, Texas
U.S.A.

Yong Liao
Department of Molecular
and Cellular Oncology
The University of Texas M.D. Anderson
Cancer Center
Houston, Texas
U.S.A.

David J. McConkey
Department of Cancer Biology
The University of Texas M.D. Anderson
Cancer Center
Houston, Texas
U.S.A.

Flavia Morales-Vasquez
Department of Breast Medical Oncology
The University of Texas M.D. Anderson
Cancer Center
Houston, Texas
U.S.A.

Kimberly A. Scata
University of Pennsylvania
Medical Center
Philadelphia, Pennsylvania
U.S.A.

Ming Tan
Department of Surgical Oncology
The University of Texas M.D. Anderson
Cancer Center
Houston, Texas
U.S.A.

Douglas Yee
Department of Pharmacology
and Medicine
University of Minnesota Cancer Center
Minneapolis, Minnesota
U.S.A.

Dihua Yu
Department of Surgical Oncology
The University of Texas M.D. Anderson
Cancer Center
Houston, Texas
U.S.A.

Xianke Zeng
Department of Pharmacology
University of Minnesota Cancer Center
Minneapolis, Minnesota
U.S.A.

Mary M. Zutter
Departments of Pathology
and Cancer Biology
Vanderbilt University Medical Center
The Vanderbilt Clinic
Nashville, Tennessee
U.S.A.

CONTENTS

Chapter 1

Overview of Resistance to Systemic Therapy in Patients with Breast Cancer

Ana Maria Gonzalez-Angulo, Flavia Morales-Vasquez and Gabriel N. Hortobagyi*

Abstract

Breast cancer is the most common cancer and the second leading cause of cancer death in American women. It was the second most common cancer in the world in 2002, with more than 1 million new cases. Despite advances in early detection and the understanding of the molecular bases of breast cancer biology, about 30% of patients with early-stage breast cancer have recurrent disease. To offer more effective and less toxic treatment, selecting therapies requires considering the patient and the clinical and molecular characteristics of the tumor. Systemic treatment of breast cancer includes cytotoxic, hormonal, and immunotherapeutic agents. These medications are used in the adjuvant, neoadjuvant, and metastatic settings. In general, systemic agents are active at the beginning of therapy in 90% of primary breast cancers and 50% of metastases. However, after a variable period of time, progression occurs. At that point, resistance to therapy is not only common but expected. Herein we review general mechanisms of drug resistance, including multidrug resistance by P-glycoprotein and the multidrug resistance protein family in association with specific agents and their metabolism, emergence of refractory tumors associated with multiple resistance mechanisms, and resistance factors unique to host-tumor-drug interactions. Important anticancer agents specific to breast cancer are described.

Breast cancer is the most common type of cancer and the second leading cause of cancer death in American women. In 2002, 209,995 new cases of breast cancer were registered, and 42,913 patients died of it.[1] In 5 years, the annual prevalence of breast cancer will reach 968,731 cases in the United States.[2] World wide, the problem is just as significant, as breast cancer is the most frequent cancer after nonmelanoma skin cancer, with more than 1 million new cases in 2002 and an expected annual prevalence of more than 4.4 million in 5 years.[1]

Breast cancer treatment currently requires the joint efforts of a multidisciplinary team. The alternatives for treatment are constantly expanding. With the use of new effective chemotherapy, hormone therapy, and biological agents and with information regarding more effective ways to integrate systemic therapy, surgery, and radiation therapy, elaborating an appropriate treatment plan is becoming more complex. Developing such a plan should be based on knowledge of the benefits and potential acute and late toxic effects of each of the therapy regimens.

Despite advances in early detection and understanding of the molecular bases of breast cancer biology, approximately 30% of all patients with early-stage breast cancer have recurrent

*Corresponding Author: Gabriel N. Hortobagyi—Department of Breast Medical Oncology, Unit 424, The University of Texas M. D. Anderson Cancer Center, 1515 Holcombe Blvd., Houston, Texas 77030, U.S.A. Email: ghortoba@mdanderson.org

Breast Cancer Chemosensitivity, edited by Dihua Yu and Mien-Chie Hung.

Table 2. General mechanisms of resistance to systemic therapy

Cellular and Biochemical Mechanisms		
	Decreased drug accumulation	
		Decreased drug influx
		Increased drug efflux
		Altered intracellular drug trafficking
	Increased inactivation of drug or toxic intermediate	
	Increased repair of or tolerance to drug-induced damage to	
		DNA
		Protein
		Membranes
	Decreased drug activation	
	Altered drug targets (quantitatively or qualitatively)	
	Altered co-factor or metabolite levels	
	Altered downstream effectors of cytotoxicity	
	Altered signaling pathway and/or apoptotic responses to drug insult	
		Altered gene expression
		DNA mutation, amplification, or deletion
		Altered transcription, posttranscription processing, or translation
		Altered stability of macromolecules
In Vivo Mechanisms		
	Pharmacological and anatomic drug barriers (tumor sanctuaries)	
	Host-drug interactions	
		Increased drug inactivation by normal tissues
		Decreased drug activation by normal tissues
		Relatively increased normal tissue drug sensitivity (toxicity)

resistance (MDR) protein 1 (MRP1) as a transporter of hydrophilic anionic and glutathione-conjugated drugs advanced the study of resistance to cancer therapy. Since then, multiple mechanisms of both in vitro and in vivo resistance have been identified. These mechanisms range from those with anatomic characteristics and pharmacological properties to those with host-drug-tumor interaction. Table 2 summarizes the different mechanisms of resistance to systemic therapy described at the molecular level and those demonstrated in vivo.

General Mechanisms of Drug Resistance

Experimental selection of drug resistance by repeated exposure to single antineoplastic agents will generally result in cross-resistance to some agents of the same class. This phenomenon is explained by shared drug transport carriers, drug-metabolizing pathways, and intracellular cytotoxic targets of these structurally and biochemically similar compounds. Generally, resistant cells retain sensitivity to drugs of different classes with alternate mechanisms of cytotoxic action. Thus, cells selected for resistance to alkylating agents or antifolates will usually remain sensitive to unrelated drugs, such as anthracyclines. Exceptions include cases with emergence of cross-resistance to multiple, apparently structurally and functionally unrelated drugs that the patient or cancer cells were never exposed to during the initial treatment.[43] Despite apparent differences within the families of drugs associated with MDR phenotypes, when the mechanisms underlying these phenotypes are identified, the involved antineoplastic agents frequently share common metabolic pathways, efflux transport systems, or sites of cytotoxic action (Table 3).

Table 3. Mechanisms of resistance for specific chemotherapeutic agents

Mechanism of Resistance	Drugs Involved	Pharmacological Defect
Decreased drug uptake	MTX	Decreased expression of folate transporter
	Arabinosylcytosine	
Decreased drug	Fludarabine	Decreased deoxycytidine kinase
activation	Cladribine	Decreased folyl- polyglutamyl synthetase
Increased drug targeting	MTX	Amplified DHFR
	5-FU	Amplified TS
	Etoposide	Altered topoisomerase II
	Doxorubicin	Altered topoisomerase II
Altered drug targeting	Etoposide	Altered topoisomerase II
	Doxorubicin	Altered topoisomerase II
	MTX	Altered DHFR
Increased detoxification	Alkylating agents	Increased GST
Enhanced DNA repair	Alkylating agents	Increased nucleotide excision repair
	Platinum derivates	Increased nucleotide excision repair
	Nitrosoureas	Increased O^6-alkyl-guanine alkyl transferase
Defective recognition	Cisplatin	Defective mismatch repair
of DNA adducts	Doxorubicin	Increased MDR expression or MDR gene amplification
	Etoposide	Increased MDR expression or MDR gene amplification
	Vinca alkaloids	Increased MDR expression or MDR gene amplification
	Paclitaxel	Increased MDR expression or MDR gene amplification
Defective checkpoint function and apoptosis	Most anticancer drugs	p53 mutations

MDR

Classic (P-Glycoprotein-Dependent) MDR

An in vitro model of MDR was described by Biedler and Riehm 3 decades ago.[44] In their studies, cultured cells selected for resistance by exposure to actinomycin D developed cross-resistance to a surprising array of structurally diverse compounds, including vinca alkaloids, puromycin, daunomycin, and mitomycin C. Induction of this pattern of cross-resistance has since been observed by numerous investigators. This resulted in initiation of the study of drug resistance. Generally, exposure of cells to drugs such as anthracyclines, vinca alkaloids, and epipodophyllotoxins is related to the classic MDR phenotype and can result in cross-resistance to all other members of the phenotype. The emergence of MDR has been associated with increased levels of expression of a membrane-bound P-glycoprotein (P-170 or MDR1 protein).

De novo and acquired cross-resistance to multiple antineoplastic agents may result from several alternative factors and processes. First, MDR patterns of cross-resistance were found to be frequently associated with decreased drug accumulation, usually because of increased drug efflux.[45] Classic MDR-associated drug resistance is mediated by P-glycoproteins. More recently, a similar but distinct MDR phenotype was attributed to the energy-dependent drug-efflux activities of MDR protein (MRP) family members.[43] An overlapping but discrete resistant MDR phenotype is associated with increased expression of the recently isolated putative efflux breast cancer resistance protein (BCRP).[46] MDR has also been described in association with overexpression of the lung resistance protein (LRP). The mechanism of LRP-associated resistance is unclear, and whether LRP alone is sufficient to confer resistance is unknown. Some have speculated that as a major vault protein, LRP is involved in nucleocytoplasmic transport and cytoplasmic sequestration of drugs.[47] Drug resistance defined by alterations in topoisomerases represents a third major category of MDR.[48,49]

MRP Family

Similar phenotypes of multiple resistance to antineoplastic agents that are associated with the expression of other membrane proteins have been described. In many of these phenotypes, resistance occurs independently of P-glycoprotein expression.[50]

A distinct gene, *mrp1* (MRP1 or MDR-associated protein 1), was isolated from a doxorubicin-selected MDR lung cancer cell line. This gene encodes a 190-kDa transmembrane protein whose structure is strikingly homologous with that of P-glycoprotein/MDR1 and other members of the ATP-binding cassette transmembrane transporter proteins.[51,52] The importance of MRP1 overexpression in clinical drug resistance is unknown. However, because MRP1 expression varies widely in tumor cells, MRP1 may be a significant mediator of drug resistance in human cancer. At least five other human MRP isoforms have been identified.[53] Among them, MRP2 (cMOAT) and MRP3 are capable of supporting efflux detoxification of cancer drugs, including epipodophyllotoxins (MRP2 and MRP3), doxorubicin (MRP2), and cisplatin (MRP2). Recent results indicated that MRP1 and MRP2 are also able to confer resistance to the polyglutamatable antifolate methotrexate (MTX).[54]

MDR Associated with Topoisomerase Poisons

Topoisomerases are nuclear enzymes that catalyze the formation of transient single- or double-stranded DNA breaks, facilitate the passage of DNA strands through these breaks, and promote rejoining of the DNA strands.[55] As a consequence of these activities, topoisomerases are thought to be critical for DNA replication, transcription, and recombination. The drugs responsible for these activities are called topoisomerase poisons and include anthracyclines, epipodophyllotoxins, and actinomycin D. Their effect is thought to depend on the DNA cleavage activities of topoisomerases. There are two classes of mammalian enzymes: topoisomerase I and topoisomerase II. Topoisomerase I catalyzes the formation of single-stranded DNA breaks, whereas topoisomerase II as well as β isoforms catalyzes both single- and double-stranded breaks.

The formation of these stabilized DNA-topoisomerase-drug complexes is thought to initiate the production of lethal DNA strand breaks. Of the chemotherapeutic drugs that affect topoisomerase activities, the topoisomerase II poisons have been found to be the most clinically important. Hence, decreased drug accumulation caused by increased expression of P-glycoprotein or MRP1 is a potential mechanism of resistance to these topoisomerase II poisons. However, a distinct pattern of topoisomerase II-related MDR that differs from the pattern of P-glycoprotein-associated MDR in several important ways has been described. For example, cells that develop topoisomerase II alterations following exposure to amsacrine may show cross-resistance to other intercalating topoisomerase II poisons but not to epipodophyllotoxins. Finally, two mammalian isozymes of topoisomerase II have been found: a 170-kDa form (topoisomerase Iα) and a 180-kDa form (topoisomerase IIβ).[56] These isozymes differ in their regulation during the cell cycle and their relative sensitivity to topoisomerase II poisons.[57] Hence, both the relative levels of the specific topoisomerase II isozymes and the total topoisomerase II activity may be significant determinants of the sensitivity of tumor cells to topoisomerase II drugs.

The molecular bases of drug resistance associated with qualitatively altered topoisomerase II expression have been described in several reports. However, the relevance of topoisomerase I for clinical drug resistance is unknown. Alternatively, altered subcellular localization of topoisomerase II isoforms[58] and altered posttranslational phosphorylation have been reported in association with some etoposide-resistant cell lines.[59,60] The cytotoxicity of topoisomerase II poisons is believed to depend on the formation of DNA strand breaks secondary to stabilization of the reversible enzyme-DNA cleavable complex.[61] A new family of drugs targeting topoisomerase II function that includes fostriecin, merbarone, aclarubicin, and bis(2,6-dioxopiperazine) derivatives (e.g., ICRF193, ICRF 187) has emerged. Also, the cytotoxic agent camptothecin has been shown to enhance topoisomerase I-mediated strand breaks. Previously, host toxicity was found to prohibit the clinical use of such topoisomerase I poisons. However, the prospect of less toxic analogues of this drug that maintain a high level of activity against topoisomerase I-rich human cancer cells has renewed interest in the clinical application of this class of compounds.[62] Consequently, the emergence of resistance to these agents may become an increasingly important consideration.

MDR Associated with Altered Expression of Drug-Metabolizing Enzymes and Drug-Conjugate Export Pumps

The manner in which cells metabolize cancer drugs and other xenobiotics is often described as three phases of detoxification. Alterations in any of these phases can influence the sensitivity and resistance to a particular drug or xenobiotic toxin. For example, phase I metabolism is mediated by cytochrome P450 mixed-function oxidases. These metabolites or the unmodified drug may then be converted to a less reactive, presumably less toxic form in phase II reactions. Phase II detoxifications include the formation of drug/xenobiotic conjugations with glutathione, glucuronic acid, or sulfate, reactions that are catalyzed by multiple isozymes of glutathione S-transferase (GST), UDP-glucuronosyl transferase, and sulfatase, respectively.[63-65] Phase III detoxification consists of exportation of the parent drug/xenobiotic or its metabolites with the use of energy-dependent transmembrane efflux pumps, including P-glycoprotein, MRP family members, and breast cancer resistance protein. Frequently, coordinated downregulation of phase I drug-activating enzymes and upregulation of specific phase II drug-conjugating enzymes are observed in cellular and animal models of drug or xenobiotic resistance.[66,67] Such a programmed cellular stress response offers a versatile, generalized protective mechanism against exposure to a variety of exogenous toxins.

Whether GST levels in tumor cells are sufficient to detoxify antineoplastic drugs to a clinically significant degree is a matter of considerable debate, and the role of GSTs in drug resistance remains uncertain because of inconsistent results from different laboratories.[68-70] Thus, the relative resistance of cells expressing drug-metabolizing enzymes may depend on cellular levels of drug conjugate transporters, including the glutathione conjugate transporters,[71] such as the MRP family proteins.[72] Indeed, recent results using model cell lines have demonstrated that combined expression of specific isozymes of GST with MRP1 is necessary to achieve full protection from the toxic effects of the cancer drug chlorambucil[41] and the carcinogen 4-nitroquinoline 1-oxide. In these studies, the expression of either GST or MRP1 alone provided little if any protection from toxic effects, a finding that illustrates the synergistic interaction of phase II and III detoxification processes in the emergence of resistance to some drugs.

Emergence of Refractory Tumors Associated with Multiple Resistance Mechanisms

The backbone of many treatment regimens designed to circumvent the proliferation of resistant tumor cells is the administration of multiple drugs with different structural properties and mechanisms of action. This approach supposes that if enough carefully selected drugs are delivered at optimal doses and intervals, individual clones of cells resistant to one class of drug will be effectively killed by another drug in the regimen. The rapid appearance of a refractory tumor despite an initially favorable cytoreductive response suggests that the emergence of tumor cell clones with resistance to multiple drugs is a common clinical occurrence. We have seen how a single genetic change such as increased P-glycoprotein or altered topoisomerase II can mediate cross-resistance to several, but not all, useful antineoplastic drugs. Although these mechanisms provide a molecular explanation for broad-spectrum resistance, it is clear that many refractory tumor clones must simultaneously develop multiple resistance mechanisms. These mechanisms may arise from multiple independent genetic changes in single-cell clones or, as suggested by Muller et al,[73] cell-to-cell transfer of genetic information.

Resistance to Free-Radical-Mediated Drug Cytotoxicity

Several antineoplastic agents form free radical intermediates that are thought to contribute to drug cytotoxicity. Anthracyclines, such as doxorubicin, are among the most important members of this class of compounds. Whereas DNA-intercalating anthracyclines can damage cells through multiple mechanisms, including inhibition of nucleic acid synthesis, induction of topoisomerase II-mediated DNA strand breaks, and perturbation of cell membranes, these quinone-hydroquinone compounds can also generate toxic free radical species that may cause cell death.[74] The semiquinone radical thus generated may either form a covalently binding free radical derivative or, in the presence of oxygen, be reoxidized to the quinone species in a reaction producing superoxide anion. Decomposition of hydrogen peroxide formed by dismutation of superoxide anion produces the highly reactive hydroxyl radical, which may directly damage DNA, lipids, and proteins. Thus, cellular factors that limit hydrogen peroxide production or repair peroxidative damage to macromolecules could theoretically confer some resistance to anthracyclines. Several pathways may contribute to protection of tumor cells from anthracycline-mediated free radical damage. First, superoxide anion formation is limited in poorly vascularized, relatively hypoxemic tissues, such as in the center of large solid tumors. Second, increased intracellular levels of catalase and glutathione peroxidase can deplete hydrogen peroxide, thus reducing the formation of toxic hydroxyl radicals.[75]

Resistance to Genotoxic Cancer Treatments Related to Suppression of Apoptotic Pathways

Chemotherapeutic drugs are cytotoxic because of their interactions with a variety of molecular targets. Despite these varied primary targets, most, if not all, cancer drugs instigate cell death, at least partially, via downstream events, especially those that converge upon pathways mediating programmed cell death or apoptosis. This process is conveniently conceptualized in three phases. First, initiation of apoptosis (e.g., secondary to chemotherapy-mediated DNA damage) is characterized by its reversibility. Second, the decision to complete the death program is irreversible. The commitment phase may involve mitochondrial changes and the release of cytochrome-*c* and apoptosis-inducing factor, which are hallmarks of apoptosis. Third, the degradation or execution phase includes downstream events, such as DNA fragmentation and morphological changes. Prior to commitment, apoptosis can be modulated by regulatory elements, such as p53 and the Bcl-2 family of proteins. Although apoptosis may be either p53 dependent or p53 independent, frequently, the cellular response to DNA damage is regulated by p53.[76] Depending on the particular cell type and damage, p53 may then initiate one of two possible pathways: apoptosis or cell cycle arrest and repair.

The mitogen-activated protein kinase-signaling cascades are involved in the regulation of cellular response to exogenous factors, including genotoxic and cytotoxic anticancer agents.[77] Additionally, the extracellular signal-regulated kinase pathway is implicated in the proliferative response to growth factors. In cells treated with potentially cytotoxic stressors, such as radiation and anticancer drugs, the p38 and stress-activated/c-Jun N-terminal protein kinase (SAPK/JNK) pathways are implicated in mediating cell cycle arrest and apoptosis. Furthermore, the Bcl-2 family of proteins comprises several important regulators of apoptosis. Although their mechanism or mechanisms of action are not completely known, the balance of expressed antiapoptotic family members (Bcl-2, Bcl-XL, Bcl-w, A1, and Mcl-1) and proapoptotic family members (Bax, Bak, Bad, Bik, and Bid) can influence the relative sensitivity of cells to toxic stressors.[78] This genomic instability may further lead to mutations that activate additional resistance mechanisms and confer more aggressive tumor behavior.[79] Thus, the expression of mutant and wild-type p53, Bcl-2 family members, mitogen-activated protein kinase (MAPK) family members, and other proteins associated with the control of apoptosis may contribute significantly to the clinical sensitivity of tumor cells. These proteins are the targets of investigational agents that may become important in future strategies for overcoming clinical drug resistance.

Resistance Factors Unique to Tumor Cells in Vivo: Host-Tumor-Drug Interactions

The failure of chemotherapy to eradicate a tumor in vivo despite exquisite sensitivity to the chemotherapeutic drug or drugs in vitro may be caused by anatomic or pharmacological sanctuaries. For example, brain and testicular barriers probably account for the relatively high frequency of acute lymphoblastic leukemia relapse at these sites.[80] In cases with a large solid tumor, failure of chemotherapy is frequently attributed to decreased drug delivery to a tumor that has overgrown its vascular supply. Additionally, development of acidosis and hypoxia in poorly perfused areas of large tumors may interfere with the cytoxicity of some drugs. Finally, altered prodrug activation by the liver or other normal tissues may profoundly influence the efficacy of drugs such as cyclophosphamide.

Mechanisms of Resistance for Agents Used to Treat Breast Cancer

Anthracyclines

Mechanism of Action

The mechanisms of action of anthracyclines are pleiotropic effects, including activation of signal transduction pathways, generation of reactive oxygen intermediates, stimulation of apoptosis, and inhibition of DNA topoisomerase II catalytic activity.

Metabolism

Anthracyclines are metabolized by reduction of a side-chain carbonyl to alcohol, resulting in some loss of cytotoxicity, and a one-electron reduction to a semiquinone free radical intermediate by flavoproteins, leading to aerobic production of superoxide anion, hydrogen peroxide, and hydroxyl radical.

Pharmacokinetics

The protein-binding rate of doxorubicin ranges from 60% to 70%, whereas its cerebro-spinal fluid (CSF)/plasma ratio is very low. Doxorubicin circulates predominantly as a parent drug, and doxorubicinol is its most common metabolite, although doxorubicin 7-deoxyaglycone and doxorubicinol 7-deoxyaglycone form in a substantial fraction of patients. In addition, substantial interpatient variation in biotransformation has been observed, and dose-related changes in clearance do not appear to be greater in men than in women. Daunorubicin metabolizes faster than an equivalent dose of doxorubicin does.

Elimination

Only 50% to 60% of the parent drug is eliminated by known routes. A substantial fraction of the parent compound is bound to DNA and cardiolipin in tissues. Although changes in anthracycline pharmacokinetics may be difficult to demonstrate in patients with mild alterations in liver function, anthracycline clearance is definitely decreased in patients with significant hyperbilirubinemia or a marked burden of metastatic tumor in the liver.

Mechanism of Resistance

The mechanism of resistance in anthracyclines is increased expression of the P-170 glycoprotein related to the enhancement of drug efflux. The evidence supporting this role includes correlation between this protein and resistance, transfer of the cloned MDR1 gene, and reversal of resistance by agents that block P-170. The in vivo cells are different from the in vitro cells. The nature of resistance that develops after a single prolonged exposure to doxorubicin was evaluated by using classic fluctuation analysis.[81] The researchers found that MDR1 expression did not occur and that the resistance arose from a spontaneous mutation with an apparent generation rate of approximately 2×10^{-6} per cell. Also, under certain circumstances, expression of the MDR1 gene clearly may be transcriptionally modulated by doxorubicin itself as well as by inhibitors of protein kinase C and calmodulin. In vivo, the resistance is more

Table 4. Approaches to overcoming or circumventing drug resistance

Prevention	
	Aggressive multiagent therapy
	Appreciation of factors that induce resistance mechanisms
	Circumvention: drug screening programs and rational drug design
	Circumvention of drug-uptake defects
Dose escalation	
	Drugs that use alternate transport mechanisms
	Agents that reverse increased efflux
	Co-factors that augment drug activation or efficacy
	Inhibition of drug inactivation
	Novel treatment modalities
Immunotherapy	
	Development of agents that target signaling and apoptotic pathways

complex, with most tumors and many normal tissues exhibiting increased expression of a gene copy.[82] Other mechanisms of resistance include a 190-kDa protein that is a member of the ATP-binding cassette transmembrane transporter superfamily. MRP expression alone, in the absence of alterations in MDR1 or topoisomerase II expression, can also produce anthracycline resistance.[83] Furthermore, altered topoisomerase II activity has been implicated in resistance to anthracyclines. Overexpression of bcl-2 can significantly diminish the toxicity of doxorubicin, as can mutations of the p53 gene.[84] Potent nuclear DNA repair systems also contribute substantially to the ability of tumor cells to withstand the cytotoxic effects of doxorubicin. For example, ADP ribosylation is a well-known posttranslational modification of topoisomerase II and plays an important role in the use of nicotinamide adenine dinucleotide (oxidized form). These results suggest that intermediary metabolism affects DNA cleavage and doxorubicin resistance.[85]

Overcoming Resistance

As discussed above, resistance to anthracyclines may occur as a consequence of P-glycoprotein overexpression or altered topoisomerase II activities. However, neither of these mechanisms will necessarily result in cross-resistance to topoisomerase II in all cases. Additionally, tumor cells resistant to classic topoisomerase II poisons frequently retain sensitivity to the cytotoxic effects of the novel class of topoisomerase II catalytic inhibitors (fostriecin, merbarone, aclarubicin, and bis(2,6-dioxopiperazine)).[86,87] This class of topoisomerase-directed drugs offers an alternative for the treatment of topoisomerase-poison-resistant tumors. Finally, structural analogues of parent topoisomerase II poisons may overcome resistance based on altered topoisomerase II.[88-89] The use of noncross-resistant agents with cytotoxic activity but different mechanisms of action after administration of anthracycline-based regimens has proven to be beneficial in patients with breast cancer.[90,91] Finally, the time and method of delivering anthracyclines affects toxicity and results. For example, continuous infusion of doxorubicin (Adriamycin) increases its therapeutic index (Table 4).[92]

Taxanes

Mechanism of Action

The mechanisms of action of taxanes are high-affinity binding to microtubules with enhanced microtubule formation at high drug concentrations and inhibition of mitosis.

Metabolism

The effects of taxanes on microtubules differ from those of the vinca alkaloids. Unlike colchicine and the vinca alkaloids, which prevent microtubule assembly, submicromolar concentrations of the taxanes decrease the lag time and shift the dynamic equilibrium between tubulin dimers and microtubule assembly and stabilize microtubules against depolymerization.[93] The metabolism and elimination of paclitaxel and docetaxel are similar. In humans, urinary excretion accounts for a small percentage of drug disposition, averaging 2%. Both hepatic metabolism and biliar excretion are also important. Approximately 80% of the administered dose is excreted in the feces within 7 days after treatment. Also, the hepatic cytochrome P450 is responsible for the bulk of drug metabolism, and the cytochrome P450 isoforms CYP3A, CYP2B, and CYP1A may play a role in biotransformation. The main metabolic pathway for taxanes consists of oxidation of a tertiary butyl group on the side chain at the C-13 position of the taxane ring as well cyclization of the side chain.[94]

Pharmacokinetics

The pharmacokinetics of taxanes consist of saturable elimination and distribution, which are particularly evident with a short (3-hour) schedule.

Elimination

Taxanes are eliminated predominantly by hepatic hydroxylation of cytochrome P450 enzymes and biliary excretion of metabolites. Less than 10% of each dose is eliminated intact in the urine.

Mechanism of Resistance

Two main mechanisms of taxane resistance have been described in cells exposed to taxanes at low concentrations for prolonged periods pf time. The first is changes in the expression of β-tubulin isotopes, mainly β-III, whereas the second is part of the MDR system. Upregulation of caveolin-1, a membrane component involved in small molecule transport and intracellular signaling, has also been found to be related to taxane resistance.[95]

Overcoming Resistance

The use of a polyoxyl compound (Cremophor) as an MDR expression modulator has been evaluated. Other modulators of MDR that have been studied include verapamil, cyclosporine A, and PC 833.[96,97] When paclitaxel is given over 3 hours at 135 to 175 mg/m^2, plasma concentrations of Cremophor are able to revert MDR in vitro.[98]

Antimetabolites

Mechanism of Action

The cytotoxic effects of antimetabolites stem from their ability to interfere with key enzymatic steps in nucleic acid metabolism. This group of agents includes three well-studied compounds: the antifolate MTX and the pyrimidine analogues 5-fluorouracil (5-FU) and arabinosylcytosine. Inhibition of dihydrofolate reductase (DHFR) leads to partial depletion of reduced folates. Polyglutamates of MTX and dihydrofolate inhibit purine and thymidylate biosynthesis.

Metabolism

The metabolism of 5-FU is complex. The best characterized mechanism of fluoropyrimidine cytotoxicity involves the inhibition of thymidylate synthase by 5-fluoro-2'-deoxyuridine monophosphate (FdUMP). Additionally, the incorporation of the metabolite 5-fluorouridine triphosphate into RNA has been correlated with cytotoxicity in some systems. Although 5-fluoro-2'-deoxyuridine triphosphate can be incorporated into DNA, the relationship between this process and the cytocidal activity of fluoropyrimidines remains undetermined.

Pharmacokinetics

Following uptake by a folate transport system, MTX can bind avidly to and inhibit DHFR, its primary enzyme target. In the presence of adequate thymidylate synthase activity, inhibition of DHFR results in depletion of the reduced folate pools essential for thymidylate and de novo purine synthesis. The cytotoxicity of MTX is significantly influenced by intracellular polyglutamation. MTX polyglutamates are retained preferentially by cells and bind more effectively to DHFR. Additionally, these polyglutamyl derivatives can inhibit other folate-dependent enzymes, including thymidylate synthase and 5-aminoimidazole-4-carboxamide ribonucleoside (AICAR) trans-formylase,[99] enzymes involved in thymidylate and de novo purine synthesis, respectively.

Elimination

Antimetabolites are eliminated primarily in the urine.

Mechanism of Resistance

Resistance to 5-FU may be conferred by alterations in enzymes involved in fluoropyrimidine metabolism, particularly those enzymes associated with the conversion of 5-FU to the thymidylate synthase inhibitor FdUMP.[100] Furthermore, changes in the thymidylate synthase level or its affinity for FdUMP have been associated with 5-FU resistance.[101]

A multifactorial process involving DHFR gene amplification, a transport defect, and a decrease in the formation of polyglutamic acid has been seen in patients with tumors resistant to MTX.[102,103] Resistance to MTX may result from a number of alternative mechanisms, including (1) reduced MTX uptake via defective folate transport systems,[104,105] such as decreased expression of the reduced folate carrier[106] or folate receptors;[107] (2) increased exportation via MRPs[108,109] or other exporters of polyglutamatable antifolates; (3) reduced polyglutamation leading to decreased drug retention as well as reduced inhibition of thymidylate synthase and AICAR transformylase;[110] (4) elevated levels of DHFR or reduced affinity of DHFR for MTX;[111,112] and (5) expression of Bcl-2 during apoptosis.

Overcoming Resistance

Strategies designed to overcome resistance to antimetabolites include dose escalation, pharmacological manipulation of drug metabolism, and rational design of new antimetabolites.[113] The rationale for the use of high-dose MTX with subsequent rescue of normal tissues by administration of the reduced folate leucovorin (N5-formyl tetrahydrofolate) in the treatment of cancers other than breast cancer was recently questioned.[114,115] Other antifolate compounds capable of inhibiting folate-dependent enzymes besides DHFR have been investigated. In particular, trimetrexate, 10-propargyl-5,8-dideazafolate, and 5,10-dideazatetra-5,6,7,8-tetrahydrofolate have shown promise in cells resistant to MTX.[116,117] Finally, the synergistic interaction between interferon and halogenated pyrimidines has been described.[118,119]

Alkylating Agents and Platinum Compounds

Mechanism of Action

All alkylating agents and platinum compounds produce alkylation of DNA through the formation of reactive intermediates that attack nucleophilic sites.

Metabolism

Cyclophosphamide is metabolized by microsomal hydroxylation and hydrolysis to phosphoramide mustard (active) and acrolein. It is excreted as inactive oxilation products. Chlorambucil undergoes chemical decomposition to active phenyl acetic acid mustard and to inert dechlorination products. Melphalan undergoes chemical decomposition to inert dechlorination products, and 20% to 35% of it is excreted unchanged in the urine. Carmustine undergoes chemical decomposition to active and inert products and enzymatic conjugation with glutathione. Finally, cisplatin covalently binds to DNA bases and disrupts DNA function. The toxicity of these agents may be related to DNA damage.

Elimination

Approximately 25% of each dose of alkylating agents and platinum compounds is excreted from the body during the first 24 hours after administration. About 90% of excretion is renal, whereas about 10% is biliary. Extensive long-term protein binding has been observed in many tissues.

Mechanism of Resistance

Resistance to alkylating agents and platinum compounds can be described by at least four broad mechanistic categories: (1) decreased alterations in transmembrane cellular drug accumulation;[120] (2) increased cytosolic drug inactivation; (3) enhanced repair of DNA damage;[121] and (4) resistance to apoptosis.[122] The correlation between the glutathione or GST level and drug resistance is variable. Indeed, some investigators have been unable to demonstrate a relationship between overexpression of multiple isozymes of GST and antineoplastic resistance.[123-124]

Aldehyde dehydrogenase is another drug-metabolizing enzyme that has been linked with resistance to cyclophosphamide derivatives in murine and human models of drug resistance.[125] These results suggest that coadministration of DNA polymerase alpha inhibitors with cisplatin is useful in overcoming cisplatin resistance. Also implicated in platinum sensitivity and resistance are alterations in mismatch repair or regulators of apoptosis, such as Bcl-2, Bax, p21, and p53.[126] Modulation of these pathways by therapeutic agents now in development represents an emerging strategy for overcoming resistance to platinum and other alkylating compounds.

Overcoming Resistance

Other results also suggest that coadministration of DNA polymerase alpha inhibitors with cisplatin is useful in overcoming cisplatin resistance. Also implicated in platinum sensitivity and resistance are alterations in mismatch repair or regulators of apoptosis, such as Bcl-2, Bax, p21, and p53.[127] Modulation of these pathways by therapeutic agents now in development represents an emerging strategy for overcoming resistance to platinum and other alkylating compounds.

Vinca Alkaloids

This group of drugs includes vincristine sulfate, vinblastine sulfate, vindesine sulfate, and vinorelbine tartrate.

Mechanism of Action

The mechanism of action of the vinca alkaloids is inhibition of polymerization of tubulin.

Metabolism

Vinca alkaloids are metabolized hepatically. Metabolites accumulate rapidly in the bile so that only 46.5% of the total biliary product is the parent compound. The specific contribution of cytochrome P450-mediated metabolism of vincristine is uncertain, although its importance is suggested by observations of enhanced clearance with phenytoin and increased toxicity with the 3A inducer itraconazole.

Pharmacokinetics

The pharmacokinetics of vinca alkaloids is characterized by large distribution volumes, high clearance rates, and long terminal half-lives. At conventional dosages, the peak plasma concentrations, which persist for only a few minutes, range from 100 to 500 nmol/L, and plasma levels remain above 1 to 2 nmol/L for relatively long durations.

Elimination

The vinca alkaloids are eliminated by biliary excretion.

Mechanism of Resistance

Resistance to vinca alkaloids arises by at least two different mechanisms and is associated with decreased drug accumulation and retention. The first mechanism is implicated by the phenomenon of pleiotropic resistance or MDR, whereas the second mechanism is one of resistance to antimicrotubule agents in vitro resulting from alterations in α- and β-tubulin proteins. An important feature of this type of resistance to the vinca alkaloids is that collateral sensitivity is conferred to the taxanes, which inhibit microtubule disassembly.

Overcoming Resistance

Studies have suggested that coadministration of DNA polymerase alpha inhibitors with vinca alkaloids is useful in overcoming resistance. Also, modulation of pathways such as Bcl, Bax, p21, and p53 by therapeutic agents now in development represents an emerging strategy for overcoming resistance to alkylating compounds.

Gemcitabine

Mechanism of Action

Gemcitabine inhibits DNA polymerase α, is incorporated into DNA, and terminates DNA-chain elongation.

Metabolism

Gemcitabine is activated to triphosphate in tumor cells, degraded to inactive uracil arabinoside by deamination, and converted to an arabinosylcytosine diphosphate choline derivative.

Elimination

Gemcitabine is eliminated by deamination in the liver, plasma, and peripheral tissues.

Mechanism of Resistance

Resistance to gemcitabine is not fully understood, although several mechanisms of resistance to gemcitabine have been described. In general, cells with deficient nucleoside transport are highly resistant to gemcitabine,[128] and the degree of resistance may vary according to the nucleoside transporter expressed on the cellular surface.[129] Also, enzymes involved in gemcitabine cell metabolism have been associated with the development of resistance to it. The initial in vitro studies suggested that deficiency in deoxycytidine kinase enzymatic activity was the most important cause of gemcitabine resistance, as gemcitabine-sensitive cell lines expressed 10 times more deoxycytidine kinase than gemcitabine-resistant ones.[130] However, experiments using KB cells from human epidermoid carcinoma suggested that the enzyme ribonucleotide reductase (RR) could play an important role. RR is specific for S phase and limits DNA synthesis. Resistant cells have 9.0 times greater expression of RR mRNA and 2.3 times greater RR activity than sensitive cells do.[131] The role of RR as a determinant of resistance to gemcitabine has been confirmed with the use of K563 erythroleukemia cell lines, in which the enzymatic activity of RR correlated with resistance to gemcitabine.[132] A cross-resistance pattern between nucleoside analogues also may have potential implications. Researchers have shown that gemcitabine has more antitumor activity than cytarabine does in sensitive (L1210 and BCLO) and resistant (LA46 and Bara C) cell lines.[133] An in vitro experiment using HL-60 promyelocytic leukemic cells made resistant to cladribine created two resistant sublines with no cross-resistance to gemcitabine.[134]

Overcoming Resistance

No strategies for overcoming gemcitabine resistance have proven to be effective. Use of combination schedules is the main approach.

Tamoxifen

Mechanism of Action

Tamoxifen binds to the estrogen receptor (ER) and induces dimerization and DNA binding to finally inactivate it.

Metabolism

Tamoxifen metabolism is mediated in the liver by cytochrome P450-dependent oxidases into 10 major metabolites.

Pharmacokinetics

After initiation of therapy, steady-state concentrations of the active metabolites of tamoxifen are achieved in 4 weeks, suggesting a half-life of 14 days.

Elimination

Metabolites and a small portion of tamoxifen are excreted in the bile as conjugates.

Mechanisms of Resistance

Several mechanisms of resistance to tamoxifen have been described. Absence of ER expression explains primary resistance in certain tumors. ER mutations may explain the variability in response to tamoxifen in patients with ER-positive tumors; however, these mutations occur in less than 1% of patients with breast cancer.[135] Alternative mRNA splicing has been identified in normal and malignant breast tissue with variants lacking one or more exons. The transcript with deleted exon 5 binds to DNA but not to estrogen and activates transcription in an estrogen-independent manner.[136] Because ER function is strongly influenced by growth factor signaling, studies have shown decreased tamoxifen response in patients whose tumors coexpress ER and HER-2.[137] Finally, the information on coactivators and coexpressors of tamoxifen resistance is limited; however, evidence of the importance of these molecules has been shown. MCF-7 tumor cells regress with the use of tamoxifen, but if tamoxifen administration is continued, they grow back in a tamoxifen-dependant manner; subsequently, withdrawal of tamoxifen causes regression.[138] N-CoR corepressor levels are suppressed in tumors stimulated by tamoxifen when compared with tumors that are sensitive to tamoxifen.[139]

Overcoming Resistance

The use of aromatase inhibitors that block ligand production is an alternative for treating tumors that are resistant to tamoxifen. Also, the use of pure antiestrogens like fulvestrant that block ER function before coactivator binding theoretically may overcome tamoxifen resistance.[140,141] Finally, the use of growth factor receptor inhibitors in the form of monoclonal antibodies and small tyrosine kinase inhibitors to reestablish tamoxifen sensitivity is being studied.

Aromatase Inhibitors, Antiestrogens, and Progestins

The mechanisms and percentages of resistance in these groups of drugs are currently being investigated.

Trastuzumab

Mechanism of Action

Trastuzumab is a humanized monoclonal antibody that selectively binds with high affinity to the extracellular domain of HER-2. It inhibits tumor-cell proliferation through antibody-dependent cellular toxicity,[142] inducing apoptosis,[143] inhibiting HER-2/neu intracellular signaling pathways,[144] and downregulating expression of HER-2 receptors.[145] It also has synergistic action in combination with chemotherapy drugs.[146,147]

Metabolism

The metabolism of trastuzumab is not clear. Clearance of it by the liver and kidneys is minimal.

Pharmacokinetics

The mean half life of trastuzumab is 21 days. Its disposition is not altered by age or renal function.

Mechanism of Resistance

Resistance to trastuzumab is an active research field. Several known mechanisms of resistance have been identified: increased production of insulin-like growth factor (insulin-like growth factor-1 or insulin-like growth factor-I receptor),[148] dysregulation of p27,[149] overexpression of epidermal growth factor receptor with activation of the AKT pathway,[150] and decreased PTEN function.[151]

Overcoming Resistance

Targeting the epidermal growth factor receptor family with monoclonal antibodies or single or multiple tyrosine kinase inhibitors to prevent or overcome trastuzumab resistance is a subject of active research. Combinations of trastuzumab with both gefitinib and erlotinib are being evaluated in phase I and II studies.[152] Several strategies for blocking insulin-like growth factor-1 signaling, including the use of monoclonal antibodies with antitumor effects in breast cancer such as αIR3[153] and antisense molecules, are being developed.[154]

Chemotherapy Sensitivity and Resistance Assays

Chemotherapy sensitivity and resistance assays are laboratory tests that pretend to select the most appropriate treatment by studying an individual's tumor behavior when exposed to certain drugs. The goal is to individualize therapy, optimize resources, and reduce toxicity. These assays are also known as chemosensitivity tests. Several of these assays have been discarded, whereas others are being studied in clinical trials. The American Society of Clinical Oncology does not recommend the use of these assays to select a therapeutic agent outside of a clinical trial, because even the assays with better potential still require more evaluation.[155]

The 3-(4,5-dimethylthyazol-2-yl)-2,5-dyphenil tetrazolium bromide assay has been studied in patients with breast cancer. Using tumor-cell suspension cultures incubated with various chemotherapeutic agents, 3-(4,5-dimethylthyazol-2-yl)-2,5-dyphenil tetrazolium bromide is added after 4 days to reduce intercellularity and generate a blue staining. The number of viable cells treated is determined according to the field intensity.[156]

In general, applying these techniques in the clinical field is significantly limited. The applicability of the results for all tumor cells, impact of the results on selecting and discarding treatments, and difficulty in accessing laboratories with the appropriate technology to apply and interpret the assays are issues that must be addressed before chemotherapy sensitivity and resistance assays are ready for prime time.

Conclusions and Future Directions

Different studies, the majority of which were performed in vitro, have identified several mechanisms of drug resistance in breast cancer. How these processes operate in vivo and their clinical impact must be further studied in controlled prospective examinations of patient tumor specimens correlated with therapeutic responses to different agents. The search for these mechanisms continues to aid the development of useful approaches to overcoming drug resistance. The use of newer technologies such as genomics and proteomics will continue to expand this field of study. For instance, recent studies using gene arrays of breast tumor tissue were able to predict response to neoadjuvant chemotherapy.[157,158]

These discoveries should impact the rationale for designing clinical trials to continue studying drug resistance and achieve the goal of being able to administer tailored therapy for breast cancer.

Acknowledgements

Funding/Support: Supported in part by the Nellie B. Connally Breast Cancer Fund.

References

1. Anonymous. Cancer Incidence, Mortality and Prevalence Worldwide, Version 1.0. GLOBOCAN: IARC Press, 2002, (http//:www-dep.iarc.fr/daba/infodata.htm).
2. Jemal A, Murray T, Samuels A et al. Cancer statistics, 2003. CA Cancer J Clin 2003; 53:5-26.
3. Pisani P, Bray F, Parkim DN. Estimates of the worldwide prevalence of cancer for 25 sites in the adult population. Int J Cancer 2002; 97:72-81.
4. Collyar DE. Breast cancer; a global prespective. J Clin Oncol 2002; 19(18 Suppl):101S-105S.
5. McGuire WL. Breast cancer prognostic factors: Evaluation guidelines. J Natl Cancer Inst 1991; 83:154-155.
6. Winer EP, Morrow M, Osborne CK et al. Malignat tumors of the breast. In: DeVita VT, Hellman S, Rosenberg S, eds. Cancer Principles and Practices of Oncology, chap 37.2. Philadephia, PA: 2001:1651-1717.
7. Singletary SE, Allred C, Ashley P et al. Revision of the American Joint Committee on Cancer staging system for breast cancer. J Clin Oncol 2002; 20:3628-3636.
8. Slamon DJ, Clark GM, Wong SG et al. Human breast cancer: Correlation of relapse and survival with amplification of the HER-2/neu oncogene. Science 1987; 235:177-182.
9. Tommasi S, Paradiso A, Mangia A et al. Biological correlation between HER-2/neu and proliferative activity in human breast cancer. Anticancer Res 1991; 11:1395-1400.
10. Fox SB, Leek RD, Smith K et al. Tumor angiogenesis in node-negative breast carcinomas - Relationship with epidermal growth factor receptor, estrogen receptor, and survival. Breast Cancer Res Treat 1994; 29:109-116.
11. Gasparini G, Boracchi P, Bevilacqua P et al. A multiparametric study on the prognostic value of epidermal growth factor receptor in operable breast carcinoma. Breast Cancer Res Treat 1994; 29:59-71.
12. Hedley DW, Clark GM, Cornelisse CJ et al. Consensus review of the clinical utility of DNA cytometry in carcinoma of the breast. Report of the DNA Cytometry Consensus Conference. Cytometry 1993; 14:482-485.
13. Makris A, Powles TJ, Dowsett M et al. p53 protein overexpression and chemosensitivity in breast cancer. The Lancet 1995; 345:1181-1182.
14. Keyomarsi K, Tucker SL, Buchholz TA et al. Cyclin E and survival in patients with breast cancer. New England Journal of Medicine Online 2002; 347:1566-1575.
15. Alkarain A, Slingerland J. Deregulation of p27 by oncogenic signaling and its prognostic significance in breast cancer. Breast Cancer Res 2004; 6:13-21.
16. Cristofanilli M, Budd GT, Ellis MJ et al. Circulating tumor cells, disease progression, and survival in metastatic breast cancer. N Engl J Med 2004; 351:781-791.
17. Funke IM, Zia A, Wild C et al. Phenotype of disseminated tumor cells in bone marrow of breast cancer patients. J Clin Oncol 2001; 7:3670S.
18. Gasparini G, Weidner N, Bevilacqua P et al. Tumor microvessel density, p53 expression, tumor size, and peritumoral lymphatic vessel invasion are relevant prognostic markers in node-negative breast carcinoma. J Clin Oncol 1994; 12:454-466.
19. Anonymous. The national institutes of health consensus development conference: Adjuvant therapy for breast cancer. Bethesda, Maryland, USA. November 1-3, 2000. Proceedings. J Natl Cancer Inst Monogr 2001; 1-152.
20. Henderson IC, Berry DA, Demetri GD et al. Improved outcomes from adding sequential Paclitaxel but not from escalating Doxorubicin dose in an adjuvant chemotherapy regimen for patients with node-positive primary breast cancer. J Clin Oncol 2003; 21:976-983.
21. Early Breast Cancer Trialists' Collaborative Group. Tamoxifen for early breast cancer: An overview of the randomised trials. Lancet 1998; 351:1451-1467.
22. Early Breast Cancer Trialists' Collaborative Group. Tamoxifen for early breast cancer. Cochrane Database Syst Rev 2001; 1:CD000486.
23. Buzdar AU. Data from the Arimidex, tamoxifen, alone or in combination (ATAC) trial: Implications for use of aromatase inhibitors in 2003. Clin Cancer Res 2004; 10:355S-361S.
24. Haq R, Zanke B. Inhibition of apoptotic signaling pathways in cancer cells as a mechanism of chemotherapy resistance. Cancer Metastasis Rev 1998; 17:233-239.
25. DeVita Jr VT. The James Ewing lecture. The relationship between tumor mass and resistance to chemotherapy. Implications for surgical adjuvant treatment of cancer. Cancer 1983; 51:1209-1220.
26. Green M, Hortobagyi GN. Neoadjuvant chemotherapy for operable breast cancer. Oncology (Huntingt) 2002; 16:871-84, (889).

27. Fisher B, Bryant J, Wolmark N et al. Effect of preoperative chemotherapy on the outcome of women with operable breast cancer. J Clin Oncol 1998; 16:2672-2685.
28. Kuerer HM, Newman LA, Buzdar AU et al. Pathologic tumor response in the breast following neoadjuvant chemotherapy predicts axillary lymph node status. Cancer Journal From Scientific American 1998; 4:230-236.
29. Fisher B, Bryant J, Wolmark N et al. Effect of preoperative chemotherapy on the outcome of women with operable breast cancer. J Clin Oncol 1998; 16:2672-2685.
30. Fisher B, Bryant J, Wolmark N et al. Effect of preoperative chemotherapy on the outcome of women with operable breast cancer. J Clin Oncol 1998; 16:2672-2685.
31. Bear HD, Anderson S, Brown A et al. The effect on tumor response of adding sequential preoperative docetaxel to preoperative doxorubicin and cyclophosphamide: Preliminary results from National Surgical Adjuvant Breast and Bowel Project Protocol B-27. J Clin Oncol 2003; 21:4165-4174.
32. Bear HD, Anderson S, Brown A et al. The effect on tumor response of adding sequential preoperative docetaxel to preoperative doxorubicin and cyclophosphamide: Preliminary results from National Surgical Adjuvant Breast and Bowel Project Protocol B-27. J Clin Oncol 2003; 21:4165-4174.
33. Heys SD, Hutcheon AW, Sarkar TK et al. Neoadjuvant docetaxel in breast cancer: 3-year survival results from the Aberdeen trial. Clin Breast Cancer 2002; 3(Suppl 2):S69-S74.
34. Buzdar AU, Hunt KK, Smith T et al. Significantly higher pathological complete remission (PCR) rate following neoadjuvant therapy with trastuzumab (H), paclitaxel (P), and anthracycline-containing chemotherapy (CT): Initial results of a randomized trial in operable breast cancer (BC) with HER/2 positive disease. Proc Am Soc Clin Oncol 2004; 22(14S).
35. Ellis MJ, Rosen E, Dressman H et al. Neoadjuvant comparisons of aromatase inhibitors and tamoxifen: Pretreatment determinants of response and on-treatment effect. J Steroid Biochem Mol Biol 2003; 86:301-307.
36. Huober J, Krainick-Strobel U, Kurek R et al. Neoadjuvant endocrine therapy in primary breast cancer. Clin Breast Cancer 2004; 5:341-347.
37. Muggia FM. Primary chemotherapy: Concepts and issues. Prog Clin Biol Res 1985; 201:377-383.
38. Parkin DM, Bray F, Ferlay J et al. Estimating the world cancer burden. Globocan 2000. Int J Cancer 2001; 94:153-156.
39. Cardoso F, Di Leo A, Lohrisch C et al. Second and subsequent lines of chemotherapy for metastatic breast cancer: What did we learn in the last two decades? Annals Oncol 2002; 3:197-207.
40. Thomas E, Holmes FA, Smith TL et al. The use of alternate, noncross-resistant adjuvant chemotherapy on the basis of pathologic response to a neoadjuvant doxorubicin-based regimen in women with operable breast cancer: Long-term results from a prospective randomized trial. J Clin Oncol 2004; 22:2294-2302.
41. Strumberg D, Nitiss JL, Rose A et al. Mutation of a conserved serine residue in a quinolone-resistant type II topoisomerase alters the enzyme-DNA and drug interactions. J Biol Chem 1999; 274:7292-7301.
42. Biedler JL, Riehm H. Cellular resistance to actinomycin D in Chinese hamster cells in vitro: Cross-resistance, radioautographic, and cytogenetic studies. Cancer Res 1970; 30:1174-1184.
43. Riordan JR, Ling V. Genetic and biochemical characterization of multidrug resistance. Pharmacol Ther 1985; 28:51-75.
44. Biedler JL, Riehm H. Cellular resistance to actinomycin D in Chinese hamster cells in vitro: Cross-resistance, radioautographic, and cytogenetic studies. Cancer Res 1970; 30:1174-1184.
45. Gros P, Croop J, Housman D. Mammalian multidrug resistance gene: Complete cDNA sequence indicates strong homology to bacterial transport proteins. Cell 1986; 47:371-380.
46. Volk EL, Rohde K, Rhee M et al. Methotrexate cross-resistance in a mitoxantrone-selected multidrug-resistant MCF7 breast cancer cell line is attributable to enhanced energy-dependent drug efflux. Cancer Res 2000; 60:3514-3521.
47. Izquierdo MA, Scheffer GL, Flens MJ et al. Relationship of LRP-human major vault protein to in vitro and clinical resistance to anticancer drugs. Cytotechnology 1996; 19:191-197.
48. Vassetzky YS, Alghisi GC, Gasser SM. DNA topoisomerase II mutations and resistance to anti-tumor drugs. Bioessays 1995; 17:767-774.
49. Fernandes DJ, Qiu J, Catapano CV. DNA topoisomerase II isozymes involved in anticancer drug action and resistance. Adv Enzyme Regul 1995; 35:265-281.
50. Chen YN, Mickley LA, Schwartz AM et al. Characterization of adriamycin-resistant human breast cancer cells which display overexpression of a novel resistance-related membrane protein. J Biol Chem 1990; 265:10073-10080.
51. Safa AR, Glover CJ, Meyers MB et al. Vinblastine photoaffinity labeling of a high molecular weight surface membrane glycoprotein specific for multidrug-resistant cells. J Biol Chem 1986; 261:6137-6140.
52. Choi KH, Chen CJ, Kriegler M et al. An altered pattern of cross-resistance in multidrug-resistant human cells results from spontaneous mutations in the mdr1 (P-glycoprotein) gene. Cell 1988; 53:519-529.

53. Lee K, Belinsky MG, Bell DW et al. Isolation of MOAT-B, a widely expressed multidrug resistance-associated protein/canalicular multispecific organic anion transporter-related transporter. Cancer Res 1998; 58:2741-2747.
54. Zhan Z, Sandor VA, Gamelin E et al. Expression of the multidrug resistance-associated protein gene in refractory lymphoma: Quantitation by a validated polymerase chain reaction assay. Blood 1997; 89:3795-3800.
55. Zhang H, D'Arpa P, Liu LF. A model for tumor cell killing by topoisomerase poisons. Cancer Cell 1990; 2:23-27.
56. Hochhauser D, Harris AL. The role of topoisomerase II alpha and beta in drug resistance. Cancer Treat Rev 1993; 19:181-194.
57. Drake FH, Hofmann GA, Bartus HF et al. Biochemical and pharmacological properties of p170 and p180 forms of topoisomerase II. Biochemistry 1989; 28:8154-8160.
58. Larsen AK, Skladanowski A. Cellular resistance to topoisomerase-targeted drugs: From drug uptake to cell death. Biochim Biophys Acta 1998; 1400:257-274.
59. Larsen AK, Skladanowski A. Cellular resistance to topoisomerase-targeted drugs: From drug uptake to cell death. Biochim Biophys Acta 1998; 1400:257-274.
60. Matsumoto Y, Takano H, Fojo T. Cellular adaptation to drug exposure: Evolution of the drug-resistant phenotype. Cancer Res 1997; 57:5086-5092.
61. Liu LF. DNA topoisomerase poisons as antitumor drugs. Annu Rev Biochem 1989; 58:351-375.
62. Giovanella BC, Stehlin JS, Wall ME et al. DNA topoisomerase I—targeted chemotherapy of human colon cancer in xenografts. Science 1989; 246:1046-1048.
63. Mannervik B, Danielson UH. Glutathione transferases—structure and catalytic activity. CRC Crit Rev Biochem 1988; 23:283-337.
64. Hayes JD, Pulford DJ. The glutathione S-transferase supergene family: Regulation of GST and the contribution of the isoenzymes to cancer chemoprotection and drug resistance. Crit Rev Biochem Mol Biol 1995; 30:445-600.
65. Brix LA, Nicoll R, Zhu X et al. Structural and functional characterisation of human sulfotransferases. Chem Biol Interact 1998; 109:123-127.
66. Ivy SP, Tulpule A, Fairchild CR et al. Altered regulation of P-450IA1 expression in a multidrug-resistant MCF-7 human breast cancer cell line. J Biol Chem 1988; 263:19119-19125.
67. Thorgeirsson SS, Huber BE, Sorrell S et al. Expression of the multidrug-resistant gene in hepatocarcinogenesis and regenerating rat liver. Science 1987; 236:1120-1122.
68. Schecter RL, Alaoui-Jamali MA, Woo A et al. Expression of a rat glutathione-S-transferase complementary DNA in rat mammary carcinoma cells: Impact upon alkylator-induced toxicity. Cancer Res 1993; 53:4900-4906.
69. Leyland-Jones BR, Townsend AJ, Tu CP et al. Antineoplastic drug sensitivity of human MCF-7 breast cancer cells stably transfected with a human alpha class glutathione S-transferase gene. Cancer Res 1991; 51:587-594.
70. Berhane K, Hao XY, Egyhazi S et al. Contribution of glutathione transferase M3-3 to 1,3-bis(2-chloroethyl)-1-nitrosourea resistance in a human nonsmall cell lung cancer cell line. Cancer Res 1993; 53:4257-4261.
71. Awasthi S, Singhal SS, Srivastava SK et al. Adenosine triphosphate-dependent transport of doxorubicin, daunomycin, and vinblastine in human tissues by a mechanism distinct from the P-glycoprotein. J Clin Invest 1994; 93:958-965.
72. Jedlitschky G, Leier I, Buchholz U et al. ATP-dependent transport of glutathione S-conjugates by the multidrug resistance-associated protein. Cancer Res 1994; 54:4833-4836.
73. Muller M, Meijer C, Zaman GJ et al. Overexpression of the gene encoding the multidrug resistance-associated protein results in increased ATP-dependent glutathione S-conjugate transport. Proc Natl Acad Sci USA 1994; 91:13033-13037.
74. Sinha BK, Mimnaugh EG, Rajagopalan S et al. Adriamycin activation and oxygen free radical formation in human breast tumor cells: Protective role of glutathione peroxidase in adriamycin resistance. Cancer Res 1989; 49:3844-3848.
75. Sinha BK. Free radicals in anticancer drug pharmacology. Chem Biol Interact 1989; 69:293-317.
76. Hall AG. Review: The role of glutathione in the regulation of apoptosis. Eur J Clin Invest 1999; 29:238-245.
77. Bellamy CO. p53 and apoptosis. Br Med Bull 1997; 53:522-538.
78. Reed JC, Miyashita T, Takayama S et al. BCL-2 family proteins: Regulators of cell death involved in the pathogenesis of cancer and resistance to therapy. Journal of Cellular Biochemistry 1996; 60:23-32.
79. Reed JC. Bcl-2 family proteins: Regulators of apoptosis and chemoresistance in hematologic malignancies. Semin Hematol 1997; 34:9-19.
80. Poplack DG, Reaman G. Acute lymphoblastic leukemia in childhood. Pediatr Clin North Am 1988; 35:903-932.

81. Chen G, Jaffrezou JP, Fleming WH et al. Prevalence of multidrug resistance related to activation of the mdr1 gene in human sarcoma mutants derived by single-step doxorubicin selection. Cancer Res 1994; 54:4980-4987.
82. Laredo J, Huynh A, Muller C et al. Effect of the protein kinase C inhibitor staurosporine on chemosensitivity to daunorubicin of normal and leukemic fresh myeloid cells. Blood 1994; 84:229-237.
83. van der Kolk DM, de Vries EG, Koning JA et al. Activity and expression of the multidrug resistance proteins MRP1 and MRP2 in acute myeloid leukemia cells, tumor cell lines, and normal hematopoietic CD34+ peripheral blood cells. Clin Cancer Res 1998; 4:1727-1736.
84. Chernov MV, Stark GR. The p53 activation and apoptosis induced by DNA damage are reversibly inhibited by salicylate. Oncogene 1997; 14:2503-2510.
85. Tanizawa A, Kubota M, Takimoto T et al. Prevention of adriamycin-induced interphase death by 3-aminobenzamide and nicotinamide in a human promyelocytic leukemia cell line. Biochem Biophys Res Commun 1987; 144:1031-1036.
86. Larsen AK, Skladanowski A. Cellular resistance to topoisomerase-targeted drugs: From drug uptake to cell death. Biochim Biophys Acta 1998; 1400:257-274.
87. Withoff S, de Jong S, de Vries EG et al. Human DNA topoisomerase II: Biochemistry and role in chemotherapy resistance (review). Anticancer Res 1996; 16:1867-1880.
88. Withoff S, de Jong S, de Vries EG et al. Human DNA topoisomerase II: Biochemistry and role in chemotherapy resistance (review). Anticancer Res 1996; 16:1867-1880.
89. Finlay GJ, Baguley BC, Snow K et al. Multiple patterns of resistance of human leukemia cell sublines to amsacrine analogues. J Natl Cancer Inst 1990; 82:662-667.
90. Seidman AD, Reichman BS, Crown JP et al. Paclitaxel as second and subsequent therapy for metastatic breast cancer: Activity independent of prior anthracycline response. J Clin Oncol 1995; 13:1152-1159.
91. Wilson WH, Berg SL, Bryant G et al. Paclitaxel in doxorubicin-refractory or mitoxantrone-refractory breast cancer: A phase I/II trial of 96-hour infusion. J Clin Oncol 1994; 12:1621-1629.
92. Anderson H, Hopwood P, Prendiville J et al. A randomised study of bolus vs continuous pump infusion of ifosfamide and doxorubicin with oral etoposide for small cell lung cancer. Br J Cancer 1993; 67:1385-1390.
93. Wilson L, Miller HP, Farrell KW et al. Taxol stabilization of microtubules in vitro: Dynamics of tubulin addition and loss at opposite microtubule ends. Biochemistry 1985; 24:5254-5262.
94. Sparreboom A, van Tellingen O, Nooijen WJ et al. Preclinical pharmacokinetics of paclitaxel and docetaxel. Anticancer Drugs 1998; 9:1-17.
95. Greenberger LM, Williams SS, Horwitz SB. Biosynthesis of heterogeneous forms of multidrug resistance-associated glycoproteins. J Biol Chem 1987; 262:13685-13689.
96. Tolcher AW, Cowan KH, Solomon D et al. Phase I crossover study of paclitaxel with r-verapamil in patients with metastatic breast cancer. J Clin Oncol 1996; 14:1173-1184.
97. Horwitz SB, Lothstein L, Manfredi JJ et al. Taxol: Mechanisms of action and resistance. Annals of the New York Academy of Sciences 1986; 466:733-744.
98. Torres K, Horwitz SB. Mechanisms of Taxol-induced cell death are concentration dependent. Cancer Research 1998; 58:3620-3626.
99. Allegra CJ, Chabner BA, Drake JC et al. Enhanced inhibition of thymidylate synthase by methotrexate polyglutamates. J Biol Chem 1985; 260:9720-9726.
100. Priest DG, Ledford BE, Doig MT. Increased thymidylate synthetase in 5-fluorodeoxyuridine resistant cultured hepatoma cells. Biochem Pharmacol 1980; 29:1549-1553.
101. Bapat AR, Zarow C, Danenberg PV. Human leukemic cells resistant to 5-fluoro-2'-deoxyuridine contain a thymidylate synthetase with lower affinity for nucleotides. J Biol Chem 1983; 258:4130-4136.
102. Hsueh CT, Dolnick BJ. Regulation of folate-binding protein gene expression by DNA methylation in methotrexate-resistant KB cells. Biochem Pharmacol 1994; 47:1019-1027.
103. Cowan KH, Jolivet J. A methotrexate-resistant human breast cancer cell line with multiple defects, including diminished formation of methotrexate polyglutamates. J Biol Chem 1984; 259:10793-10800.
104. Grant SC, Kris MG, Young CW et al. Edatrexate, an antifolate with antitumor activity: A review. [Review]. Cancer Invest 1993; 11:36-45.
105. Sirotnak FM, Moccio DM, Kelleher LE et al. Relative frequency and kinetic properties of transport-defective phenotypes among methotrexate-resistant L1210 clonal cell lines derived in vivo. Cancer Res 1981; 41:4447-4452.
106. Dixon KH, Lanpher BC, Chiu J et al. A novel cDNA restores reduced folate carrier activity and methotrexate sensitivity to transport deficient cells. J Biol Chem 1994; 269:17-20.

107. Chung KN, Saikawa Y, Paik TH et al. Stable transfectants of human MCF-7 breast cancer cells with increased levels of the human folate receptor exhibit an increased sensitivity to antifolates. J Clin Invest 1993; 91:1289-1294.
108. Kool M, van der LM, de Haas M et al. MRP3, an organic anion transporter able to transport anti-cancer drugs. Proc Natl Acad Sci USA 1999; 96:6914-6919.
109. Hooijberg JH, Broxterman HJ, Scheffer GL et al. Potent interaction of flavopiridol with MRP1. British Journal of Cancer 1999; 81:269-276.
110. Cowan KH, Jolivet J. A methotrexate-resistant human breast cancer cell line with multiple defects, including diminished formation of methotrexate polyglutamates. J Biol Chem 1984; 259:10793-10800.
111. Volk EL, Rohde K, Rhee M et al. Methotrexate cross-resistance in a mitoxantrone-selected multidrug-resistant MCF7 breast cancer cell line is attributable to enhanced energy-dependent drug efflux. Cancer Res 2000; 60:3514-3521.
112. Rhee MS, Wang Y, Nair MG et al. Acquisition of resistance to antifolates caused by enhanced gamma-glutamyl hydrolase activity. Cancer Res 1993; 53:2227-2230.
113. Spears CP. Clinical resistance to antimetabolites. Hematol Oncol Clin North Am 1995; 9:397-413.
114. Ackland SP, Schilsky RL. High-dose methotrexate: A critical reappraisal. J Clin Oncol 1987; 5:2017-2031.
115. Kamen BA, Winick NJ. High dose methotrexate therapy: Insecure rationale? Biochem Pharmacol 1988; 37:2713-2715.
116. Jackson RC, Jackman AL, Calvert AH. Biochemical effects of a quinazoline inhibitor of thymidylate synthetase, N-(4-(N-((2-amino-4-hydroxy-6-quinazolinyl)methyl)prop-2-ynylamino) benzoyl)-L-glutamic acid (CB3717), on human lymphoblastoid cells. Biochem Pharmacol 1983; 32:3783-3790.
117. Beardsley GP, Moroson BA, Taylor EC et al. A new folate antimetabolite, 5,10-dideaza-5, 6,7,8-tetrahydrofolate is a potent inhibitor of de novo purine synthesis. J Biol Chem 1989; 264:328-333.
118. Elias L, Crissman HA. Interferon effects upon the adenocarcinoma 38 and HL-60 cell lines: Antiproliferative responses and synergistic interactions with halogenated pyrimidine antimetabolites. Cancer Res 1988; 48:4868-4873.
119. Auerbach M, Elias EG, Orford J. Experience with methotrexate, 5-fluorouracil, and leucovorin (MFL): A first line effective, minimally toxic regimen for metastatic breast cancer. Cancer Invest 2002; 20:24-28.
120. Klatt O, Stehlin Jr JS, McBride C et al. The effect of nitrogen mustard treatment on the deoxyribonucleic acid of sensitive and resistant Ehrlich tumor cells. Cancer Res 1969; 29:286-290.
121. Nogae I, Kohno K, Kikuchi J et al. Analysis of structural features of dihydropyridine analogs needed to reverse multidrug resistance and to inhibit photoaffinity labeling of P-glycoprotein. Biochem Pharmacol 1989; 38:519-527.
122. Zamble DB, Lippard SJ. Cisplatin and DNA repair in cancer chemotherapy. Trends Biochem Sci 1995; 20:435-439.
123. Fairchild CR, Moscow JA, O'Brien EE et al. Multidrug resistance in cells transfected with human genes encoding a variant P-glycoprotein and glutathione S-transferase-pi. Mol Pharmacol 1990; 37:801-809.
124. Townsend AJ, Tu CP, Cowan KH. Expression of human mu or alpha class glutathione S-transferases in stably transfected human MCF-7 breast cancer cells: Effect on cellular sensitivity to cytotoxic agents. Mol Pharmacol 1992; 41:230-236.
125. Hilton J. Deoxyribonucleic acid crosslinking by 4-hydroperoxycyclophosphamide in cyclophosphamide-sensitive and -Resistant L1210 cells. Biochem Pharmacol 1984; 33:1867-1872.
126. Perez RP. Cellular and molecular determinants of cisplatin resistance. Eur J Cancer 1998; 34:1535-1542.
127. Perez RP. Cellular and molecular determinants of cisplatin resistance. Eur J Cancer 1998; 34:1535-1542.
128. Mackey JR, Mani RS, Selner M et al. Functional nucleoside transporters are required for gemcitabine influx and manifestation of toxicity in cancer cell lines. Cancer Res 1998; 58:4349-4357.
129. Mackey JR, Mani RS, Selner M et al. Functional nucleoside transporters are required for gemcitabine influx and manifestation of toxicity in cancer cell lines. Cancer Res 1998; 58:4349-4357.
130. Ruiz VHV, Veerman G, Eriksson S et al. Development and molecular characterization of a 2',2'-difluorodeoxycytidine-resistant variant of the human ovarian carcinoma cell line A2780. Cancer Res 1994; 54:4138-4143.
131. Goan YG, Zhou B, Hu E et al. Overexpression of ribonucleotide reductase as a mechanism of resistance to 2,2-difluorodeoxycytidine in the human KB cancer cell line. Cancer Res 1999; 59:4204-4207.

132. Bergman AM, Pinedo HM, Jongsma AP et al. Decreased resistance to gemcitabine (2',2'-difluorodeoxycitidine) of cytosine arabinoside-resistant myeloblastic murine and rat leukemia cell lines: Role of altered activity and substrate specificity of deoxycytidine kinase. Biochem Pharmacol 1999; 57:397-406.
133. Schirmer M, Stegmann AP, Geisen F et al. Lack of cross-resistance with gemcitabine and cytarabine in cladribine-resistant HL60 cells with elevated 5'-nucleotidase activity. Exp Hematol 1998; 26:1223-1228.
134. Abbruzzese JL, Grunewald R, Weeks EA et al. A phase I clinical, plasma, and cellular pharmacology study of gemcitabine. J Clin Oncol 1991; 9:491-498.
135. Roodi N, Bailey LR, Kao WY et al. Estrogen receptor gene analysis in estrogen receptor-positive and receptor-negative primary breast cancer. J Natl Cancer Inst 1995; 87:446-451.
136. Gottardis MM, Jordan VC. Development of tamoxifen-stimulated growth of MCF-7 tumors in athymic mice after long-term antiestrogen administration. Cancer Res 1988; 48:5183-5187.
137. Pietras RJ, Arboleda J, Reese DM et al. Her-2 tyrosine kinase pathway targets estrogen receptor and promotes hormone-independent growth in human breast cancer cells. Oncogene 1995; 10:2435-2446.
138. Gottardis MM, Martin MK, Jordan VC. Long-term tamoxifen therapy to control transplanted human breast tumor growth in athymic mice. In: Salmon SE, ed. Adjuvant Therapy of Cancer V. Orlando: 1987:447-453.
139. Lavinsky RM, Jepsen K, Heinzel T et al. Diverse signaling pathways modulate nuclear receptor recruitment of N-CoR and SMRT complexes. Proc Natl Acad Sci USA 1998; 95:2920-2925.
140. Dauvois S, White R, Parker MG. The antiestrogen ICI 182780 disrupts estrogen receptor nucleocytoplasmic shuttling. J Cell Sci 1993; 106(Pt 4):1377-1388.
141. Parker MG, Arbuckle N, Dauvois S et al. Structure and function of the estrogen receptor. Ann NY Acad Sci 1993; 684:119-126.
142. Baselga J, Norton L, Albanell J et al. Recombinant humanized anti-HER2 antibody (Herceptin) enhances the antitumor activity of paclitaxel and doxorubicin against HER2/neu overexpressing human breast cancer xenografts. Cancer Research 1998; 58:2825-2831.
143. Sliwkowski MX, Lofgren JA, Lewis GD et al. Nonclinical studies addressing the mechanism of action of trastuzumab (Herceptin). Semin Oncol 1999; 26:60-70.
144. Lewis GD, Figari I, Fendly B et al. Differential responses of human tumor cell lines to anti-p185HER2 monoclonal antibodies. Cancer Immunol Immunother 1993; 37:255-263.
145. Pegram MD, Baly D, Wirth C et al. Antibody dependent cell-mediated cytotoxicity in breast cancer patients in Phase III clinical trials of a humanized anti-HER2 antibody. Proceedings of the American Association for Cancer Research 1997; 38:602.
146. Pegram M, Hsu S, Lewis G et al. Inhibitory effects of combinations of HER-2/neu antibody and chemotherapeutic agents used for treatment of human breast cancers. Oncogene 1999; 18:2241-2251.
147. Argiris A, Wang CX, Whalen SG et al. Synergistic interactions between tamoxifen and trastuzumab (Herceptin). Clin Cancer Res 2004; 10:1409-1420.
148. Sachdev D, Yee D. The IGF system and breast cancer. <[11] Journal Name> 2001; 8:197-209.
149. Miller KD. The role of ErbB inhibitors in trastuzumab resistance. Oncologist 2004; 9(Suppl 3):16-19.
150. Moulder SL, Yakes FM, Muthuswamy SK et al. Epidermal growth factor receptor (HER1) tyrosine kinase inhibitor ZD1839 (Iressa) inhibits HER2/neu (erbB2)-overexpressing breast cancer cells in vitro and in vivo. Cancer Res 2001; 61:8887-8895.
151. Esteva FJ. Monoclonal antibodies, small molecules, and vaccines in the treatment of breast cancer. Oncologist 2004; 9(Suppl 3):4-9.
152. Chernicky CL, Tan H, Yi L et al. Treatment of murine breast cancer cells with antisense RNA to the type I insulin-like growth factor receptor decreases the level of plasminogen activator transcripts, inhibits cell growth in vitro, and reduces tumorigenesis in vivo. Mol Pathol 2002; 55:102-109.
153. Lu YH, Zi XL, Zhao YH et al. Insulin-like growth factor-I receptor signaling and resistance to trastuzumab (Herceptin). J Natl Cancer Inst 2001; 93:1852-1857.
154. Nagata Y, Lan KH, Zhou X et al. PTEN activation contributes to tumor inhibition by trastuzumab, and loss of PTEN predicts trastuzumab resistance in patients. Cancer Cell 2004; 6:117-127.
155. Schrag D, Garewal HS, Burstein HJ et al. American society of clinical oncology technology assessment: Chemotherapy sensitivity and resistance assays. J Clin Oncol 2004; 22:3631-3638.
156. Xu JM, Song ST, Tang ZM et al. Predictive chemotherapy of advanced breast cancer directed by MTT assay in vitro. Breast Cancer Res Treat 1999; 53:77-85.
157. Symmans WF, Ayers M, Clark EA et al. Total RNA yield and microarray gene expression profiles from fine-needle aspiration biopsy and coreneedle biopsy samples of breast carcinoma. Cancer 2003; 97:2960-71.
158. Chang JC, Wooten EC, Tsimelzon A et al. Gene expression profiling for the prediction of therapeutic response to docetaxel in patients with breast cancer. Lancet 2003; 2(362):362-9.

Chapter 2

Roles of Multidrug Resistance Genes in Breast Cancer Chemoresistance

M. Tien Kuo*

Abstract

ATP binding cassette (ABC)-containing drug efflux transporters play important roles in regulating intracellular drug concentrations that determine cell sensitivity to chemotherapeutic agents. Of particular relevance to cancer chemotherapy are the transporters P-glycoprotein (Pgp) encoded by multidrug resistance 1 gene, multidrug resistance protein (MRP), and breast cancer resistance protein (BCRP). More than 80% of currently used antitumor agents can be transported by these three transporters, and overexpression of these transporters renders multidrug resistance to a broad spectrum of antitumor agents. Elevated expression of these transporters is frequently found in breast cancers and correlations with elevated expression of Pgp or MRP1 to chemotherapeutic outcomes have been observed in some cases, suggesting that these transporters may contribute to chemoresistance in breast cancers. However, attempts to modulate the activities of these transporters using reversal agents have met with limited success. Future studies should focus on better understanding of the upregulation mechanisms of ABC transporter genes in breast cancers, and of the pharmacologic mechanisms of transporter-reversal agent interactions. These studies may lead to novel strategies for improving chemotherapeutic efficacies through targeted interventions of these ABC transporters.

Introduction

Breast cancer is a major health threat to women worldwide. One in every ten new cancers diagnosed each year is female breast cancer. It is also the principal cause of cancer-related death in women.[1] In United States alone, breast cancer is estimated to account for 32% (215,900) of all new cancer cases among women in year 2004, making breast cancer the leader among the 10 top cancer types.[2] Primary breast cancers without distant spread are highly curable with local or regional treatment. However, most women with primary breast cancer have subclinical metastases and eventually develop distant metastases that complicates the curability of the disease.

Over the past several decades, breast cancer survival rates have significantly improved.[3] While many factors are credited, including the development of early detection methods, this improvement can be attributed to the development of new treatment modalities and new drugs. Regimens based on anthracyclines (doxorubicin, daunomycin, and epirubicin)[4] and taxanes (paclitaxel and docetaxel)[5,6] are the most frequently used combination therapy for breast cancers. However, the response rates remain suboptimal. Moreover, few effective

*M. Tien Kuo—Department of Molecular Pathology, Unit 89, The University of Texas M.D. Anderson Cancer Center, 1515 Holcombe Blvd., Houston, Texas 77030, U.S.A.
Email: tkuo@mdanderson.org

Breast Cancer Chemosensitivity, edited by Dihua Yu and Mien-Chie Hung.

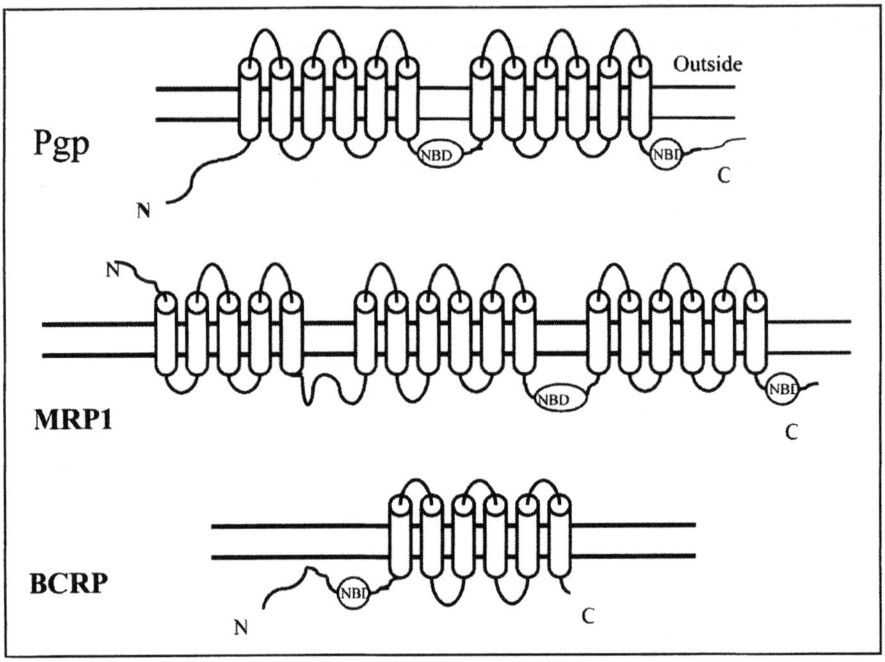

Figure 1. Predicted secondary structures of ABC transporters. Three classes of transporters are presented here each contains multiple transmembrane domain (TMD)(cylinders) and intracellularly located nucleotide-binding domain (NBD). Pgp, MRP1 and BCRP have 12, 17, and 6 TMD, respectively. MDR1/Pgp and MRP1 have two NBD whereas BCRP has only one. N and C refer to the amino- and carboxyl-terminal ends of the molecules, respectively.

therapeutic regimens are available to treat those had been exposed to anthracyclines or failed to anthracycline treatments.[7,8] These observations underscore the importance of multidrug resistance in breast cancer chemotherapy.

One important strategy by which cancer cells acquire drug resistance is the overexpressing drug transporters through which intracellular drug contents reduce to sublethal levels. Of particular importance are ATP-binding cassette (ABC) transporters.

Forty-eight ABC proteins, grouped into seven subfamilies ranging from A to G, are encoded by the human genome (see http://nutrigene.4t.com/humanabc.htm), but only about a dozen are associated with resistance to chemotherapeutic agents. The first ABC transporter known to be associated with multidrug resistance to chemotherapeutic agents was identified about three decades ago as P-glycoprotein (MDR1/Pgp, ABCB1).[9] The later realization that MDR1/Pgp alone could not account for all the MDR in many independently established multidrug resistance cells led to the discoveries of other drug resistance-related transporters, notably multidrug resistance (-acssociated) protein (MRP1, ABCC1)[10] and breast cancer resistance protein (BCRP, ABCG2).[11] These ABC transporters contain multiple transmembrane domains (TMD) and intracellularly localized nucleotide-binding domains (NBD) (Fig. 1). These transporters function as efflux pumps by eliminating a diverse array of structurally dissimilar compounds. Because many antitumor agents used in current breast cancer chemotherapy are substrates of these ABC transporters and because these ABC transporters are frequently overexpressed in breast cancers, it is relevant to discuss their roles in breast cancer chemoresistance. Because of space limitation, this review can only briefly describes MDR1, MRP1 and BCRP, and evaluates their roles in breast cancer.

Table I. Properties of some ABC transporters

ABC Transporter	Resistance Spectrum*	Reversal Agents*	Important Endogenous* Substrates	Major Tissue Expression*
MDR1 (ABCB1)	¶Anth. Vinc,	Vera, Cycl,	Phospholipid	blood-brain barrier,
MDR2 (ABCB2)	Etop, Taxa, Colc	GF120918,		adrenal gland, liver
MRP1 (ABCC1)	Anth. Vinc. Etop. Topo,		LTC4	ubiquitous
MRP2 (ABCC2)	Anth. Vinc. Etop. Camp.		Bilirubin glucuronide	liver, kidney
MRP3 (ABCC3)	Etop, MTX		$E_2$17G, LTC4	liver, intestine, pancreas, kidney
BCRP (ABCG2)	Dox, Mitoxantrone, Anth, ST1571, Topo	GF120918, Grfitnib	Porphyrin, heme	placenta, hemapoietic stem cells, kidney liver

* The list is not meant to be complete, just representatives are included. ¶ Anth, anthracyclines; Camp, Camptotechin; Colc, Colcine; Cycl, cylosporin A; Etop, Etoposine; MTX, methotrexate; Mito, Mitoxantrone; Topo, Topotechan; Vera, Verpamil; Vinc, Vinca;

General Descriptions of MDR1/Pgp, MRP1, and BRCP

MDR1/Pgp

The *MDR1*-encoded Pgp is responsible for multidrug resistance in cultured cells exposed to antitumor agents, including doxorubicin, vincristine, and taxanes, etoposide, teniposide, Actinomycin D.[12-14] Many of these agents are used to treat breast cancers. Although structurally dissimilar, they are generally hydrophobic and therefore readily to interact with cytoplasmic membrane. It is believed that these agents enter the cells through passive diffusion and subsequently evicted by Pgp through a drug concentration gradient across the membrane. How Pgp transports such structurally diverse substrates has been a challenging topic to structural biologists and pharmacologists alike. While X-ray crystallographic information of Pgp is not available, crystallographic determinations of a bacterial homolog of multidrug transporter MsbA have been instrumental in elucidating the transport mechanism of Pgp.[15] This information, together with biochemical studies which have identified several drug binding sites on various TMD of Pgp,[14] suggest that the initial event of Pgp-mediated drug transport is substrate binding, resulting in conformational changes that bring the two NBD into cross proximity to facilitate ATP binding. Mutation analyses have demonstrated that both NBD are required for transporter activity. Nucleotide binding and subsequent ATP hydrolysis provide the needed energy for releasing the substrate outward through the multi-TMD forming pore.[14,16] However, much of the complex dynamic and vectorial processes involved in the Pgp-mediated drug transport remains to be learned.

Humans have two *MDR* genes, *MDR1* and *MDR2.* Only *MDR1*-encoded Pgp functions as drug transporter. *MDR2*-encoded Pgp functions as a phospholipid transporter. MDR1/Pgp is expressed in many normal tissues, including liver, kidney, small intestine, colon, adrenal gland, and blood-brain barrier, whereas MDR2/Pgp is expressed mainly in the liver (Table 1). Mice have three *mdr* genes, two of which (*mdr1a* and *mdr1b*) are drug transporters, whereas the third (*mdr2*) has a similar function as human *MDR1*. The endogenous substrates for MDR1, mdr1a, mdr1b are not known. Mice without *mdr1a* (-/-) or both *mdr1a* (-/-) *mdr1b*(-/-) alleles generated by knockout strategies are viable and fertile, suggesting that *mdr1a* and *mdr1b* are not essential for cell viability.[17,18] However, these animals exhibit elevated sensitive to cytotoxic effects upon challenged by cytotoxic agents. While the endogenous substrates for P-gp remains unknown, it is generally accepted that animals utilize this efflux pump to prevent xenotoxins from entering the body (intestine, colon) and to remove cytotoxic compounds once inside the body (liver, kidney, bone marrow, and brain).

MDR1/Pgp-mediated transport can be inhibited by the so-called MDR-reversal agents or Pgp blockers. Some inhibitors have been in clinical applications in attempts to block MDR in chemotherapy-resistant tumors that express elevated levels of Pgp. Agents, e.g., calcium channel blocker verapamil and immunosuppressive agent cyclosporin A, are themselves Pgp substrates. They act as competitive inhibitors to Pgp-mediated transport. Other inhibitors, e.g., PSC-833, GF120918, and LY335979 have been used in various stages of clinical trials.[13] The development of clinically applicable reversal agents are an important avenue in combating MDR in cancer chemotherapy.

If Pgp plays a role in cancer chemotherapy, its expression in tumor cells most likely is elevated. Indeed, Pgp expression levels are frequently elevated in many types of cancer. Understanding the upregulation mechanisms is of importance for modulating its expression. Most of our understanding on MDR1 regulation mechanisms are from cultured cell studies.[19] Upregulation of MDR1 in cultured cells can be at the transcriptional and/or posttranscriptional levels. Transcriptional regulation involves a host of basal transcriptional factors, e.g., NF-Y, SP1, Egr1, and ets-1. Moreover, MDR1 expression can be induced by various stress conditions, including UV, inflammation, carcinogens, hypoxia, and chemotherapeutic agents. We have demonstrated that induction of MDR1 by the carcinogen (2-acetylaminofluorene) is mediated by DNA sequence located at -6092 bp which contains a NF-kappaB binding site, through upstream signaling via phosphoinositide 3-kinase- Rac1-and NAD(P)H oxidase-AKT pathway.[20] Upregulation of MDR1 expression by chemotherapeutics is in part by posttranscriptional (enhanced mRNA stability). Posttranslational regulation phosphorylation is associated with enhanced MDR activity. Pgp phosphorylation can be regulated by PKCα which in turn is regulated by wild-type p53, a tumor suppressor protein.[21]

MRP

MRP1 was first identified in doxorubicin-resistant cells that did not express elevated levels of Pgp.[10] Like Pgp1, MRP1 contains two intracellularly localized NBD. However, unlike Pgp1, it contains 17 TMD[14,23,24] (Fig. 1). The function of the five extra TMD is not clear, they are apparently not essential for catalytic function, because deleting this domain did not compromise its activity.[22] Overexpression of MRP1 conferred resistance to a spectrum of antitumor agents that is similar, but not identical, to that of Pgp1. For example, while taxanes are good substrate for MDR1 Pgp but are poor for MRP1. Additionally, unlike Pgp1, MRP1-mediated efflux requires cofactors, glutathione (GSH), glucuronic acid or sulfate. Mice lacking *mrp*1 are viable and fertile but have a deficient imflammatory response to its mediator leucotriene LTC4, which is an endogenous substrate of MRP1.[25]

Since the discovery of MRP1, eight related sequences have been identified, i.e, MRP2 to MRP9.[23,24] MRP1 and MRP2 have similar substrate selectivity but the tissue expression profiles are quite different: MRP1 expression is rather ubiquitous, whereas MRP2 expression is restricted to liver and kidney. Hepatic MRP2 is involved in the hepatobiliary extrusion of bilirubin glucuronide and defected MRP2 is associated with Dubin-Johnson syndrom.[26]

While the structural organization of MRP3 is similar to those of MRP1 and MRP2, namely, it also possesses 17 TMD, but the substrate specificity of MRP3 is quite different from those of MRP1 and MRP2. MRP3-mediated transport does not require intracellular GSH. Etoposide appears to be transported by MRP3 in unmodified form,[27] whereas vincristine and doxorubicin which are transported by MRP1 and MRP2 through GSH conjugates are not transported by MRP3. It is important to note that cancer cells overexpress MRP3 are not resistant to anthracyclines which are important antitumor agents in breast cancer treatment.

MRP4 and MRP5 contain 12 TMD rather than 17 TMD, making them structurally more like MDR1 than does MRP1. MRP4[28] and MRP5 transport cyclic nucleotides and nucleotide analogs which are not transported by MRP1, MRP2, or MRP3. The roles of MRP4 and MRP5 in breast cancer chemoresistance are not known. MRP6, MRP7, MRP8, and MRP9 are newly cloned MRP gene family members. The substrate specificities and pharmacologic properties of these ABC transporters remain to be determined.

Like MDR1, expression of MRP1 in cultured cells can be induced by a variety of cytotoxic agents including prooxidants, heavy metals, antitumor agents, and nitric oxides. Because MRP1-mediated efflux requires GSH, intracellular GSH levels may play important roles in regulating the expression of MRP1. Biosynthesis of GSH is regulated by the rate-limiting enzyme γ-glutamylcysteine synthetase (γ-GCS), which consists of one heavy catalytic subunit (γ-GCSh) and one light (regulatory) subunit (γ-GCSl). Our laboratory has demonstrated that expression of γ-GCSh can be induced by many cytotoxic agents that also upregulate MRP1.[29] Moreover, expression of MRP1 and γ-GCSh is frequently upregualted in colorectal cancers.[30] These observations suggest that both genes may be regulated by the same mechanisms. Transcriptional regulation of *γ-GCSh* gene expression is mediated by an oxidative stress response element (ORE) located at -3802 bp which interacts with the leucine zipper transcription factor complex Nrf2/Maf.[31] However, no ORE element has yet been identified in the promoter of MRP1. Expression of *MRP3* can also be induced by prooxidants.[32] While MDR1, MPR1 and MRP3, like γ-GCSh, may be considered as a stress inducible ABC transporters, but because many different signaling pathways can be associated with stress-induced gene expression, regulation mechanisms of these genes may not be the same.

BCRP

The ABC transporter BCRP was first cloned from the doxorubicin-resistant MCF-7 breast cancer cell line,[11] but it is not implied that the expression is associated with breast cancer. This tranporter encodes only 655 amino acids, about one half of the sizes of MDR1 and MRP transporter (Fig. 1). It is likely that two half-molecules form a homodimer to function as a drug transporter.[33-35] Cell lines selected for resistance to many antitumor agents, including mitoxantrone, topotecan, doxorubicin, SN-38 exhibit MDR phenotype and overexpressed BCRP, suggesting an important role of BCRP in MDR development. BCRP-mediated transport apparently does not require GSH cofactor. Like *mdr1a*(-/-), *mdr1b*(-/-), and *mrp*(-/-) mice, *bcrp*(-/-) animals are fertile with no apparent phenotypic alterations as compared with those in the wild-type animals, suggesting that murine *bcrp* is not essential for normal animal physiology.[36] The fact that these individual knockout animals fail to display normal physiological abnormality also suggest that there are functional redundancy among these transporters.

Roles of Pgp, MRP1, and BCRP in Breast Cancer Chemotherapy

For an ABC transporter to play a role in reducing cancer chemoresistance, its expression levels should be inversely correlated with the chemosensitivity of antitumor agents that are known to be substrates of the transporter. In addition, an enhanced response to chemotherapy should be observed when inhibitors or reversal agents are used. These criteria are discussed here in the context of Pgp, MRP1 and BCRP in breast cancer chemotherapy. A review describing similar issues has recently been published.[37]

Expression of MDR1/Pgp in nonneoplastic breast tissue and in breast cancer tissues has been extensively investigated at mRNA and protein levels. MDR1 mRNA levels were mainly determined by using RT-PCR method and proteins levels were by immunohistochemical (IHC), flow cytometry, and western blot analyses. Agreement between IHC analyses and RT-PCR results were found in many studies, although inconsistent results were also found, perhaps because the levels of MDR1/Pgp regulation (transcriptional vs. prostranslcriptional regulation) vary in different patient population. The disparate results may also reflect differences in analytic methodologies, including tissue sampling (IHC analysis detects expression in tumor cells whereas RT-PCR may use a heterogeneous pool of cell types), the use of different probes (RT-PCR can be gene-specific, whereas some antibodies used for IHC can cross-react both

MDR1- and *MDR2*-encoded Pgp). It is therefore, careful evaluation of experimental designs is needed before results can be compared.[38]

Many studies aimed at determining the correlation between expression levels of MDR1/Pgp in various tumor types and responses to chemotherapy with antitumor agents that are substrates of MDR1/Pgp have been published. For breast cancer treatment, while some studies showed positive correlations between reduced Pgp expression levels and improved response rates[39,40] whereas other failed to find such a correlation.[40]

Expression of MRP1 is frequently observed in breast cancer even before chemotherapy, and chemotherapy has been reported to increased MRP1 expression.[41] A correlation between MRP1 expression and patient survival rates after chemotherapy has been noted in some studies.[41,42] whereas other reports showed no correlation between MRP1 expression and prognosis.[43-45]

Evaluation of BCRP expression in human cancers has most been performed in leukemia. And several studies have also been published for the expression of BCRP in breast cancers, mostly determined by using RT-PCR method. While levels of BCRP in AML patients are variable in some studies and the expression levels are increased in associated with relapsed/reflactory,[46] whereas other studies did not show correlation.[47] In breast cancer, expression levels of BCRP are low.[48] The role of BCRP in the chemoresistance of breast cancer remains to be investigated.

The fact that ABC transporter expression levels and resistance to chemotherapy are positively correlated in some, but not all, breast cancers may reflect differences in analytic methods, patient population, or the chemotherapeutic drugs used. The use of reversal agents for combating MDR1/Pgp-related clinical drug resistance began soon after the discovery of Pgp inhibitors in many types of cancer, including breast cancer. Verapamil was one of the very early discovered MDR1 reversal agent used in clinical trials. From a pool of four studies, verapamil appears to resensitize 15% of advanced breast cancer patients refractory to anthracycline-containing regimens.[18] However, this may not be beneficial because the response rate of the same patients to alternative second-line chemotherapy could achieve a better response. Reversal agents such as quinidine and biricodar in clinical trials have not shown evidence of benefits. These results suggest that expression levels of MDR1/Pgp1 levels are not readily for prognostic evaluation of drug sensitivities and thus for pharmacologic intervention for improving chemotherapeutic efficacies remain to be further developed.

The difficulties associated with clinical trials using reversal agents may be explained as follows: Reliably assessing the contribution of the overexpressed ABC transporter to drug resistance is difficult, even the transporters are overexpressed in the tumors. Not all the Pgp^- tumor respond to chemotherapeutics; and not all the Pgp^+ tumors are resistant to chemotherapy. Aside from the technical aspects in measuring expression levels, no studies have convincingly shown that high levels of transporter expression translate into high transporter activities. Another difficulty is that there is no direct evidence showing that reversal agents indeed downregulate the transporter activity at tumor sites. A third consideration is that multiple ABC transporters can pump the same antitumor substrate, and in many cases, overexpression of multiple transporters is found in tumors. Thus, inhibition of one or a few ABC transporters may not be sufficient to bring down drug resistance to therapeutic achievable levels. The functional redundancy of ABC transporter family proteins may then encourage the development of reversal agents that can simultaneously inhibit multiple transporters. For instance, some Pgp inhibitors such as cyclosporin A and PSC 833 also inhibit the function of MRP, albeit less effective, and another Pgp inhibitor GF129018 can also suppress the function of BCRP. Last but not the least, in clinical settings where combination chemotherapy is often used, multiple mechanisms may contribute to the overall response to chemotherapeutic agents. Inhibition of drug transport alone may be insufficient to overcome the overall drug resistance.

Conclusion

The discovery of ABC transporters associated with MDR phenotype in cultured cells revealed an important mechanism bywhich cancer cells acquire resistance to many chemotherapeutic agents. Much has been learned about how expression of these transporters, notably MDR1,

MRP1, and BCRP, in cultured cells confer resistance to antitumor agents. This knowledge holds great implications for clinical drug resistance. Overexpression of Pgp and MRP1 in some breast cancers has been correlated with chemoresistance in clinical setting in some studies but not in others. The disappointing results of clinical applications in using reversal agents suggest that more investigation is needed for translational gains in breast cancer chemotherapy. Future studies should focus on the molecular basis of how the expression of these transporters is regulated in normal breast cells and in their malignant counterparts. These studies may lead to novel strategies of controlling MDR through gene regulation. Another area of research may involve developing strategies for modulating transporter activities through better understanding pharmacodynamic and pharmacogenetic behaviors of reversal agents. In combination of advancing imaging systems, suppression of transporter activities in tumor sites can be measured. These studies may eventually lead to effective evaluation on the roles of these ABC transporters in breast cancer chemoresistance and the development of strategies of circumventing it.

Acknowledgements

Research in author's laboratory was supported in part by grants CA72404 and CA79085 from the National Cancer Institute.

References

1. Bray F, McCarron P, Parkin DM. The changing global patterns of female breast cancer incidence and mortality. Breast Cancer Res 2004; 6:229-239.
2. Jemal A, Tiwari RC, Murray T et al. Cancer statistics, 2004. CA cancer J Clin 2004; 54:8-29.
3. Giordano SH, Buzdar AU, Smith TL et al. Is breast cancer survival improving? Cancer 2004; 100:44-52.
4. Hortobagyi GN. Treatment of breast cancer. N Engl J Med 1998; 339:974-984.
5. Piccart M. The roles of taxenes in the adjuvant treatment of early breast cancers. Breast Cancer Res Treat 2003; 79:S25-S34.
6. Nabholtz JMA. Docetaxel-anthracycline combinations in metastatic brease cancers. Breast Cancer Res Treat 2003; 79:S3-S9.
7. Valero V, Holmes FA, Walters RS et al. Phase II trial of docetaxel: A new, highly effective antineoplastic agent in the management of patients with anthracycline-resistant metastatic breast cancer. J Clin Oncol 1995; 13:23886-2894.
8. Nabholtz JM, Gelmon K, Bontenbal M et al. Multicenter, randomized comparative study of two doses of paclitaxel in patients with metastatic breast cancer. J Clin Onco 1996; 14:1858-1867.
9. Juliano RL, Ling V. A surface glycoprotein modulating drug permeability in Chinese hamster ovary cell mutants. Biophys Acta Biochim 1975; 445:152-162.
10. Cole SP, Bhardway JH, Gerlach JE et al. Overexpression of a transporter gene in a multidrug-resistant human lung cancer cell line. Science 1992; 258:1650-1654.
11. Doyle LA, Yang W, Abruzzo LV et al. A multidrug resistance transporter from human MCF-7 breast cancer cells. Proc Natl Acad Sci USA 1998; 5:15665-15670.
12. Deng L, Tatebe S, Lin-Lee YC et al. MDR and MRP gene families as cellular determinant factors for resistance to clinical anticancer agents. Cancer Treat Res 2002; 112:49-66.
13. Schinkel AH, Jonker JW. Mammalian drug efflux transporters of the ATP binding cassette (ABC) family: An overview. Adv Drug Deliv Rev 2003; 55:3-29.
14. Borst P, Elferink RO. Mammalian ABC transporters in health and disease. Annu Rev Biochem 2002; 71:537-592.
15. Chang G, Roth CB. Structure of MsbA from E. coli: A homolog of the multidrug resistance ATP binding cassette (ABC) transporters. Science 2001; 293:1793-1800.
16. Higgins CF, Linton KJ. The ATP switch model for ABC transporter. Nature Struct Mol Biol 2004; 11:918-926.
17. Schinkel AH, Smit JJM, Van Tellingen O et al. Disruption of the mouse mdr1a P-glycoprotein gene leads to a deficiency in the blood-brain barrier and to increased sensitivity to drugs. Cell 1994; 77:491-502.
18. Schinkel AH, Mayer U, Wagenaar E et al. Normal viability and altered pharmacokinetics in mice lacking mdr1-type (drug-transporting) P-glycoproteins. Proc Natl Acad Sci USA 1997; 94:4028-4033.
19. Scotto KW. Transcriptional regulation of ABC drug transporters. Oncogene 2003; 22:7496-7511.
20. Kuo MT, Liu Z, Wei Y et al. Induction of human MDR1 gene expression by 2-acetylaminofluorene is mediated by effectors of the phosphoinositide 3-kinase pathway that activate NF-kappaB signaling. Oncogene 2002; 21:1945-1954.

21. Zhan M, Yu D, Lin J et al. Transcriptional repression of protein kinase Cα via Sp1 by wild-type p53 is involved in inhibition of MDR1 P-glycoprotein phosphorylation. J Biol Chem 2005; 280:4825-4833.
22. Bakos E, Evers R, Szakacs G et al. Functional multidrug resistance protein (MRP1) lacking the N-terminal transmembrane domain. J Biol Chem 1998; 273:32167-32175.
23. Kruh GD, Belinsky MG. The MRP family of drug efflux pumps. Oncogene 2003; 22:7537-7552.
24. Heimeur A, Conseil G, Deeley RG et al. The MRP-related and BCRP/ABCG2 multidrug resistance proteins: Biology, substrate specificity and regulation. Curr Drug Metab 2004; 5:21-53.
25. Wijnholds J, Evers R, van Leusden MR et al. Increased sensitivity to anticancer drugs and decreased inflammatory response in mice lacking the multidrug resistance-associated protein. Nature Med 1997; 3:1275-1279.
26. Keppler D, Kartenbeck J. The canalicular conjugate export pump encoded by the cmrp/cmoat gene. Prog Liver Dis 1996; 14:55-67.
27. Zelcer N, Saeki T, Reid G et al. Characterization of drug transport by the human mulltidrug resistance protein 3 (ABCC3). J Biol Chem 2001; 276:46400-46407.
28. Schuetz JD, Connelly MC, Sun D et al. MRP4: A previously unidentified factor in resistance to nucleoside-based antiviral drugs. Naure Med 1999; 5:1048-1051.
29. Yamane Y, Furuichi M, Song R et al. Expression of multidrug resistance protein/GS-X pump and gamma-glutamylcysteine synthetase genes is regulated by oxidative stress. J Biol Chem 1998; 273:31075-31085.
30. Kuo MT, Bao JJ, Curley SA et al. Frequent coordinated overexpression of the MRP/GS-X pump and gamma-glutamylcysteine synthetase genes in human colorectal cancers. Cancer Res 1996; 56:3642-3644.
31. Zipper LM, Mulcahy RT. The Keap1 BTB/POZ dimerization function is required to sequester Nrf2 in cytoplasm. J Biol Chem 2002; 277:36544-36552.
32. Lin-Lee YC, Tatebe S, Savaraj N et al. Differential sensitivities of the MRP gene family and gamma-glutamylcysteine synthetase to prooxidants in human colorectal carcinoma cell lines with different p53 status. Biochem Pharmacol 2001; 61:555-563.
33. Allen JD, Schinkel AH. Multidrug resistance and pharmacological protein mediated by the breast cancer resistance protein (BCRP/ABCG2). Mol Cancer Therap 2002; 1:427-434.
34. Doyle LA, Ross DD. Multidrug resistance mediated by the breast cancer resistance protein BCRP (ABCG2). Oncogene 2003; 22:7340-7358.
35. Bates SE, Robey R, Miyake K et al. The roles of half-transporters in multidrug resistance. J Bioenergetics and Biomem 2001; 33:503-517.
36. Jonker JW, Buitelaar M, Wagenaar E et al The breast cancer resistance protein protects against a major chlorophyll-derived dietary phototoxin and protoporphyria. Proc Natl Acad Sci USA 2002; 99:15649-15654.
37. Leonessa F, Clarke R. ATP binding cassette transporters and drug resistance in brease cancer. Endocrine-related Cancer 2003; 10:43-73.
38. Beck WT, Grogan TM, Willman CL et al. Methods to detect P-glycoprotein-associated multidrug resistance in patients' tumors: Consensus recommendations. Cancer Res 1996; 56:3010-3020.
39. Verrelle P, Meissonnier F, Fonck Y et al. Clinical relevance of immunohistochemical detection of multidrug resistance P-glycoprotein in breast carcinoma. J Nat Cancer Inst 1991; 83:111-116.
40. Chevillard S, Pouillart P, Beldjord C et al. Sequential assessment of multidrug resistance phenotype and measurement of S-phase fraction as predictive markers of breast cancer response to neoadjuvant chemotherapy. Cancer 1996; 77:292-300.
41. Rudas M, Filipits M, Taucher S et al. Expression of MRP1, LRP and Pgp in breast carcinoma patients treated with preoperative chemotherapy. Breast Cancer Res Treat 2003; 81:149-157.
42. Filipits M, Malayeri R, Suchomel RW et al. Expression of the multidrug resistance protein (MRP1) in breast cancer. Anticancer Res 1999; 19:5043-5049.
43. Ferrero JM, Etienne MC, Formento JL et al. Application of an original RT-PCR-ELISA multiplex assay for MDR1 and MRP, along with p53 determination in node-positive breast cancer patients. Br J Cancer 2000; 82:171-177.
44. Faneyte IF, Kristel PM, van de Vijver MJ. Multidurg resistance associated genes MRP1, MRP2 and MRP3 in primary and anthracycline exposed breast cancer. Anticancer Res 2004; 24:2931-2939.
45. Kanzaki A, Toi M, Nakayama K et al. Expression of multidrug resistance-related transporters in human breast carcinoma. Jpn J Cancer Res 2001; 92:452-458.
46. Van den Heuvel-Eibrink MM, Wiemer EA, Prins A et al. Leukemia. 2002; 16:833-839.
47. Sargent JM, Williamson CJ, Maliepaard M et al. Breast cancer resistance protein expression and resistance to daunorubicin in blast cells from patients with acute myeloid leukaemia. Brit J Haematol 2001; 115:257-262.
48. Faneyte IF, Kristel PM, Maliepaard M et al. Expression of the breast cancer resistance protein in breast cancer. Clinic Cancer Res 2002; 8:1068-1074.

CHAPTER 3

Therapy-Induced Apoptosis in Primary Tumors

David J. McConkey*

Abstract

An enormous body of literature has accumulated over the past 15 years implicating apoptosis (programmed cell death) in breast cancer cell death induced by conventional and investigational cancer therapies in preclinical models. As a result, new therapeutic approaches that directly target key components of apoptotic pathways are either entering or will soon enter clinical trials in patients, raising hopes that the information gained from the preclinical studies can be translated to improve patient care. However, there is a new appreciation for the fact that apoptosis is not the only relevant pathway that mediates physiological cell death, and many investigators are challenging the notion that targeting apoptosis is the best means of optimizing therapeutic efficacy in primary tumors. Here I will review some of the basic concepts that have emerged from the study of apoptosis in preclinical models, the evidence that apoptosis does or does not mediate the effects of current front line therapies in patients, and the new strategies that are emerging that are designed to more directly target apoptotic pathways.

Introduction

Components of the Core Apoptotic Machinery

Kerr, Wyllie, and Currie first coined the term "apoptosis" to describe a series of stereotyped morphological alterations they observed in cells undergoing diverse examples of physiological cell death.[1,2] A more biochemical definition of the response was advanced in 1980 with Wyllie's observation that the chromatin within apoptotic cells is fragmented in a regular pattern to produce the so-called "DNA ladders" that are usually associated this form of cell death.[3] However, it was not until Horvitz and his coworkers began to define the genes required for programmed cell death in the nematode *Caenorhabditis elegans*[4-8] that a true understanding of the molecular regulation of the process began to emerge. Their first studies established that two genes, termed ced-3 and ced-4, were required for all of the 131 cell deaths that occur during embryonic development of the organism,[4] and in subsequent work they identified two more (ced-9 and egl-1) that function upstream of ced-3 and ced-4 to control their activation.[5,7] These genes all have structural and functional homologues in higher organisms (Fig. 1), and a good deal is known about how they promote cell death. Ced-3 is the founding member of a family of aspartate-specific cysteine proteases termed caspases,[8,9] and ced-4 is an adaptor protein homologous to human amino-terminal activation function (AF-1) domain (for

*David J. McConkey—Department of Cancer Biology-173, University of Texas M.D. Anderson Cancer Center, 1515 Holcombe Boulevard, Houston, Texas 77030, U.S.A. Email: dmcconke@mdanderson.org

Breast Cancer Chemosensitivity, edited by Dihua Yu and Mien-Chie Hung.

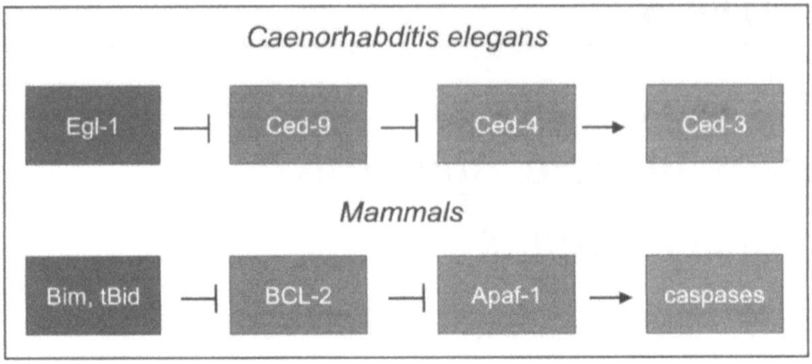

Figure 1. Control of caspase activation by Bcl-2 family polypeptides in the nematode *Caenorhabditis elegans* and mammals. In the nematode, activation of the cysteine protease ced-3 is facilitated by the adaptor protein ced-4, which is normally held in check by the ced-9 protein. Induction of EGL-1 leads to inhibition of ced-9, releasing ced-4 to promote ced-3 interaction. This pathway is structurally and functionally conserved in mammals, where caspases are the cyteine proteases, Apaf-1 is the caspase-activating adaptor protein, anti-apoptotic members of the Bcl-2 family inhibit Apaf-1 activation, and the BH3-only proteins trigger cell death in part by inhibiting Bcl-2 and its homologues.

"apoptosis protease activating factor-1")[10-12] that promotes ced-3 activation.[6] Ced-9 is homologous to human Bcl-2, and like Bcl-2 it inhibits caspase activation.[7] Conversely, Egl-1 is homologous to the BH3-only Bcl-2-like protein subfamily, and like mammalian BH3-only proteins it inhibits Ced-9 to promote caspase activation (Fig. 1).[5] Thus, it appears that the positive and negative regulators of apoptosis function to promote or inhibit caspase activation, respectively, leading most investigators to conclude that apoptosis is best defined by its dependency on caspase activation.[13] Studies of the effects of targeting Apaf-1[14,15] or key downstream caspases (3, 7 or 9)[16-19] in mice or chicken cells have largely supported this conclusion.

Investigations into the molecular control of apoptosis by viruses also converged with genetic studies in *Drosophila* to produce important information about another family of cell death regulators termed inhibitor of apoptosis proteins (IAPs). First identified by Lois Miller's group in baculoviruses,[20,21] IAPs contain structural elements termed baculovirus inhibitor of apoptosis repeats (BIR domains) that directly bind to caspases 3, 7, and 9 and inhibit their enzymatic activation and/or interaction with substrates.[22-24] Parallel studies in Drosophila identified three genes (reaper, hid, and grim) that play central roles in the regulation of programmed cell death in the organism[25-27] (Fig. 2), and subsequent functional studies demonstrated that all three contain short N-terminal peptide motifs that enable them to bind to a Drosophila IAP (DIAP1) to inhibit its functions.[24,28] Again, a structural homologue of these proteins has been identified in mammalian cells (termed second mitochondrial activator of caspases/SMAC in human cells[29,30] or DIABLO in murine cells) that also binds to IAPs (particularly XIAP) and promotes caspase activation when it is released from mitochondria during apoptosis (Fig. 2).[29-31]

Studies in primary tumor tissues have confirmed that alterations in the expression and/or function of key apoptotic regulatory proteins accumulate in human tumors. The best example of the role of Bcl-2 in tumorigenesis can be found in nonHodgkins B cell lymphomas, where the t(14;18) translocation juxtaposes the Eμ enhancer with the *bcl-2* coding sequence,[32] driving high level expression of Bcl-2 in B cell precursors.[33] Bcl-2 is also overexpressed in chronic lymphocytic leukemia (CLL)[34] and androgen-independent prostate cancer,[35] but its role in breast cancer is less clear. High level expression of Bcl-2 has been linked to therapeutic resistance,[36] but several studies have concluded that in untreated tumors Bcl-2 is expressed by well-differentiated, ER^+PR^+, $EGFR^-$, HER^-2^- tumors,[37-41] strongly suggesting that other molecular defect(s) in the control of apoptosis play more important roles in the later stages of tumor progression.[42-44]

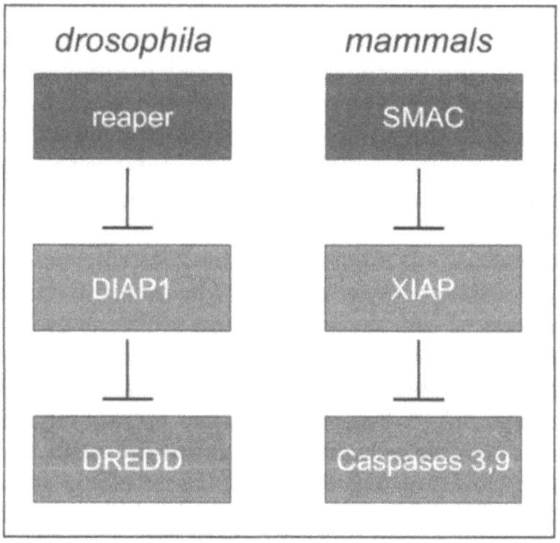

Figure 2. Control of caspase activation by inhibitor of apoptosis proteins (IAPs) in Drosophila and mammals. In Drosophila, developmental cell death is dependent upon induction of REAPER, a small polypeptide that contains a critical N-terminal motif that enables it to bind to and inhibit Drosophila inhibitor of apoptosis protein-1 (DIAP1). This releases the Drosophila caspase, DREDD, from its interaction with DIAP1 and allows it to trigger cell death. In mammals, caspases 3 and 9 are held in check by mammalian IAPs (particularly XIAP). This interaction is disrupted by SMAC (second mitochondrial activator of caspases), a mammalian REAPER homologue that is released from mitochondria with cytochrome c in response to Bid cleavage or other signals.

Alterations in the expression of IAPs may also promote breast cancer progression and resistance to adjuvant therapy. The first IAP to be implicated in tumor progression was survivin, which with one BIR domain is the smallest member of the family. Survivin is expressed at high levels in many human tumor cell lines[45] and its expression correlates with drug sensitivity in the NCI's panel of 60 human cancer cell lines.[46] Furthermore, one report demonstrated that four common human breast cancer cell lines displayed relatively high levels of caspase-3-like activity and processed (active) caspase-3 as compared to nontumorigenic normal mammary epithelial cells and that the viability of these cells was dependent on their coexpression of high levels of surviving and XIAP.[47] However, survivin expression has been correlated with expression of Bcl-2[48] and appears to be associated with a good prognosis.[49] On the other hand, baseline levels of XIAP have been linked to the levels of cleaved (active) caspases 3 and 6,[50] consistent with the results obtained with cell lines.[47] Therefore, a subset of human breast cancers may rely on continuous expression of XIAP for survival.

Current Approaches to Target Apoptosis

Effects of Conventional Chemotherapy and Radiation on Apoptosis in Preclinical Models

Some of the first evidence that cancer chemotherapy might work by inducing apoptosis in tumor target cells came from studies of the effects of glucocorticoids in primary hematological tumor cells and leukemic cell lines.[51,52] Soon after Wyllie demonstrated that glucocorticoids induce apoptosis in immature thymocytes, others showed that they also stimulate apoptosis in primary ALL and CLL cells as well as in ALL and lymphoma cell lines (CEM-C7, WEHI) in vitro.[51-55] This was followed by a wave of studies demonstrating that cancer chemotherapeutic agents such as cisplatin, etoposide, and nucleoside analogues induce apoptosis in human solid and hematological tumor cell lines in vitro.[56-58]

growth of MCF-7 tumors in vivo, and some of the early work established that the effects are associated with increased tumor cell death.[86] Nonetheless, their effects on the growth of experimental tumors are largely cytostatic,[88] which contrasts with the effects of conventional chemo- or radiotherapy, where tumor involution associated with large increases in apoptosis can be observed. There are certainly limitations associated with generating general conclusions from preclinical studies that primarily employ one cell line, and MCF-7 cells are unique among solid tumor lines because they express an inactive mutant form of caspase-3.[89] Clearly, some estrogen-dependent tumors may retain a dependency upon estrogen for their survival, and in these tumors ER antagonists like tamoxifen or aromatase inhibitors would be expected to induce tumor cell death and cause frank regression. However, it is also possible that most ER-positive tumor cells retain their sensitivity to estrogen for cell cycle progression but no longer depend on it for survival, and that the therapeutic benefit obtained from adjuvant hormonal therapy is mediated by cell cycle arrest.

The data emerging from neoadjuvant studies with SERMs and aromatase inhibitors strongly supports this conclusion. For example, in a recent study Dowsett and coworkers studied the effects of neoadjuvant anastrozole or tamoxifen alone or combined in a clinical trial involving 330 patients (the IMPACT trial).[90] Biopsies were collected before and 2 or 12 weeks after initiation of therapy for analysis of proliferation (by Ki-67 immunohistochemistry) and apoptosis (by TUNEL staining). Overall, the vast majority of tumors displayed reductions in proliferation at both 2 and 12 weeks. Furthermore, the effects of therapy on proliferation correlated well with the clinical responses observed within each arm of a larger adjuvant trial of the same design (the ATAC trial), where single-agent anastrozole outperformed either single-agent tamoxifen or the tamoxifen plus anastrozole combination.[91,92] In contrast, apoptosis was not increased in any of the treatment arms.[90] Instead, mean levels of apoptosis appeared to decrease slightly in all of the treatment arms at 2 weeks and in the anastrozole arm at 12 weeks, but only the latter reached statistical significance.[90]

These results may have important implications for best exploiting the biological effects of these drugs. Use of a cytostatic drug in the setting of bulky disease is unlikely to produce substantial tumor regression but would be expected to be quite effective in blocking the expansion of micrometastases and/or prevent recurrence. Furthermore, given the relationship between cell proliferation and apoptosis sensitivity discussed above, combining a cytostatic drug with a cytotoxic one may not be the best means of exploiting the pro-apoptotic effects of the latter. This may be especially true in the case of combinations involving taxanes, which appear to kill cells via a cyclin-dependent kinase-dependent mechanism only after they have been arrested in mitosis. It would probably be better to combine them with agent(s) such as TRAIL that promote apoptosis most effectively in growth-arrested cells, as will be discussed below.

Effects of Growth Factor Receptor Inhibitors on Apoptosis

Recent studies have implicated members of the erbB family of growth factor receptors in breast cancer progression and the development of resistance to SERMs and aromatase inhibitors.[93] The gene encoding erbB2/HER-2 is amplified in approximately one third of tumors and is associated with a poor disease-free and overall survival and response to therapy.[94] Preclinical studies of the biology of HER-2 indicate that it inhibits apoptosis induced by taxanes via a p21-dependent mechanism[95,96] and that it promotes tumor metastasis by promoting invasion and angiogenesis.[97-99] A blocking anti-HER-2 antibody (herceptin) inhibits the growth of HER-2-expressing breast cancer cells in vitro and tumor xenografts in vivo,[100,101] and exposure to combinations of herceptin and conventional therapies (particularly taxanes) lead to further growth inhibition and enhanced clinical activity.[102,103]

The receptor for epidermal growth factor (EGFR, also erbB1) also appears to play an important role in breast cancer progression.[104-106] Although the EGFR can function as a homodimer, it preferentially forms heterodimers with HER-2 and probably serves as the primary ligand-binding component of these heterodimers.[106] Furthermore, recent work indicates that EGFR is expressed on the vast majority of primary tumors that express activated (phosphorylated) HER-2 and that patients with these tumors have a very poor prognosis.[106] Breast cancers constitutively express EGFR ligands (EGF, TGF-α, HB-EGF)[107,108] and it appears that the ligands promote autocrine and paracrine proliferation. Several small molecule and antibody-based inhibitors of the EGFR have been developed,[109,110] and preclinical studies with these agents have demonstrated that they inhibit tumor cell growth in vitro and in vivo.[110]

HER-2 amplification is more prevalent in ER^- as compared to ER^+ tumors, and HER-2 expression is increased in tumors that develop resistance to SERMs. Recent studies have provided a biological explanation for this observation by showing that HER-2 and the EGFR interact directly with the ER pathway to enhance cell proliferation.[111] ErbB-mediated phosphorylation of the ER appears to promote its sensitivity to ligand-dependent and possibly ligand-independent activation, and erbB family proteins alter tumor cell expression of transcriptional coactivators and corepressors that control the inhibitory activity of the SERMs. Thus, combination therapy with tamoxifen plus herceptin leads to synergistic inhibition of tumor growth in preclinical models, and clinical trials are underway to test the effects of EGFR or HER-2 blockade on tumor response to SERMs or aromatase inhibitors in patients.[111,112]

Although ErbB antagonists have displayed very promising activity in preclinical models, it appears that, like the SERMs, their effects may be largely cytostatic rather than cytotoxic. Thus, in xenograft studies erbB antagonists consistently inhibit tumor growth (completely at high doses), but they do not appear to induce marked tumor regression. This may explain why herceptin (like tamoxifen) is most effective when used to prevent recurrence in the adjuvant setting and why studies of single-agent EGFR antagonists have yielded disappointing results in patients with bulky metastatic disease. Indeed, preclinincal studies of single-agent herceptin or various EGFR antagonists (ZD1839/Iressa, OSI-774/Tarceva, C225/Erbitux) have demonstrated that they consistently reduce tumor cell proliferation (as measured by Ki-67, PCNA, or BrdU staining) but have more variable effects on apoptosis.[110] Similarly, the preliminary results that have been obtained to date in neoadjuvant trials with these agents have demonstrated that they tend to inhibit proliferation without inducing substantial increases in apoptosis.[112]

It has recently been appreciated that these effects on tumor cell proliferation can have important (negative) implications for erbB-based conbination therapy with cytotoxic agents. Some of this concern was spawned by the disappointing results of combinations of EGFR antagonists plus taxanes or other cell cycle-active agents in large clinical trials in patients with nonsmall cell lung cancer, where the combinations performed no better than single-agent chemotherapy.[113,114] Because taxanes induce apoptosis via a cdc2-dependent mechanism in cells that are first arrested in M phase,[95] the p27-dependent growth arrest induced by erbB antagonists in cancer cells at the G_1/S restriction point might actually interfere with apoptosis induced by taxanes and any other agent that is most active at other phases of the cell cycle. Indeed, a recent study concluded that intermittent, high-dose therapy with Iressa given two days before docetaxel was more effective than continuous exposure to both agents, in part because higher doses of Iressa could be delivered and in part because cell cycle interference was avoided.[115]

Targeting Core Apoptotic Pathway Components

TRAIL and Agonistic Anti-TRAIL Receptor Antibodies

TNF-related apoptosis-inducing ligand (TRAIL) is a cell surface-associated cytokine that was isolated based on its homology to tumor necrosis factor (TNF).[116] However, unlike TNF or TNF's other cell death-associated homologue, FasL, systemic administration of TRAIL at physiologically relevant doses does not result in detectable toxicity in rodents or primates.[117] TRAIL stimulates apoptosis via the so-called "extrinsic", or death receptor-mediated pathway. Specifically, upon engagement by trimeric ligand, TRAIL's two receptors [known as TRAIL receptors 1 and 2 (TR1, TR2) or death receptors 4 and 5 (DR4, DR5)][118,119] bind to a cytosolic adaptor protein, Fas-associated death domain (FADD),[120] which recruits two of the caspases

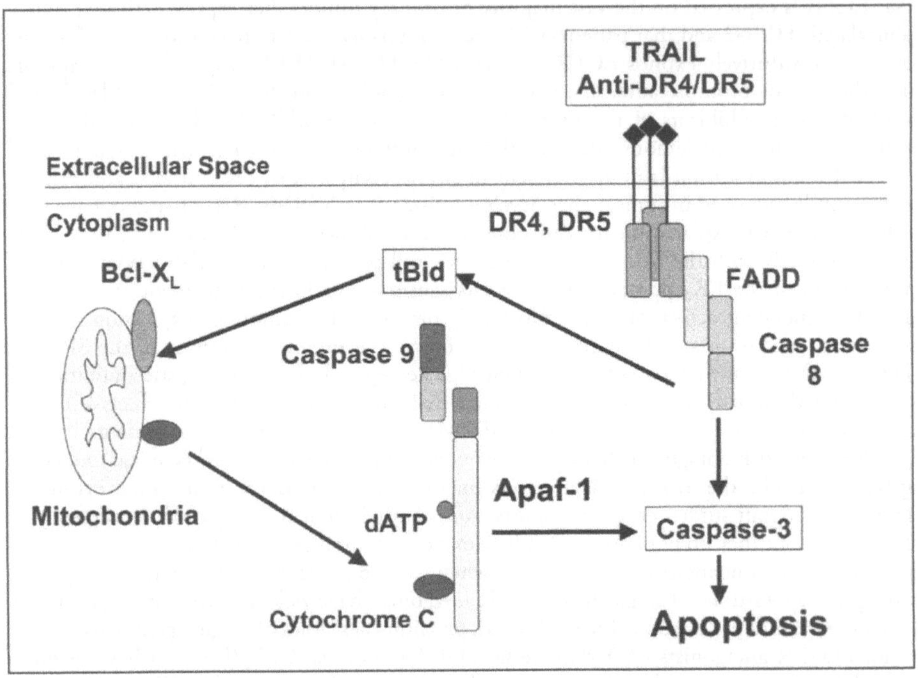

Figure 3. Downstream effectors of TRAIL receptor-mediated cell death. TRAIL and agonistic anti-TRAIL receptor antibodies oligomerize TRAIL's receptors (death receptors 4 and 5), leading to recruitment of the adaptor protein, FADD (Fas-associated death domain) and procaspases 8 and 10 and caspase activation. Active caspase-8 can then either directly promote activation of caspase-3 (Type I cells) or can acdtivate the so-called "intrinsic" pathway by cleaving Bid, which then translocates to mitochondria to promote cytochrome c release. Inhibitors of apoptosis proteins (IAPs), and in particular XIAP, appear to dampen caspase-8-mediated activation of caspase-3. Thus, inhibitors of XIAP and SMAC peptides and mimetics are particularly potent TRAIL-sensitizing agents because they may lower the threshold for direct caspase-8-mediated activation of caspase-3.

(8 and 10)[121,122] to the activated receptor (Fig. 3). This complex of aggregated death receptor(s), FADD, and caspase-8/10 is known as the death-inducing signaling complex, or DISC,[123] and it functions to promote caspase activation via aggregation. Once activated, caspases 8/10 can either directly activate caspases 3 and 7 to stimulate cell death (in so-called Type I cells),[124] or they interact with the mitochondrial (intrinsic) pathway of apoptosis (in Type II cells)[124] by promoting cytochrome c release to activate caspase-9 and stimulate cell death. The latter is accomplished by caspase-mediated cleavage of the BH3 protein, Bid,[125,126] forming a truncated product that is capable of directly activating Bax and Bak and restructuring mitochondria to induce cytochrome c release.[127] Although many different properties might contribute to make cells Type I or Type II, most solid tumors appear to be Type II cells, and recent work suggests that direct caspase-8/10-mediated caspase-3/7 activation is limited in these cells by the expression of polypeptides (XIAP, α/β-crystallin) that function to attenuate caspase-3 activation.[128-131]

Approximately half of human breast cancer cell lines are highly sensitive to TRAIL receptor-mediated apoptosis at baseline,[129,132] and apoptosis sensitivity can be augmented in many of the rest by combining TRAIL with conventional or investigational anti-cancer therapies.[132] These observations prompted parallel efforts by the University of Alabama SPORE in Breast Cancer[132] ("TRA-8") and Human Genome Sciences, Inc. (HGS)[133] ("HGS-ETR1",

"HGS-ETR-2", and "HGS-TR2J") to develop antibodies that are capable of crosslinking TRAIL's receptors to stimulate cell death. Subsequent in vivo studies have confirmed that recombinant TRAIL and these agonistic anti-DR antibodies display potent growth inhibitory activity in human breast cancer xenografts without detectable toxicity.[132] Furthermore, their in vivo activities are greatly enhanced by combining them with conventional chemotherapy.[132] Phase I and II clinical trials with the HGS antibodies have been open since 2003, and a consortium of biotechnology companies (Amgen, Genentech, and Immunex) cooperated to open Phase I trials employing recombinant TRAIL in 2005. There are advantages and disadvantages associated with the antibodies and rhTRAIL that make these trials far from redundant. The antibodies have very long serum half lives (> 1 week), which enables them to be delivered less frequently and in a more sustained fashion. However, each antibody only engages one of TRAIL's two receptors, and it is possible that sustained exposure of normal cells to death receptor signaling will produce more substantial toxicity than transient exposure. On the other hand, recombinant TRAIL is cleared from the serum very rapidly, necessitating a more frequent (daily) dosing schedule. Perhaps because of this, preclinical studies demonstrated that mice and primates tolerate very high doses of rhTRAIL (100 mg/kg) without toxicity, and in fact no MTD has been reached to date. Some evidence is emerging that less frequent dosing with higher doses of rhTRAIL are more efficacious than chronic exposure to lower doses.

Although there is considerable enthusiasm for TRAIL and the agonistic antibodies as candidate therapies for refractory breast cancer, it seems unlikely that single-agent TRAIL will display substantial clinical activity in this population of heavily pretreated patients. Thus, identifying the most promising combination regimens and evaluating their potential toxicities in preclinical studies remains a high research priority. A wide array of agents are capable of promoting additive or synergistic enhancement of apoptosis in human breast cancer cells in vitro, and in some cases xenograft studies have been performed which demonstrate that these combination regimens are tolerated in vivo (Table 1). Importantly, most of these TRAIL-sensitizing agents are cytostatic, and in fact evidence has been advanced which suggests that tumor cell sensitivity to TRAIL may be highest in the G_1 phase of the cell cycle.[134,135] It may be possible to evaluate all of these combinations in clinical trials in patients, but this will certainly prove to be expensive, so prioritizing them in terms of their promise should be attempted. Furthermore, some (or perhaps all) of these combinations will unmask toxicity in normal tissues, and understanding these toxicity profiles will be important to fully exploit TRAIL's anti-cancer activity. For example, we have found the proteasome inhibitor bortezomib (Velcade, PS-341) to be an extremely potent TRAIL-sensitizing agent in vitro and in vivo, but simultaneously combining TRAIL with bortezomib (at its MTD) results in rapid lethality. Although a detailed characterization of the target tissue(s) involved is underway, this toxicity can be avoided by decreasing the dose of bortezomib or by staggering dosing (bortezomib followed by TRAIL 16 h later).

Table 1. TRAIL-sensitizing agents

Compound	References
Chemotherapy	184-190
HDAC inhibitors	191-198
Retinoids	199, 200
PPARγ agonists	201-203
Flavopiridol	204
Herceptin	205
Velcade	133, 135, 206-215

Bcl-2 Family Proteins

As discussed earlier, members of the Bcl-2 family usually control commitment to apoptotic cell death, and as such they make extremely attractive therapeutic targets. As the oldest member of the family, most previous efforts have been directed towards targeting Bcl-2 itself, although as discussed above it is not clear that Bcl-2 is the most relevant anti-apoptotic family member in breast cancer. The first attempts to target Bcl-2 as a therapeutic approach in cancer involved the use of specific antisense oligonucleotides, usually identified through a combination of computer-based predictive modeling and empiricism, that partially downregulate expression of *bcl-2* mRNA in whole cells.[136-138] One such reagent (Genasense/oblimersen, developed by Genta, Inc)[137,139] is currently being evaluated for single agent efficacy in Phase III clinical trials in patients with CLL, and it is being combined with various therapeutic modalities in Phase II trials in a variety of different disease sites, including breast cancer. Preliminary findings in prostate cancer and multiple myeloma suggest that the compound has biological and clinical activity.[140,141] Another antisense product (SPC-2996), developed by Santaris Pharma, is being evaluated in Phase II and II trials in patients with CLL in Europe (see www.santaris.com). Unfortunately, Genasense failed to reach the primary survival endpoint in a randomized, open label Phase III trial in melanoma when given with dacarbazine compared to dacarbazine alone,[142] dampening enthusiasm for the approach somewhat. However, whether or not Bcl-2 plays an obligate role in the progression of melanoma has not been established, so the negative trial results may be more a problem of an incomplete understanding of tumor biology rather than to a poor therapeutic approach.

Clinical experience with Genasense raises several other issues that are important to the further development of antagonists of Bcl-2 and other anti-apoptotic members of the family. First, considering that these proteins function by inhibiting the actions of pro-apoptotic members of the family, it seems likely that effective inhibition of their expression will sensitize most cancer cells to cytotoxic stimuli but will not directly result in apoptosis. Second, it will be important to develop pharmacodynamic assays that are capable of measuring their effects on their targets, either in biopsies or via noninvasive strategies, to ensure that target inhibition was efficient. Third, and most importantly, there is probably considerable redundancy among members of the family, such that strategies designed to be highly specific might fail because other members compensate for loss of one family member's expression. Evidence in support of this concept can be found in preclinical studies with Bcl-2 or Bcl-X_L transgenics, where modulation of one of the proteins in thymocytes led to compensatory changes in the other.[143] Finally, as touched upon above, a better understanding of the biological relevance of each family member to disease progression and therapeutic resistance in breast cancer must be obtained. Although preclinical studies can aid in this effort, well-designed neoadjuvant trials with these agents in patients could provide the most valuable information about the importance of inter-patient heterogeneity and the factor(s) that dictate reliance upon one family member versus another.

A newer, more sophisticated approach involves using gene therapy to drive expression of proapoptotic Bcl-2 family proteins.[144] Such strategies have been developed and evaluated for Bax,[145] Bak,[146] and Bik[147] in preclinical models, where effective gene transfer has almost invariably produced very efficient induction of cell death. The major concern associated with these approaches is that overexpression of Bax and Bak can be lethal to packaging cells[144] and to normal tissues. To address this concern, various groups are employing cancer-specific promoters to selectively drive high-level transgene expression.[148] Although concerns about promoter "leakiness" and gene transfer efficiency remain, preclinical studies provide strong evidence for the feasibility of this approach.[148] With improvements in viral delivery systems systemic gene transfer may be possible within the context of metastatic disease. Again, because Bax and Bak probably need to be "activated" by a proapoptotic signal to be optimally effective, it is likely that the biological effects of Bcl-2 family-directed gene therapy will be best exploited by combining these agents with cytotoxic stimuli (conventional chemotherapy, TRAIL, etc).

A third emerging approach to target Bcl-2 family proteins involves the development of so-called "BH3 mimetics".[149,150] These peptide-based compounds and small molecules are designed to either disrupt the interactions between anti-apoptotic members of the Bcl-2 family and pro-apoptotic members of the family or directly activate Bax and Bak. Proof-of-principle for this approach came from studies with isolated BH3 domains themselves, and this work demonstrated that some BH3 domains, such as those found in Bid or Bim, function as direct activators of Bax and Bak, triggering mitochondrial pore formation and release of cytochrome C and SMAC via a BH3-dependent mechanism.[149,150] Other BH3 proteins, such as Bad or Bik, cannot directly activate Bax or Bak. Rather, they function as facilitators of apoptosis by virtue of their abilities to bind to Bcl-2 and other anti-apoptotic members of the Bcl-2 family, thereby displacing the direct activators such as Bid or Bax so that they can directly promote cytochrome c release.[149,150] There are clearly conceptual strengths and weaknesses associated with mimicking the direct activators or the facilitators. Although the direct activators would be expected to trigger cell death independently of any other signal, this could produce significant systemic toxicity that might be most severe in cells that possess the lowest thresholds for apoptosis, which might include normal cells. Conversely, while the Bad or Bik mimetics might require a second signal for efficient cell death induction, one could imagine developing agents that selectively target only the anti-apoptotic Bcl-2 family protein that is required for survival in a given tumor. Thus, given the functional redundancy among members of the family, this might represent a strategy to induce selective tumor cell killing.

Several groups have developed modified (cell-permeant) BH3 peptides that display promising activity in preclinical models.[151-153] In the most prominent example of the feasibility of such an approach, Korsmeyer's group synthesized Bid BH3 analogs that contained α,α-disubstituted unnatural amino acids to stabilize the critical α-helix found within the Bid BH3 domain, thereby making it more protease resistant and cell permeant.[153] These so-called "stapled" BH3 peptides directly induced apoptosis in whole cells and inhibited the growth of leukemia xenografts in vivo, prolonging survival.[153] Importantly, this was accomplished without systemic toxicity, addressing one of the concerns raised above. Others have shown that other strategies can enhance the membrane permeability of BH3 peptides, allowing them to enter cells and induce apoptosis.[151,152]

Past experience with nonantibody proteins and peptides in vivo has been mixed. Liposomal encapsulation and other approaches can enhance their serum half lives, but overall they tend to display relatively poor pharmacokinetic properties and low bioavailability. Not withstanding the preclinical successes described above, it is clear that the development of small molecule approaches to target Bcl-2 family proteins would receive significantly more enthusiasm from industry. High throughput screening strategies have already identified several small molecules (gossypol,[154-156] antimycin A,[157] chelerythrine,[158] and others) that bind to Bcl-2, Bcl-X_L, and Bcl-w with low micromolar affinity. These compounds can induce apoptosis in tumor cells in vitro and in some cases in vivo, but their clinical applicability will probably be limited because they do not inhibit their targets in the nanomolar range.

Recently, a more potent Bcl-2 family antagonist was developed by Abbott laboratories in collaboration with a large group of very experienced academic investigators.[159] Called ABT-737, the compound binds to Bcl-2, Bcl-X_L, and Bcl-w with subnanomolar affinity.[159] It appears to function as a Bad mimetic (i.e., facilitating rather than directly causing apoptosis), and it exhibits synergistic cytotoxicity with chemotherapeutic agents and radiation in vitro.[159] Importantly, it also displayed single-agent activity in lymphoma and leukemia cell lines, primary patient-derived cells, and small cell lung cancer (SCLC) cell lines, all of which probably depend on the drug target(s) for their survival.[159] Finally, dose-intensive therapy (100 mg/kg/day for 3 weeks) with ABT-787 in mice bearing subcutaneous human SCLC xenografts produced complete regressions and cures in a high percentage of mice, and lower doses of the drug (25-75 mg/kg/day) still displayed significant anti-tumor activity.[159] Clinical trials with ABT-787 in lymphoma, leukemia, and SCLC are scheduled to open soon. Together with the ongoing

121. Muzio M, Chinnaiyan AM, Kischkel FC et al. FLICE, a novel FADD-homologous ICE/CED-3-like protease, is recruited to the CD95 (Fas/APO-1) death—inducing signaling complex. Cell 1996; 85:817-827.
122. Fernandes-Alnemri T, Armstrong RC, Krebs J et al. In vitro activation of CPP32 and Mch3 by Mch4, a novel human apoptotic cysteine protease containing two FADD-like domains. Proc Natl Acad Sci USA 1996; 93:7464-7469.
123. Kischkel FC, Hellbardt S, Behrmann I et al. Cytotoxicity-dependent APO-1 (Fas/CD95)-associated proteins form a death-inducing signaling complex (DISC) with the receptor. EMBO J 1995; 14:5579-5588.
124. Scaffidi C, Fulda S, Srinivasan A et al. Two CD95 (APO-1/Fas) signaling pathways. EMBO J 1998; 17:1675-1687.
125. Li H, Zhu H, Xu CJ et al. Cleavage of BID by caspase 8 mediates the mitochondrial damage in the Fas pathway of apoptosis. Cell 1998; 94:491-501.
126. Luo X, Budihardjo I, Zou H et al. Bid, a Bcl2 interacting protein, mediates cytochrome c release from mitochondria in response to activation of cell surface death receptors. Cell 1998; 94:481-490.
127. Scorrano L, Ashiya M, Buttle K et al. A distinct pathway remodels mitochondrial cristae and mobilizes cytochrome c during apoptosis. Dev Cell 2002; 2:55-67.
128. Wilkinson JC, Cepero E, Boise LH et al. Upstream regulatory role for XIAP in receptor-mediated apoptosis. Mol Cell Biol 2004; 24:7003-7014.
129. Kamradt MC, Lu M, Werner ME et al. The small heat shock protein alpha B-crystallin is a novel inhibitor of TRAIL-induced apoptosis that suppresses the activation of caspase-3. J Biol Chem 2005; 280:11059-11066.
130. Kamradt MC, Chen F, Sam S et al. The small heat shock protein alpha B-crystallin negatively regulates apoptosis during myogenic differentiation by inhibiting caspase-3 activation. J Biol Chem 2002; 277:38731-38736.
131. Kamradt MC, Chen F, Cryns VL. The small heat shock protein alpha B-crystallin negatively regulates cytochrome c- and caspase-8-dependent activation of caspase-3 by inhibiting its autoproteolytic maturation. J Biol Chem 2001; 276:16059-16063.
132. Buchsbaum DJ, Zhou T, Grizzle WE et al. Antitumor efficacy of TRA-8 anti-DR5 monoclonal antibody alone or in combination with chemotherapy and/or radiation therapy in a human breast cancer model. Clin Cancer Res 2003; 9:3731-3741.
133. Georgakis GV, Li Y, Humphreys R et al. Activity of selective fully human agonistic antibodies to the TRAIL death receptors TRAIL-R1 and TRAIL-R2 in primary and cultured lymphoma cells: Induction of apoptosis and enhancement of doxorubicin- and bortezomib-induced cell death. Br J Haematol 2005; 130:501-510.
134. Jin Z, Dicker DT, El-Deiry WS. Enhanced sensitivity of G1 arrested human cancer cells suggests a novel therapeutic strategy using a combination of simvastatin and TRAIL. Cell Cycle 2002; 1:82-89.
135. Lashinger LM, Zhu K, Williams SA et al. Bortezomib abolishes tumor necrosis factor-related apoptosis-inducing ligand resistance via a p21-dependent mechanism in human bladder and prostate cancer cells. Cancer Res 2005; 65:4902-4908.
136. Cotter FE, Waters J, Cunningham D. Human Bcl-2 antisense therapy for lymphomas. Biochim Biophys Acta 1999; 1489:97-106.
137. Klasa RJ, Gillum AM, Klem RE et al. Oblimersen Bcl-2 antisense: Facilitating apoptosis in anticancer treatment. Antisense Nucleic Acid Drug Dev 2002; 12:193-213.
138. Kim R, Tanabe K, Emi M et al. Modulation of tamoxifen sensitivity by antisense Bcl-2 and trastuzumab in breast carcinoma cells. Cancer 2005; 103:2199-2207.
139. Banerjee D. Genasense (Genta Inc). Curr Opin Investig Drugs 2001; 2:574-580.
140. Tolcher AW, Chi K, Kuhn J et al. A phase II, pharmacokinetic, and biological correlative study of oblimersen sodium and docetaxel in patients with hormone-refractory prostate cancer. Clin Cancer Res 2005; 11:3854-3861.
141. Badros AZ, Goloubeva O, Rapoport AP et al. Phase II study of G3139, a Bcl-2 antisense oligonucleotide, in combination with dexamethasone and thalidomide in relapsed multiple myeloma patients. J Clin Oncol 2005; 23:4089-4099.
142. Frantz S. Lessons learnt from Genasense's failure. Nat Rev Drug Discov 2004; 3:542-543.
143. Chao DT, Linette GP, Boise LH et al. Bcl-XL and Bcl-2 repress a common pathway of cell death. J Exp Med 1995; 182:821-828.
144. Kagawa S, Pearson SA, Ji L et al. A binary adenoviral vector system for expressing high levels of the proapoptotic gene bax. Gene Ther 2000; 7:75-79.
145. Li X, Marani M, Yu J et al. Adenovirus-mediated Bax overexpression for the induction of therapeutic apoptosis in prostate cancer. Cancer Res 2001; 61:186-191.

146. Pataer A, Fang B, Yu R et al. Adenoviral Bak overexpression mediates caspase-dependent tumor killing. Cancer Res 2000; 60:788-792.
147. Zou Y, Peng H, Zhou B et al. Systemic tumor suppression by the proapoptotic gene bik. Cancer Res 2002; 62:8-12.
148. Chen JS, Liu JC, Shen L et al. Cancer-specific activation of the survivin promoter and its potential use in gene therapy. Cancer Gene Ther 2004; 11:740-747.
149. Letai A, Bassik MC, Walensky LD et al. Distinct BH3 domains either sensitize or activate mitochondrial apoptosis, serving as prototype cancer therapeutics. Cancer Cell 2002; 2:183-192.
150. Kuwana T, Bouchier-Hayes L, Chipuk JE et al. BH3 domains of BH3-only proteins differentially regulate Bax-mediated mitochondrial membrane permeabilization both directly and indirectly. Mol Cell 2005; 17:525-535.
151. Holinger EP, Chittenden T, Lutz RJ. Bak BH3 peptides antagonize Bcl-xL function and induce apoptosis through cytochrome c-independent activation of caspases. J Biol Chem 1999; 274:13298-13304.
152. Wang JL, Zhang ZJ, Choksi S et al. Cell permeable Bcl-2 binding peptides: A chemical approach to apoptosis induction in tumor cells. Cancer Res 2000; 60:1498-1502.
153. Walensky LD, Kung AL, Escher I et al. Activation of apoptosis in vivo by a hydrocarbon-stapled BH3 helix. Science 2004; 305:1466-1470.
154. Kitada S, Leone M, Sareth S et al. Discovery, characterization, and structureactivity relationships studies of proapoptotic polyphenols targeting B-cell lymphocyte/leukemia-2 proteins. J Med Chem 2003; 46:4259-4264.
155. Mohammad RM, Wang S, Aboukameel A et al. Preclinical studies of a nonpeptidic small-molecule inhibitor of Bcl-2 and Bcl-X(L) [(-)-gossypol] against diffuse large cell lymphoma. Mol Cancer Ther 2005; 4:13-21.
156. Oliver CL, Miranda MB, Shangary S et al. (-)-Gossypol acts directly on the mitochondria to overcome Bcl-2- and Bcl-X(L)-mediated apoptosis resistance. Mol Cancer Ther 2005; 4:23-31.
157. Tzung SP, Kim KM, Basanez G et al. Antimycin A mimics a cell-death-inducing Bcl-2 homology domain 3. Nat Cell Biol 2001; 3:183-191.
158. Chan SL, Lee MC, Tan KO et al. Identification of chelerythrine as an inhibitor of BclXL function. J Biol Chem 2003; 278:20453-20456.
159. Oltersdorf T, Elmore SW, Shoemaker AR et al. An inhibitor of Bcl-2 family proteins induces regression of solid tumours. Nature 2005; 435:677-681.
160. Deveraux QL, Takahashi R, Salvesen GS et al. X-linked IAP is a direct inhibitor of cell-death proteases. Nature 1997; 388:300-304.
161. Arkin M. Protein-protein interactions and cancer: Small molecules going in for the kill. Curr Opin Chem Biol 2005; 9:317-324.
162. Takahashi R, Deveraux Q, Tamm I et al. A single BIR domain of XIAP sufficient for inhibiting caspases. J Biol Chem 1998; 273:7787-7790.
163. Sun C, Cai M, Gunasekera AH et al. NMR structure and mutagenesis of the inhibitor-of-apoptosis protein XIAP. Nature 1999; 401:818-822.
164. Riedl SJ, Renatus M, Schwarzenbacher R et al. Structural basis for the inhibition of caspase-3 by XIAP. Cell 2001; 104:791-800.
165. Sun C, Cai M, Meadows RP et al. NMR structure and mutagenesis of the third Bir domain of the inhibitor of apoptosis protein XIAP. J Biol Chem 2000; 275:33777-33781.
166. Shiozaki EN, Chai J, Rigotti DJ et al. Mechanism of XIAP-mediated inhibition of caspase-9. Mol Cell 2003; 11:519-527.
167. Wu G, Chai J, Suber TL et al. Structural basis of IAP recognition by Smac/DIABLO. Nature 2000; 408:1008-1012.
168. Liu Z, Sun C, Olejniczak ET et al. Structural basis for binding of Smac/DIABLO to the XIAP BIR3 domain. Nature 2000; 408:1004-1008.
169. Altieri DC. Survivin, versatile modulation of cell division and apoptosis in cancer. Oncogene 2003; 22:8581-8589.
170. Li F, Ambrosini G, Chu EY et al. Control of apoptosis and mitotic spindle checkpoint by survivin. Nature 1998; 396:580-584.
171. Hu Y, Cherton-Horvat G, Dragowska V et al. Antisense oligonucleotides targeting XIAP induce apoptosis and enhance chemotherapeutic activity against human lung cancer cells in vitro and in vivo. Clin Cancer Res 2003; 9:2826-2836.
172. Arnt CR, Chiorean MV, Heldebrant MP et al. Synthetic Smac/DIABLO peptides enhance the effects of chemotherapeutic agents by binding XIAP and cIAP1 in situ. J Biol Chem 2002; 277:44236-44243.

173. Yang L, Mashima T, Sato S et al. Predominant suppression of apoptosome by inhibitor of apoptosis protein in nonsmall cell lung cancer H460 cells: Therapeutic effect of a novel polyarginine-conjugated Smac peptide. Cancer Res 2003; 63:831-837.
174. Sun H, Nikolovska-Coleska Z, Yang CY et al. Structurebased design, synthesis, and evaluation of conformationally constrained mimetics of the second mitochondria-derived activator of caspase that target the X-linked inhibitor of apoptosis protein/caspase-9 interaction site. J Med Chem 2004; 47:4147-4150.
175. Sun H, Nikolovska-Coleska Z, Yang CY et al. Structurebased design of potent, conformationally constrained Smac mimetics. J Am Chem Soc 2004; 126:16686-16687.
176. Oost TK, Sun C, Armstrong RC et al. Discovery of potent antagonists of the antiapoptotic protein XIAP for the treatment of cancer. J Med Chem 2004; 47:4417-4426.
177. Fulda S, Wick W, Weller M et al. Smac agonists sensitize for Apo2L/TRAIL- or anticancer drug-induced apoptosis and induce regression of malignant glioma in vivo. Nat Med 2002; 8:808-815.
178. Wu TY, Wagner KW, Bursulaya B et al. Development and characterization of nonpeptidic small molecule inhibitors of the XIAP/caspase-3 interaction. Chem Biol 2003; 10:759-767.
179. Nikolovska-Coleska Z, Xu L, Hu Z et al. Discovery of embelin as a cell-permeable, small-molecular weight inhibitor of XIAP through structurebased computational screening of a traditional herbal medicine three-dimensional structure database. J Med Chem 2004; 47:2430-2440.
180. Li L, Thomas RM, Suzuki H et al. A small molecule Smac mimic potentiates TRAIL- and TNFalpha-mediated cell death. Science 2004; 305:1471-1474.
181. Lowe SW, Cepero E, Evan G. Intrinsic tumour suppression. Nature 2004; 432:307-315.
182. Sansal I, Sellers WR. The biology and clinical relevance of the PTEN tumor suppressor pathway. J Clin Oncol 2004; 22:2954-2963.
183. Thompson JE, Thompson CB. Putting the rap on Akt. J Clin Oncol 2004; 22:4217-4226.
184. Keane MM, Ettenberg SA, Nau MM et al. Chemotherapy augments TRAIL-induced apoptosis in breast cell lines. Cancer Res 1999; 59:734-741.
185. Chinnaiyan AM, Prasad U, Shankar S et al. Combined effect of tumor necrosis factor-related apoptosis-inducing ligand and ionizing radiation in breast cancer therapy. Proc Natl Acad Sci USA 2000; 97:1754-1759.
186. Nagane M, Pan G, Weddle JJ et al. Increased death receptor 5 expression by chemotherapeutic agents in human gliomas causes synergistic cytotoxicity with tumor necrosis factor-related apoptosis-inducing ligand in vitro and in vivo. Cancer Res 2000; 60:847-853.
187. Ohtsuka T, Buchsbaum D, Oliver P et al. Synergistic induction of tumor cell apoptosis by death receptor antibody and chemotherapy agent through JNK/p38 and mitochondrial death pathway. Oncogene 2003; 22:2034-2044.
188. Xu ZW, Kleeff J, Friess H et al. Synergistic cytotoxic effect of TRAIL and gemcitabine in pancreatic cancer cells. Anticancer Res 2003; 23:251-258.
189. Singh TR, Shankar S, Chen X et al. Synergistic interactions of chemotherapeutic drugs and tumor necrosis factor-related apoptosis-inducing ligand/Apo-2 ligand on apoptosis and on regression of breast carcinoma in vivo. Cancer Res 2003; 63:5390-5400.
190. Ravi R, Jain AJ, Schulick RD et al. Elimination of hepatic metastases of colon cancer cells via p53-independent cross-talk between irinotecan and Apo2 ligand/TRAIL. Cancer Res 2004; 64:9105-9114.
191. Zhang XD, Gillespie SK, Borrow JM et al. The histone deacetylase inhibitor suberic bishydroxamate: A potential sensitizer of melanoma to TNF-related apoptosis-inducing ligand (TRAIL) induced apoptosis. Biochem Pharmacol 2003; 66:1537-1545.
192. Rosato RR, Almenara JA, Dai Y et al. Simultaneous activation of the intrinsic and extrinsic pathways by histone deacetylase (HDAC) inhibitors and tumor necrosis factor-related apoptosis-inducing ligand (TRAIL) synergistically induces mitochondrial damage and apoptosis in human leukemia cells. Mol Cancer Ther 2003; 2:1273-1284.
193. Guo F, Sigua C, Tao J et al. Cotreatment with histone deacetylase inhibitor LAQ824 enhances Apo-2L/tumor necrosis factor-related apoptosis inducing ligand-induced death inducing signaling complex activity and apoptosis of human acute leukemia cells. Cancer Res 2004; 64:2580-2589.
194. Nakata S, Yoshida T, Horinaka M et al. Histone deacetylase inhibitors upregulate death receptor 5/TRAIL-R2 and sensitize apoptosis induced by TRAIL/APO2-L in human malignant tumor cells. Oncogene 2004; 23:6261-6271.
195. Chopin V, Slomianny C, Hondermarck H et al. Synergistic induction of apoptosis in breast cancer cells by cotreatment with butyrate and TNF-alpha, TRAIL, or anti-Fas agonist antibody involves enhancement of death receptors' signaling and requires P21(waf1). Exp Cell Res 2004; 298:560-573.

196. Sonnemann J, Gange J, Kumar KS et al. Histone deacetylase inhibitors interact synergistically with tumor necrosis factor-related apoptosis-inducing ligand (TRAIL) to induce apoptosis in carcinoma cell lines. Invest New Drugs 2005; 23:99-109.
197. Singh TR, Shankar S, Srivastava RK. HDAC inhibitors enhance the apoptosis-inducing potential of TRAIL in breast carcinoma. Oncogene 2005; 24:4609-4623.
198. Vanoosten RL, Moore JM, Karacay B et al. Histone deacetylase inhibitors modulate renal cell carcinoma sensitivity to TRAIL/Apo-2L-induced apoptosis by enhancing TRAIL-R2 expression. Cancer Biol Ther 2005; 4.
199. Sun SY, Yue P, Hong WK et al. Augmentation of tumor necrosis factor-related apoptosis-inducing ligand (TRAIL)-induced apoptosis by the synthetic retinoid 6-[3-(1-adamantyl)-4-hydroxyphenyl]-2-naphthalene carboxylic acid (CD437) through up-regulation of TRAIL receptors in human lung cancer cells. Cancer Res 2000; 60:7149-7155.
200. Cuello M, Coats AO, Darko I et al. N-(4-hydroxyphenyl) retinamide (4HPR) enhances TRAIL-mediated apoptosis through enhancement of a mitochondrial-dependent amplification loop in ovarian cancer cell lines. Cell Death Differ 2004; 11:527-541.
201. Goke R, Goke A, Goke B et al. Pioglitazone inhibits growth of carcinoid cells and promotes TRAIL-induced apoptosis by induction of p21waf1/cip1. Digestion 2001; 64:75-80.
202. Kim Y, Suh N, Sporn M et al. An inducible pathway for degradation of FLIP protein sensitizes tumor cells to TRAIL-induced apoptosis. J Biol Chem 2002; 277:22320-22329.
203. Lu M, Kwan T, Yu C et al. Peroxisome proliferator-activated receptor gamma agonists promote TRAIL-induced apoptosis by reducing survivin levels via cyclin D3 repression and cell cycle arrest. J Biol Chem 2005; 280:6742-6751.
204. Kim DM, Koo SY, Jeon K et al. Rapid induction of apoptosis by combination of flavopiridol and tumor necrosis factor (TNF)-alpha or TNF-related apoptosis-inducing ligand in human cancer cell lines. Cancer Res 2003; 63:621-626.
205. Cuello M, Ettenberg SA, Clark AS et al. Down-regulation of the erbB-2 receptor by trastuzumab (herceptin) enhances tumor necrosis factor-related apoptosis-inducing ligand-mediated apoptosis in breast and ovarian cancer cell lines that overexpress erbB-2. Cancer Res 2001; 61:4892-4900.
206. Sayers TJ, Brooks AD, Koh CY et al. The proteasome inhibitor PS-341 sensitizes neoplastic cells to TRAIL-mediated apoptosis by reducing levels of c-FLIP. Blood 2003; 102:303-310.
207. Johnson TR, Stone K, Nikrad M et al. The proteasome inhibitor PS-341 overcomes TRAIL resistance in Bax and caspase 9-negative or Bcl-xL overexpressing cells. Oncogene 2003; 22:4953-4963.
208. An J, Sun YP, Adams J et al. Drug interactions between the proteasome inhibitor bortezomib and cytotoxic chemotherapy, tumor necrosis factor (TNF) alpha, and TNF-related apoptosis-inducing ligand in prostate cancer. Clin Cancer Res 2003; 9:4537-4545.
209. Yin D, Zhou H, Kumagai T et al. Proteasome inhibitor PS-341 causes cell growth arrest and apoptosis in human glioblastoma multiforme (GBM). Oncogene 2005; 24:344-354.
210. Papageorgiou A, Lashinger L, Millikan R et al. Role of tumor necrosis factor-related apoptosis-inducing ligand in interferon-induced apoptosis in human bladder cancer cells. Cancer Res 2004; 64:8973-8979.
211. Matta H, Chaudhary PM. The proteasome inhibitor bortezomib (PS-341) inhibits growth and induces apoptosis in primary effusion lymphoma cells. Cancer Biol Ther 2005; 4:77-82.
212. Nikrad M, Johnson T, Puthalalath H et al. The proteasome inhibitor bortezomib sensitizes cells to killing by death receptor ligand TRAIL via BH3-only proteins Bik and Bim. Mol Cancer Ther 2005; 4:443-449.
213. Bai J, Sui J, Demirjian A et al. Predominant Bcl-XL knockdown disables antiapoptotic mechanisms: Tumor necrosis factor-related apoptosis-inducing ligand-based triple chemotherapy overcomes chemoresistance in pancreatic cancer cells in vitro. Cancer Res 2005; 65:2344-2352.
214. Nencioni A, Wille L, Dal Bello G et al. Cooperative cytotoxicity of proteasome inhibitors and tumor necrosis factor-related apoptosis-inducing ligand in chemoresistant Bcl-2-overexpressing cells. Clin Cancer Res 2005; 11:4259-4265.
215. Zhu H, Guo W, Zhang L et al. Proteasome inhibitors-mediated TRAIL resensitization and Bik accumulation. Cancer Biol Ther 2005; 4.

CHAPTER 4

Cell Cycle Deregulation in Breast Cancer: Insurmountable Chemoresistance or Achilles' Heel?

Laura Lambert and Khandan Keyomarsi*

Abstract

Deregulation of the G1 cyclin, cyclin E, has been shown to be both the most powerful predictor of prognosis in early stage breast cancer as well as a significant determinant of tumor aggressiveness.[1,2] It may also contribute to treatment failure due to chemoresistance. Because some form of cell cycle deregulation is present in all malignant cells,[3] increasing understanding of these processes is starting to provide new opportunities to overcome the cells' resistance mechanisms.

One particular form of cyclin E deregulation, the generation of hyperactive low molecular weight isoforms, is especially intriguing. Because only tumor cells contain the machinery necessary to generate these isoforms,[4] they not only provide a mechanism of targeting critical cell cycle events, but their presence may also provide both a means of increased specificity for targeting malignant cells, as well as an objective measure of response.

This review describes the mechanisms of resistance to commonly used systemic therapies for the treatment of breast cancer, with particular respect to the role of the cell cycle. The mechanisms and effects of the deregulation of cyclin E in breast cancer are reviewed and novel approaches to circumventing chemoresistance through abrogation of the malignant cell cycle are proposed.

Introduction

Tumor resistance to systemic antineoplastic therapy is the main cause of failure of breast cancer treatment. For early stage breast cancer, adjuvant endocrine and cytotoxic agents have resulted in only an 8-37% reduction in mortality.[5,6] For patients with more advanced disease the success rate is even lower. Investigation into the means by which tumor cells resist cytotoxic therapies have revealed multiple mechanisms of drug resistance and efforts to devise ways of circumventing resistance are currently underway.

The cytotoxic mechanisms of most conventional chemotherapeutic agents used in the current treatment of breast cancer (doxorubicin, cyclophospamide, 5-flourouracil, methotrexate and the taxanes) are attributable to their damaging or inhibitory effects on DNA. However, as illustrated by the high rate of resistance, this approach is limited in a number of ways. First it is highly nonspecific. Second, these agents rely upon a relative rate of cell division to establish a cytotoxic threshold to distinguish between rapidly dividing malignant cells and normal cells. Another limitation is the nonlethality of the effect of the drug on the DNA with the ultimate outcome (susceptibility versus resistance) dependent upon the status of the cell's mechanisms of DNA repair and apoptosis. Because of the redundancy of the cell salvage pathways, continuing to use conventional approaches only prolongs the inevitable occurrence of drug resistance (Table 1).

*Corresponding Author: Khandan Keyomarsi—Departments of Surgical Oncology and Experimental Radiation Oncology at University of Texas, MD Anderson Cancer Center, 1515 Holcombe Blvd., Houston, Texas 77030, U.S.A. Email: kkeyomar@mdanderson.org

Breast Cancer Chemosensitivity, edited by Dihua Yu and Mien-Chie Hung.

Table 1. Response rates and possible mechanisms of resistance in neoadjuvant chemotherapy and endocrine regimens for breast cancer

Neoadjuvant Chemotherapy/ Endocrine Regimen	Response	Possible Mechanism(s) of Resistance	Ref.
Adriamycin (doxorubicin) and **cyclophosphamide** (AC)	Pathologic complete 10% Objective clinical 70%	**Adriamycin**: Increased cellular efflux Alterations in topoisomerases Aberrant intracellular localization **Cyclophosphomide**: Intracellular inactivation Increased conjugation	19-27, 31-42, 117
Adriamycin and **Taxol** (paclitaxel) (AT)	Pathologic complete 16% Objective clinical 89%	**Taxol**: Increased cellular efflux Impaired microtubule polymerization Microtubule instability	87-92, 117
Flourouracil, **Adriamycin**, and **cyclophosphamide** (FAC)	Pathologic - complete 24% - partial 55% Clinical - complete 18% - partial 82%	**Flourouracil**: Reduced anabolism Increased catabolism Reduced FdUMP affinity Increased thymidylate synthase Mode of administration	43,44, 50-56, 118
Taxol	Pathologic - complete 24% - partial 55% Clinical - complete 18% - partial 82%	(See above)	118
Tamoxifen	Objective clinical 17-36%	**Tamoxifen**: Her2 over-expression ER-negative tumor	87-92, 119-124
Aromatase inhibitors Letrozole Anastrozole Exemestane	Objective clinical Letrozole 30-55% Anastrozole 21-43% Exemestane 41%	**Aromatase inhibitors**: Lack of estrogen-response	114, 119-124
Trastuzumab (Herceptin)*	Objective - complete 6% - partial 20%	**Trastuzumab**: Decreased PTEN	116,125

*Used for treatment of metastatic breast cancer.

The *sine quo non* of the malignant phenotype is deregulation of the cell cycle.[3] However, while deregulation of the tightly controlled cell cycle events clearly leads to malignant transformation, it also provides intriguing targets for alternative therapeutic approaches to overcome the problem of chemoresistance. One target of particular interest for this approach is the cyclinE/cyclin-dependent kinase 2 (Cdk2) complex and the G1/S transition of the cell cycle.

The G1/S transition is regulated through the cooperation of two essential, parallel cell cycle pathways, RB and Myc, which converge on the control of the G1 cyclin-dependent kinase

- Cdk2.[7-12] Cdk2 activity in the G1/S transition is both rate-limiting and necessary for cell replication, and it is dependent upon appropriate interaction with the G1 cyclin, cyclin E.[13,14] A number of recent studies have suggested that deregulation of cyclin E plays a significant role in the aggressiveness of breast cancer and other malignancies.[1,2,15-18] In fact, a form of cyclin E deregulation caused by the generation of recently identified hyperactive low molecular weight (LMW) isoforms has been shown to be the most powerful predictor of outcome in patients with early stage breast cancer.[2] Because only tumor cells possess the machinery to generate these forms,[4] they provide both a potential means of identifying malignant versus normal cells as well as a multi-leveled target within an essential cell cycle pathway. For these reasons therapies designed to take advantage of the deregulation of cyclin E and the G1/S transition are appealing. This review describes the mechanisms of resistance to commonly used systemic therapies for the treatment of breast cancer, with particular respect to the role of the cell cycle. The mechanisms and effects of the deregulation of cyclin E in breast cancer are reviewed and novel approaches to circumventing chemoresistance through abrogation of the malignant cell cycle are proposed.

Conventional Chemotherapies of Breast Cancer

Anthracyclines

Anthracycline-based chemotherapy is the current standard of care in breast cancer treatment. Anthracyclines (doxorubicin, epirubicin) are intercalating, topoisomerase II poisons that bind to double-stranded DNA causing structural changes which interfere with DNA and RNA synthesis. Multiple forms of resistance to these drugs have been identified. Because many of these agents are natural products, resistance by cellular efflux mechanisms, such as the *mdr1*, *mrp1* and *mrp2* gene product members of the ATP-binding cassette (ABC) family, have been demonstrated.[19-21] In addition, alterations in topoisomerases, including point mutations as well as defects in phosphorylation, have been described in some drug-resistant cell lines.[22,23] Furthermore, aberrant intracellular localization (cytoplasmic) has been implicated by decreasing the potential for DNA binding.[24-27] Finally, although not yet clearly demonstrated, because these agents function by causing structural DNA damage which should ultimately lead to apoptosis, alterations in the apoptotic proteins of the cell (e.g., p53 and the Bcl-2 family), have been suggested to confer drug resistance.[28]

Alkylating Agents

The alkylating agent cyclophosphamide is frequently used in anthracycline-based chemotherapy regimens for breast cancer. A member of the nitrogen mustard family, cyclophosphamide activation requires cytochrome P450-mediated oxidation in the liver to produce 4-hydroxycyclophosphamide. Relatively nonpolar, 4-hydroxycyclophosphamide readily diffuses into target cells where its tautomer, aldophosphamide, decomposes to the active alkylating agent, phosphoramide mustard.[29] At least three mechanisms of resistance to cyclophosphamide have been identified. Because cyclophosphamide enters the cell through diffusion, it is not a known substrate for the multiple-drug-resistance (MDR) export systems.[30] Intracellular inactivation of cyclophosphamide by its natural detoxifier, aldehyde dehydrogenase, has been shown not only to protect normal cells from the cytotoxicity of this agent, but also to confer resistance in tumor cells.[31-36] In addition, increased 4-hydroxycyclophosphamide glutathione conjugation, either spontaneous or through enhanced transcription of glutathione S-transferase, has been shown to contribute to cyclophosphamide resistance.[37-41] Finally, resistance related to the cell's ability to either repair DNA interstrand cross-links or to arrest in the G2 phase of the cell cycle in response to the alkylating damage has also been demonstrated.[42]

Antimetabolites

The pyrimidine analog 5-flourouracil (5-FU) is used in the management of many epithelial malignancies, including breast cancer. Potential mechanisms of cytotoxicity caused by 5-FU include RNA incorporation,[43,44] dTTP depletion by thymidylate synthase inhibition,[45] DNA incorporation, or DNA damage due to excision of uracil or 5-FU.[46-49] Resistance to 5-FU therapy has been demonstrated in the form of reduced anabolism of the analog to the nucleotide form either through altered condensation with pyrophosphorylribose-5-PO4 (PRPP) or the pyrimidine salvage pathway.[43,44] In addition, increased catabolism of 5-FU due to elevated dihydropyrimidine dehydrogenase (DPD) activity can lead to decreased sensitivity and has been shown to be a predictor of decreased response in some tumor types.[50,51] Other mechanisms of resistance have been related to changes in thymidylate synthase (reduced affinity for FdUMP,[52] increased rate of synthesis or activity[53]), and the mode of exposure to the drug (enteral versus parenteral).[54-56]

Folate Antagonists

Another important agent in the management of breast cancer is the folate antagonist, methotrexate (MTX). MTX stoichiometrically inhibits the enzyme dihydrofolate reductase (DHFR) leading to decreased availability of thymidine, decreased DNA synthesis and ultimately cell death.[57] Resistance to MTX can be either intrinsic or acquired. A significant mechanism of intrinsic resistance to MTX is reduced formation of long-chain MTX polyglutamates due to decreased folylpolyglutamate synthetase (FPGS) activity which can lead to both decreased affinity for DHFR as well as increased cell efflux.[58-62] Other mechanisms of intrinsic resistance to MTX include impaired transport through the reduced folate carrier (RFC),[63,64] and increased DHFR levels due to increased levels of the transcription factor E2F which occur in the absence of the tumor-suppressor retinoblastoma protein.[65-67] Acquired mechanisms of resistance to MTX include increased DHFR activity due to amplification of its gene,[68-74] altered binding of MTX to DHFR due to DHFR mutations,[75-79] decreased MTX uptake secondary to decreased long-chain polyglutamate formation, and decreased influx through the RFC.[80]

Microtubule-Targeting Agents

Recently added to the breast cancer chemotherapy armamentarium are the taxanes (paclitaxel, docetaxel) which are naturally-occurring antimicrotubule agents. Taxanes have been shown to prevent depolymerization of the microtubule by binding and stabilizing the molecular conformation of the protofilament of the microtubule.[81] This stabilization causes a mitotic arrest at the metaphase/anaphase juncture.[82] The mechanisms of cell death caused by the taxanes include apoptosis through the activation of caspase 3 and 8 as well as a noncaspase activated mechanism of DNA fragmentation that causes apoptosis.[83-86] Multiple possible mechanisms of resistance to taxane therapy exist including increased expression of the *mdr1* gene and Pgp efflux pump,[87] structural alterations in the α- and β- tubulins which impair microtubule polymerization,[87-92] and dynamic instability of the microtubule caused by increased expression of the β_{III} isotype of β tubulin.[90-92]

Hormonal and Targeted Therapies

Because of the important role of estrogen in the development of breast cancer, endocrine therapy, either in the form of anti-estrogens or estrogen deprivation, plays a significant role in the medical treatment of breast cancer. With respect to the cell cycle, estrogen has been shown to have a regulatory role of the molecules involved in the G1/S phase progression, including the expression and function of c-Myc[93-95] and cyclin D1.[96,97] Furthermore, other studies have demonstrated estrogen-mediated inhibition of the generation of the cyclin-dependent kinase inhibitor (CKI) p21, resulting in increased cyclinE/Cdk2 complex activity.[96,98,99] Deregulation of any of these cell cycle regulators may contribute to increased anti-estrogen resistance.

In addition, increasing evidence suggests that breast cancer growth may also be influenced by the coordinated actions of the estrogen receptor (ER) and the HER2 growth factor receptor signaling pathway.[100] Estrogen binding of the ER induces a series of both membrane-bound (G-protein-coupled receptor activation[101]) and nuclear events (phosphorylation of the receptor, conformational alteration, receptor dimerization, receptor complex-promoter binding, and recruitment of coactivators).[102] The nuclear events ultimately lead to the transcriptional regulation of the ER target genes.[103,104] The membrane-bound events have been shown to lead to the paracrine or autocrine activation of the HER2 signaling pathway through the release of epidermal growth factor (EGF).[105] Activation of the HER2 signaling pathway initiates a kinase signaling cascade which has been shown to augment the transcriptional activation potential of ER resulting in enhanced cell proliferation and survival.[105-107] This "crosstalk" between the ER and the HER2 signaling pathway may also be one of the major mechanisms for resistance to endocrine therapy in breast cancer treatment.[105,108,109]

The current mainstay of anti-estrogen therapy, tamoxifen, is known to display partial agonist-antagonist activities in different tissues and cells, depending upon the various ER coactivators and corepressors present.[110] Like estrogen, tamoxifen also has both nuclear and membrane-bound effects.[105] In addition to preventing the binding of estrogen to the ER, under favorable conditions such as negative or very low levels of HER2, tamoxifen's effects are primarily antagonistic and nuclear. In this setting, the ER conformation induced by the binding of tamoxifen leads to the recruitment of corepressors and deacetylases which inhibit transcriptional activity. On the other hand, in the setting of abundant HER2, evidence suggests that agonist effects of tamoxifen may predominate through membrane-bound events which lead to HER2 signaling activation, tumor growth and resistance to anti-estrogen therapy.[105,110]

Options to overcome anti-estrogen therapy resistance in breast cancer patients include two currently used therapies: estrogen deprivation through aromatase inhibition and inhibition of HER2 signaling by the monoclonal antibody receptor tyrosine kinase inhibitor—trastuzumab (Herceptin). Aromatase inhibitors (AIs) are a group of agents that inhibit the steroid hydroxylations involved in the conversion of androstenedione to estrone, thereby lowering both the circulating and intratumoral amounts of estrogen available to bind the ER.[111] In theory, these agents should be able to abrogate both the membrane-bound HER2 activating ER events, as well as the nuclear steroid signaling events. In support of this theory, clinical trials have demonstrated the superiority of AIs over tamoxifen in both HER2-overexpressing breast cancers as well as ER-positive/PR-negative tumors.[112,113] Resistance to AIs is thought not to be due to failure of these agents to suppress estradiol, but rather through resistance to the hormone itself.[114]

Trastuzumab is a humanized monoclonal antibody that specifically binds to the extracellular domain of the HER2/neu tyrosine kinase receptor. Down-regulation and inactivation of the receptor by the antibody occur through multiple mechanisms including accelerated degradation, interference with the hetrodimerization of the receptor, and targeting of the immune system to HER2 overexpressing cells.[115] In addition, trastuzumab has been shown to stabilize and activate the PTEN tumor suppressor leading to down-regulation of the P13K-Akt signaling pathway and initiating cell cycle arrest.[116] Recently, Nagata et al demonstrated that when the expression of PTEN is reduced, the antitumoral effects of trastuzumab are impaired. Based on these findings, the authors predicted and confirmed that clinical resistance to trastuzumab correlated with low levels of PTEN.[116]

The Cell Cycle as a Therapeutic Target in Breast Cancer

Deregulation of G1/S Transition

Cell division is a complex and orderly process divided into four phases involving cell growth and monitoring (G1 and G2 phase), DNA synthesis (S phase), and mitosis (M phase).[126] In the settings of favorable cellular and tissue environments, cells can initiate their own division and enter a mitogen-dependent growth phase (early G1). Upon entering the cell cycle, the order and quality of the cell cycle events are monitored and ensured by a series of checkpoints.[127] Commitment to genome replication and eventual cell division occurs late in the G1 phase at a period defined as the restriction point.[128] Recent studies have suggested that this molecular "point of no return" revolves around the activity of Cdk2 and its G1-associated cyclin, cyclin E, which is also the point of convergence of the RB (p16-Cdk4/6-cyclin D-pRb) and Myc proto-oncogene pathways.[7-12]

Cdk2 belongs to a family of serine and threonine protein kinases whose substrates include intracellular, cell cycle-regulatory proteins that control the major cell cycle events: DNA replication, mitosis and cytokinesis. One of the most important functions of Cdk2 is the mid-late G1 phase phosphorylation and inactivation of the tumor suppressor pRb which, in normal cells, is essential to cell cycle progression. Like all Cdks, Cdk2 activity is governed by an array of enzymes and proteins, the most prominent of which are cyclins. Unlike Cdk levels which normally remain constant throughout the cell cycle, cyclins, as the name implies, undergo a tightly regulated cycle of synthesis and degradation resulting in the cyclic assembly and activation of cyclin-Cdk complexes.[129,130] In each phase of the cell cycle, Cdk activity is dependent upon binding to the appropriate cyclin protein and it is this activation that propels the cell through the cell cycle. In late G1 phase, cyclin E complexes with Cdk2 to control the transition into S-phase.[131]

In normally dividing cells, the G1-synthesis and S phase-degradation of cyclin E are tightly regulated.[132] In late G1, cyclin E transcription is activated when pRb is hyperphosphorylated by cyclin D/Cdk4/Cdk6 complexes, relieving repression of the cyclin E gene. This event causes a G1 arrest allowing further accumulation of cyclin E protein. This accumulation continues to a level where cyclin E/Cdk2 itself phosphorylates pRb, relieving the repression of the S-phase cyclin, cyclin A, and Cdk1, and allowing the cell cycle to progress to mitosis.[133] Concomitant activation of cyclin E-Cdk2 kinase also occurs through the Myc proto-oncogene pathway.[12] c-Myc proto-oncogene is a mitogen-induced transcription factor of the helix-loop-helix/leucine zipper protein family whose role in cyclin E activation includes both direct mechanisms (transcriptional effects) and indirect mechanisms (sequestration or enhanced degradation of the cyclinE/Cdk2 inhibitor p27).[11,12,134-136] Deregulation of any of these cell cycle components can lead to the unscheduled expression of cyclin E that is often seen in cancer (Fig. 1).

Multiple mechanisms of malignant deregulation of cyclin E have been identified including gene amplification,[137,138] overexpression,[139,140] downregulation of inhibitory proteins such as p27,[141] faulty degradation[139,140,142] and the generation of LMW isoforms of cyclin E.[4,143] Of these cyclin E alterations, the most profound is the generation of the LMW isoforms which have been associated with poor clinical outcomes in breast cancer and other malignancies. In fact, in a retrospective study of 395 breast cancer patients, the presence of the LMW isoforms of cyclin E was found to be eight times more predictive of poor prognosis than nodal status.[2] Significant biochemical and functional differences between the full-length and LMW isoforms of cyclin E are thought to explain the correlation between this type of deregulation and increased breast cancer mortality.[144]

Six cyclin E isoforms (EL1-6) have been identified (Fig. 2).[1] The predominant, full-length (50-kDa) isoform (EL1) is the only isoform found in normal cells. The LMW isoforms (EL2-6) are generated either by alternative translation (EL4) or proteolytic processing of the full-length protein by an elastase-like protease which creates two paired-isoforms (EL2/3 and EL5/6). Only tumor cells are capable of processing cyclin E into its LMW forms which are nuclear and functionally hyperactive.[143]

Tumorigenic properties associated with the LMW cyclin E isoforms involve both aberrant control of both the cell cycle as well as many aspects of DNA replication. In normal cells, direct binding of chromatin by cyclin E initiates DNA replication and also potentially blocks rereplication.[145] Cyclin E has been shown to induce histone gene transcription at the beginning of S phase through the phosphorylation of NPAT[146,147] and control centrosome

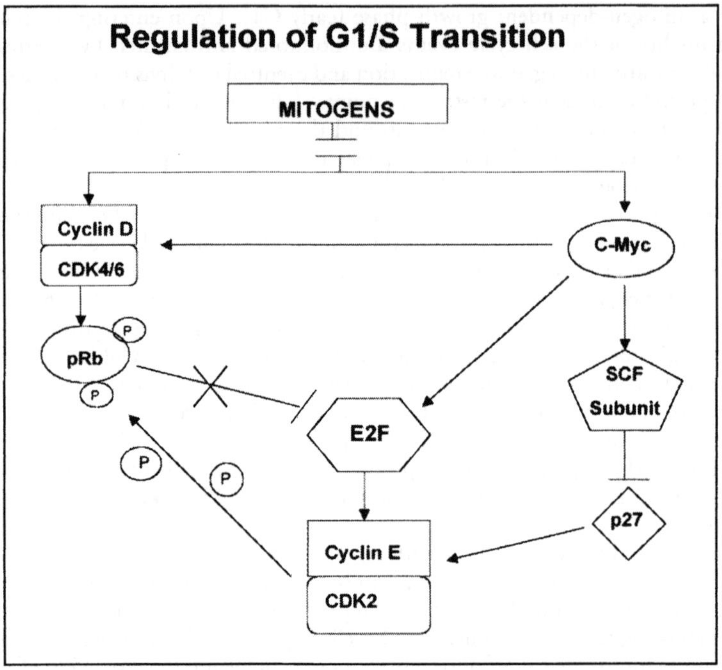

Figure 1. Regulation of the G1/S transition by the cyclin-dependent kinase (Cdk) 2 and its G1-associated cyclin, cyclin E, at the point of convergence of the RB (p16-Cdk4/6-cyclin D-pRb) and Myc proto-oncogene pathways.

duplication through the phosphorylation of nucleophosmin B23[148,149] and stabilization of the Mmps1p-like kinase.[148] Additional cyclin E substrates involved in other DNA replication processes such as transcriptional regulation (SWI/SNF),[150] pre-mRNA splicing (spliceosomal protein)[151,152] and modulation of transcription factors (Id2, Id3)[153,154] have also been identified. Deregulation of cyclin E impacts many of these aspects of DNA replication, often conferring a growth advantage to tumor cells.

With respect to the cell cycle, the LMW forms of cyclin E have been shown to result in decreased cell doubling times, decreased cell size and loss of growth factor requirements for proliferation.[131,155] These effects are due to both the increased biochemical and biological activity of the LMW forms as compared with the full-length cyclin E. Specifically, because of the increased affinity for Cdk2 of the LMW cyclin E, there appears to be at least a two-fold increase in associated Cdk2 kinase activity and a three- to five-fold increase in resistance to the Cdk inhibitors p21 and p27 in cells with these forms.[156] Through this increased activity deregulated cyclin E has been shown to independently and sufficiently phosphorylate pRb, enough to induce aberrant cell cycle progression.[4]

Targeting the G1/S Transition Therapeutically

The central role of cyclinE/Cdk2 in the regulation of the G1/S transition makes this complex an attractive target for novel cancer therapy. First, differential expression of the tumor-specific LMW cyclin E provides a unique means of both identifying and targeting tumor cells only, potentially increasing selective lethality of the therapy. In addition the same target may also act as a more objective measure of both the degree of tumor aggressiveness as well as therapeutic response. Elucidation of the mechanisms of this differential expression have helped identify opportunities for therapeutic exploitation.

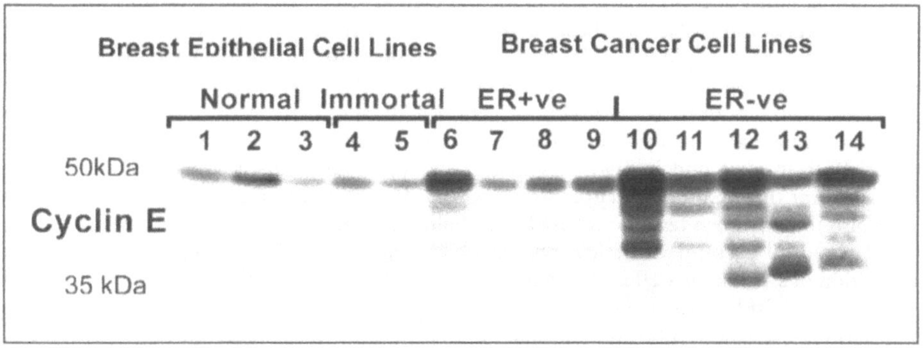

Figure 2. Western blot analysis of cyclin E in normal and immortalized breast epithelial cell lines and estrogen receptor positive (ER+ve) and negative (ER-ve) breast cancer cell lines. Deregulated cyclin E caused by the proteolytic generation of hyperactive, low molecular weight isoforms (35-50 kDa), is seen only in the breast cancer cell lines.

Proteolytic processing of the full length cyclin E has recently been identified as the mechanism responsible for the generation of the hyperactive LMW forms of cyclin E seen in some tumor cells.[4,143] Two proteolytically sensitive domains in cyclin E have been identified and four of the five LMW forms are accounted for by proteolysis at these two sites, with post-translational modification creating two closely migrating doublets—EL2/3 and EL5/6. Sequence analysis of the proteolytically cleaved regions of cyclin E have identified an elastase-like serine protease as responsible for generating these LMW forms.[4]

The differential expression of the LMW forms of cyclin E in tumor versus normal cells may be due to either increased elastase-like activity in tumor cells, increased elastase inhibitor levels in normal cells, or decreased elastase-inhibitor levels in tumor cells. Each of these possible mechanisms presents a potential target for cancer therapy. Recent studies looking at the neutrophil (elastase) inhibitor, CE-2072, demonstrated partial abrogation of some of the LMW forms of cyclin E in the breast cancer cell line MDA-MB-157, a cell line that expresses all 6 isoforms of cyclin E. In comparison, CE-2072 treatment of MCF-10A breast cancer cells, which do not express the LMW isoforms, did not affect the expression of cyclin E in these cells. In addition, treatment with CE-2027 was found to cause partial arrest in the G1 phase of the cell cycle in tumor cells, but not normal cells. These results suggest a cause and effect relationship between the disappearance of some of the LMW forms of cyclin E in tumor cells and partial growth arrest of these cells.[156] Although elastase inhibitors are not used in the clinic for the treatment of cancer at this time, some reports have suggested that the use of these agents for chemotherapy may provide a high therapeutic index. Following identification of the specific protease of the elastase class which cleaves cyclin E into the LMW forms, cyclin E-specific protease inhibitors may then be engineered.

The differential expression of the LMW forms of cyclin E in tumor versus normal cells may also occur through a relative decrease in the presence or function of an endogenous elastase inhibitor—elafin.[156] Thus an alternative approach to elastase inhibition could be to increase intracellular levels of functional elafin. Potential mechanisms for this approach include increased elafin expression through adenoviral gene therapy or by the administration of the elafin protein in target-specific, trigger-specific liposomes. While effective results through this means of drug delivery remain on the horizon,[157] it is possible that someday liposomes targeted to breast cancer-specific membrane receptors (e.g., ER, HER2) could deliver the relatively small elafin protein (9 kDa) intracellularly where a tumor-specific enzyme (elastase) could release the liposomal payload. Finally, as on-going studies better elucidate the mechanisms by which elafin is down-regulated in tumor cells, other approaches to increasing elafin expression will become available.

Another target at this nodal point in the cell cycle is Cdk2. Because of their central role in cell cycle regulation, Cdks have been targeted for both drug and small molecule therapy. The two basic schemes employed to inhibit Cdks include either direct blockade of their kinase activity or targeting of their major regulators (indirect). Over 50 direct chemical Cdk inhibitors have been described with varying degrees of Cdk specificity. Most of these compounds modulate kinase activity by interacting specifically with the ATP-binding pocket of the enzyme. Both in vitro and in vivo Cdk-specific cell cycle and anti-tumoral effects have been described for three Cdk modulators—flavopiridol, R-roscovitine, and BMS-387032—which have also recently been tested in phase I and II clinical trials.

Flavopiridol is a semisynthetic flavonoid which appears to induce cell cycle arrest by direct inhibition of all Cdks as well as through transcriptional repression of cyclin D1.[158-160] Phase I trials for flavopiridol have demonstrated tolerable toxicity with some objective responses across a spectrum of advanced solid and nonsolid tumors.[161,162] Furthermore, a Phase II trial in metastatic lung cancer showed a median overall survival consistent with both a randomized trial of four platinum-based chemotherapy regimens and with the survival observed with the approved EGFR inhibitor gefitinab (Iressa).[163-165] R-roscovitine (CYC202) is an olomoucine analogue and a potent inhibitor of Cdk1, Cdk2, and Cdk5.[166] Preclinical studies in multiple xenograft models have shown antitumoral effects in the forms of both cell cycle arrest as well as evidence of apoptosis.[167] Two phase I clinical trials of oral CYC202 have demonstrated tolerable toxicity[168,169] and both single agent and combination chemotherapy phase II clinical trials are being planned. BMS-387032 is an aminothiazole Cdk2 inhibitor with a 10-100-fold selectivity for Cdk2 over Cdk1, Cdk4 and other kinases.[170] In vitro and in vivo antiproliferative effects of this class of compounds include cell cycle arrest with loss of pRb phosphorylation and some evidence of apoptosis. Three phase I trials have shown tolerable toxicity and some objective responses.[171-173] Phase II and combination phase I trials are planned.

One nonspecific chemical Cdk modulator, UCN-01, has also been tested in clinical trials. In addition to anti-Cdk activity, UCN-01 also exhibits a number of other cell cycle and non-cell cycle molecular effects. With respect to the cell cycle, UCN-01 has been shown to abrogate both the G1[174-181] and G2 checkpoints through inappropriate cdc2 activation[182] and chk1 inhibition,[183-185] and also appears to possess increased cytotoxicity in cells with p53 mutations.[182] Important non-cell cycle effects include potent inhibition of protein kinase C isoenzymes and modulation of the PI3 kinase/Akt survival pathway.[186,187] UCN-01 has been evaluated in both phase I and II trials with tolerable toxicity and some objective responses.[188,189] Synergistic effects of UCN-01 have been observed with many chemotherapeutic agents in preclinical models[174,190-193] and clinical trials of combination chemotherapies are underway.

While the Cdk modulation approach is certainly intriguing, one major limitation of the current agents under investigation is their lack of true cytotoxicity. Although most of the agents being tested in clinical trials have shown some preclinical evidence of inducing apoptosis, a G1 or G2 cell cycle arrest is the predominant result. For this reason, results of the combination chemotherapy trials are eagerly awaited.

Another limitation shared by both these agents and other conventional chemotherapies is a lack of tumor-specificity. Once again, cell cycle deregulation in the form of LMW cyclin E isoforms may help overcome this lack of specificity. Indole-3-carbinol (I3C) is an indirect Cdk2 inhibitor which has recently been shown to induce a G1 arrest in breast cancer cells by inhibiting Cdk2 activity associated with the LMW forms of cyclin E.[194] In a study by, Garcia et al, MCF-7 breast cancer cells treated with I3C demonstrated a shift in the size distribution of the Cdk2 protein complex from an enzymatically active 90kDa protein to a larger, 200kDa protein, with reduced kinase activity. In addition, the treated cells appeared to have lost their association with the 35 kDa LMW isoform of cyclin E as compared with nontreated cells. Furthermore, I3C treatment was also associated with a subcellular cytoplasmic localization of the Cdk2-cyclin E complex. These changes were felt to be indole-specific as treatment with the I3C natural dimerization product, DIM, or the anti-estrogen, tamoxifen, did not produce similar results. No changes in CKI (p21 or p27) levels were seen with I3C treatment. While compelling, this study is not without some limitations. Whether the effects of I3C on MCF-7 breast cancer cells are tumor-specific has not been determined as they were not compared to normal breast epithelial cells. Nor was the generalizability of the I3C treatment effects assessed in other cancer cell lines that express the proteolytic generated LMW isoforms of cyclin E (e.g., MDA-MB-157, MDA-MB-436, and Ovcar).

Other potential indirect modulators of Cdk2 activity worth considering include the CKIs p27 and p21. With respect to breast cancer, increasing the expression of p21 may provide an additional means of overcoming some anti-estrogen resistance as well as increase anti-estrogen sensitivity in ER-negative breast cancers. In a study by Chen et al,[195] after demonstrating a strong association between p21 and ER expression, the investigators proceeded to induce the ER and estrogen receptor element promoters in an estrogen responsive manner through over-expression of p21 in a p21-negative, ER-negative breast cancer cell line. These cells were sensitive to both the growth inhibitor effects of anti-estrogen treatment as well as the growth stimulatory effects of 17β-estradiol. These findings suggest that p21 may play a significant role in the estrogen-signaling pathway and raise the possibility that anti-estrogen therapy may be effective in p21-positive, ER-negative breast cancers. Furthermore, a number of commonly used breast cancer chemotherapeutic agents have also been shown to induce p21, including paclitaxel,[196,197] doxorubicin,[198] and vinorelbine,[199] raising the potential of treatment strategies that combine chemotherapy and anti-hormonal therapy in ER-negative breast cancers induced to express p21.

Other possible strategies for targeting CKIs include increased protein expression through gene therapy or administration of tumor-targeted peptidomimetics of CKIs or other peptides that inhibit CDK activity. Because both p21 and p27 are substrates for ubiquitination and proteosome-dependent degradation, strategies designed to decrease the turnover of these CKIs through inhibition of ubiquitin-mediated proteolysis by the proteosome should also be considered. In fact, induction of both p21 and p27 in MDA-MB-157 cells through inhibition of the proteosome by treatment with the HMG-CoA reductase inhibitor, lovastatin, has been demonstrated to cause a G1 arrest.[200] In this study, the mechanism of p21 and p27 accumulation was clearly shown to be due to unique inhibitory effects of the closed-ring prodrug form of lovastatin on the proteosome, and not related to the HMG-CoA reductase inhibition of the open-ring form of the drug. With respect to breast cancer, as low levels of p27 have also been correlated with poor prognosis in young breast cancer patients,[16] efforts geared towards increasing the levels of both p27 and p21, for previously described reasons, may be particularly helpful in overcoming cell cycle-related drug resistance. Currently investigations with other proteosome inhibitors such as farnesyl transferase inhibitors are also on-going.

Summary

As some facet of cell cycle deregulation is present in all tumors, it is reasonable to consider cancer a disease of the cell cycle. In addition to driving the malignant transformation of normal cells, cell cycle deregulation also contributes to the chemotherapy resistance of cancer cells, as these agents often rely on the presence of normal cell cycle checkpoints to cause cell death. However, while this cell cycle-driven resistance often seems insurmountable, it may ultimately prove to be the Achilles' heel of cancer cell survival.

As illustrated in this review, the deregulated cell cycle provides multiple opportunities for tumor targeted therapies to either break the cycle by reregulation or to target it in combination with more conventional chemotherapies in ways that result in mitotic catastrophe (e.g., DNA damage plus G1 and G2 checkpoint abrogation.) However, in order for these cell cycle-directed strategies to work, there are some basic requirements that need to be met. First, specificity through differential expression of the target in normal versus tumor cells must be present. Second, the mechanism of the differential expression needs to be understood. Finally, the mechanism needs to be exploited therapeutically. Deregulation of cyclin E through the proteolytic generation of hyperactive LMW isoforms meets these criteria and means of exploiting this potential Achilles' heel are underway.

References

1. Keyomarsi K, O'Leary N, Molnar G et al. Cyclin E, a potential prognostic marker for breast cancer. Cancer Res 1994; 54(2):380-385.
2. Keyomarsi K, Tucker SL, Buchholz TA et al. Cyclin E and survival in patients with breast cancer. N Engl J Med 2002; 347(20):1566-1575.
3. Pardee AB. G1 events and regulation of cell proliferation. Science 1989; 246(4930):603-608.
4. Porter DC, Zhang N, Danes C et al. Tumor-specific proteolytic processing of cyclin E generates hyperactive lower-molecular-weight forms. Mol Cell Biol 2001; 21(18):6254-6269.
5. Tamoxifen for early breast cancer: An overview of the randomised trials. Early Breast Cancer Trialists' Collaborative Group. Lancet 1998; 351(9114):1451-1467.
6. Polychemotherapy for early breast cancer: An overview of the randomised trials. Early Breast Cancer Trialists' Collaborative Group. Lancet 1998; 352(9132):930-942.
7. Berns K, Hijmans EM, Bernards R. Repression of c-Myc responsive genes in cycling cells causes G1 arrest through reduction of cyclin E/CDK2 kinase activity. Oncogene 1997; 15(11):1347-1356.
8. Berns K, Martins C, Dannenberg JH et al. p27kip1-independent cell cycle regulation by MYC. Oncogene 2000; 19(42):4822-4827.
9. Beier R, Burgin A, Kiermaier A et al. Induction of cyclin E-cdk2 kinase activity, E2F-dependent transcription and cell growth by Myc are genetically separable events. Embo J 2000; 19(21):5813-5823.
10. Roussel MF, Theodoras AM, Pagano M et al. Rescue of defective mitogenic signaling by D-type cyclins. Proc Natl Acad Sci USA 1995; 92(15):6837-6841.
11. Leone G, DeGregori J, Sears R et al. Myc and Ras collaborate in inducing accumulation of active cyclin E/Cdk2 and E2F. Nature 1997; 387(6631):422-426.
12. Santoni-Rugiu E, Falck J, Mailand N et al. Involvement of Myc activity in a G(1)/S-promoting mechanism parallel to the pRb/E2F pathway. Mol Cell Biol 2000; 20(10):3497-3509.
13. Bartek J, Bartkova J, Lukas J. The retinoblastoma protein pathway and the restriction point. Curr Opin Cell Biol 1996; 8(6):805-814.
14. Sherr CJ, Roberts JM. CDK inhibitors: Positive and negative regulators of G1-phase progression. Genes Dev 1999; 13(12):1501-1512.
15. Muller-Tidow C, Metzger R, Kugler K et al. Cyclin E is the only cyclin-dependent kinase 2-associated cyclin that predicts metastasis and survival in early stage nonsmall cell lung cancer. Cancer Res 2001; 61(2):647-653.
16. Porter PL, Malone KE, Heagerty PJ et al. Expression of cell-cycle regulators p27Kip1 and cyclin E, alone and in combination, correlate with survival in young breast cancer patients. Nat Med 1997; 3(2):222-225.
17. Richter J, Wagner U, Kononen J et al. High-throughput tissue microarray analysis of cyclin E gene amplification and overexpression in urinary bladder cancer. Am J Pathol 2000; 157(3):787-794.
18. Sui L, Dong Y, Ohno M et al. Implication of malignancy and prognosis of p27(kip1), Cyclin E, and Cdk2 expression in epithelial ovarian tumors. Gynecol Oncol 2001; 83(1):56-63.
19. Gros P, Ben Neriah YB, Croop JM et al. Isolation and expression of a complementary DNA that confers multidrug resistance. Nature 1986; 323(6090):728-731.
20. Koike K, Kawabe T, Tanaka T et al. A canalicular multispecific organic anion transporter (cMOAT) antisense cDNA enhances drug sensitivity in human hepatic cancer cells. Cancer Res 1997; 57(24):5475-5479.
21. Cole SP, Chanda ER, Dicke FP et al. NonP-glycoprotein-mediated multidrug resistance in a small cell lung cancer cell line: Evidence for decreased susceptibility to drug-induced DNA damage and reduced levels of topoisomerase II. Cancer Res 1991; 51(13):3345-3352.
22. DeVore RF, Corbett AH, Osheroff N. Phosphorylation of topoisomerase II by casein kinase II and protein kinase C: Effects on enzyme-mediated DNA cleavage/religation and sensitivity to the antineoplastic drugs etoposide and 4'-(9-acridinylamino)methane-sulfon-m-anisidide. Cancer Res 1992; 52(8):2156-2161.
23. Takano H, Kohno K, Ono M et al. Increased phosphorylation of DNA topoisomerase II in etoposide-resistant mutants of human cancer KB cells. Cancer Res 1991; 51(15):3951-3957.
24. Feldhoff PW, Mirski SE, Cole SP et al. Altered subcellular distribution of topoisomerase II alpha in a drug-resistant human small cell lung cancer cell line. Cancer Res 1994; 54(3):756-762.
25. Mirski SE, Cole SP. Cytoplasmic localization of a mutant M(r) 160,000 topoisomerase II alpha is associated with the loss of putative bipartite nuclear localization signals in a drug-resistant human lung cancer cell line. Cancer Res 1995; 55(10):2129-2134.
26. Harker WG, Slade DL, Parr RL et al. Selective use of an alternative stop codon and polyadenylation signal within intron sequences leads to a truncated topoisomerase II alpha messenger RNA and protein in human HL-60 leukemia cells selected for resistance to mitoxantrone. Cancer Res 1995; 55(21):4962-4971.

27. Wessel I, Jensen PB, Falck J et al. Loss of amino acids 1490Lys-Ser-Lys1492 in the COOH-terminal region of topoisomerase IIalpha in human small cell lung cancer cells selected for resistance to etoposide results in an extranuclear enzyme localization. Cancer Res 1997; 57(20):4451-4454.
28. Rubin E, Halt W. Anthracyclines and DNA Intercalators/Epipodophyllotoxins/Camptothecins/DNA Topoisomerases. In: Kufe DW, Pollock RE, Weichselbaum RR, Bast RC, Gansler TS, Holland JF, Frei E, eds. Cancer Medicine 6. Vol 1. Hamilton: BC Decker, 2003:783.
29. Fenselau C, Kan MN, Rao SS et al. Identification of aldophosphamide as a metabolite of cyclophosphamide in vitro and in vivo in humans. Cancer Res 1977; 37(8 Pt 1):2538-2543.
30. Colvin M. Alkylating agents and platinum antitumor compounds. In: Kufe DW, Pollock RE, Weichselbaum RR, Bast RC, Gansler TS, Holland JF, Frei E, eds. Cancer Medicine 6. Vol 1. Hamilton: BC Decker, 2003:762.
31. Colvin M, Russo JE, Hilton J et al. Enzymatic mechanisms of resistance to alkylating agents in tumor cells and normal tissues. Adv Enzyme Regul 1988; 27:211-221.
32. Hilton J. Role of aldehyde dehydrogenase in cyclophosphamide-resistant L1210 leukemia. Cancer Res 1984; 44(11):5156-5160.
33. Koelling TM, Yeager AM, Hilton J et al. Development and characterization of a cyclophosphamide-resistant subline of acute myeloid leukemia in the Lewis x Brown Norway hybrid rat. Blood 1990; 76(6):1209-1213.
34. Parsons PG, Lean J, Kable EP et al. Relationships between resistance to cross-linking agents and glutathione metabolism, aldehyde dehydrogenase isozymes and adenovirus replication in human tumour cell lines. Biochem Pharmacol 1990; 40(12):2641-2649.
35. Rekha GK, Sreerama L, Sladek NE. Intrinsic cellular resistance to oxazaphosphorines exhibited by a human colon carcinoma cell line expressing relatively large amounts of a class-3 aldehyde dehydrogenase. Biochem Pharmacol 1994; 48(10):1943-1952.
36. Sreerama L, Sladek NE. Identification of a methylcholanthrene-induced aldehyde dehydrogenase in a human breast adenocarcinoma cell line exhibiting oxazaphosphorine-specific acquired resistance. Cancer Res 1994; 54(8):2176-2185.
37. Buller AL, Clapper ML, Tew KD. Glutathione S-transferases in nitrogen mustard-resistant and -sensitive cell lines. Mol Pharmacol 1987; 31(6):575-578.
38. Nakagawa K, Saijo N, Tsuchida S et al. Glutathione-S-transferase pi as a determinant of drug resistance in transfectant cell lines. J Biol Chem 1990; 265(8):4296-4301.
39. Pallante SL, Lisek CA, Dulik DM et al. Glutathione conjugates. Immobilized enzyme synthesis and characterization by fast atom bombardment mass spectrometry. Drug Metab Dispos 1986; 14(3):313-318.
40. Puchalski RB, Fahl WE. Expression of recombinant glutathione S-transferase pi, Ya, or Yb1 confers resistance to alkylating agents. Proc Natl Acad Sci USA 1990; 87(7):2443-2447.
41. Tew KD, Bomber AM, Hoffman SJ. Ethacrynic acid and piriprost as enhancers of cytotoxicity in drug resistant and sensitive cell lines. Cancer Res 1988; 48(13):3622-3625.
42. O'Connor PM, Ferris DK, White GA et al. Relationships between cdc2 kinase, DNA cross-linking, and cell cycle perturbations induced by nitrogen mustard. Cell Growth Differ 1992; 3(1):43-52.
43. Mulkins MA, Heidelberger C. Biochemical characterization of fluoropyrimidine-resistant murine leukemic cell lines. Cancer Res 1982; 42(3):965-973.
44. Mulkins MA, Heidelberger C. Isolation of fluoropyrimidine-resistant murine leukemic cell lines by one-step mutation and selection. Cancer Res 1982; 42(3):956-964.
45. Berger SH, Hakala MT. Relationship of dUMP and free FdUMP pools to inhibition of thymidylate synthase by 5-fluorouracil. Mol Pharmacol 1984; 25(2):303-309.
46. Houghton JA, Houghton PJ. Elucidation of pathways of 5-fluorouracil metabolism in xenografts of human colorectal adenocarcinoma. Eur J Cancer Clin Oncol 1983; 19(6):807-815.
47. Kufe DW, Major PP, Egan EM et al. 5-Fluoro-2'-deoxyuridine incorporation in L1210 DNA. J Biol Chem 1981; 256(17):8885-8888.
48. Ingraham HA, Tseng BY, Goulian M. Mechanism for exclusion of 5-fluorouracil from DNA. Cancer Res 1980; 40(4):998-1001.
49. Morikawa K, Fan D, Denkins YM et al. Mechanisms of combined effects of gamma-interferon and 5-fluorouracil on human colon cancers implanted into nude mice. Cancer Res 1989; 49(4):799-805.
50. Jiang W, Lu Z, He Y et al. Dihydropyrimidine dehydrogenase activity in hepatocellular carcinoma: Implication in 5-fluorouracil-based chemotherapy. Clin Cancer Res 1997; 3(3):395-399.
51. Etienne MC, Cheradame S, Fischel JL et al. Response to fluorouracil therapy in cancer patients: The role of tumoral dihydropyrimidine dehydrogenase activity. J Clin Oncol 1995; 13(7):1663-1670.
52. Bapat AR, Zarow C, Danenberg PV. Human leukemic cells resistant to 5-fluoro-2'-deoxyuridine contain a thymidylate synthetase with lower affinity for nucleotides. J Biol Chem 1983; 258(7):4130-4136.
53. Berger SH, Jenh CH, Johnson LF et al. Thymidylate synthase overproduction and gene amplification in fluorodeoxyuridine-resistant human cells. Mol Pharmacol 1985; 28(5):461-467.

54. Aschele C, Sobrero A, Faderan MA et al. Novel mechanism(s) of resistance to 5-fluorouracil in human colon cancer (HCT-8) sublines following exposure to two different clinically relevant dose schedules. Cancer Res 1992; 52(7):1855-1864.
55. Pizzorno G, Handschumacher RE. Effect of clinically modeled regimens on the growth response and development of resistance in human colon carcinoma cell lines. Biochem Pharmacol 1995; 49(4):559-565.
56. Sobrero AF, Aschele C, Guglielmi AP et al. Synergism and lack of cross-resistance between short-term and continuous exposure to fluorouracil in human colon adenocarcinoma cells. J Natl Cancer Inst 1993; 85(23):1937-1944.
57. Pizzorno G, Diasio R, Cheng Y. Pyrimidine and Purine Antimetabolites. In: Kufe DW, Pollock RE, Weichselbaum RR, Bast RC, Gansler TS, Holland JF, Frei E, eds. Cancer Medicine 6. Vol 1. Hamilton: BC Decker, 2003:748.
58. Curt GA, Jolivet J, Carney DN et al. Determinants of the sensitivity of human small-cell lung cancer cell lines to methotrexate. J Clin Invest 1985; 76(4):1323-1329.
59. Li WW, Waltham M, Tong W et al. Increased activity of gamma-glutamyl hydrolase in human sarcoma cell lines: A novel mechanism of intrinsic resistance to methotrexate (MTX). Adv Exp Med Biol 1993; 338:635-638.
60. Longo GS, Gorlick R, Tong WP et al. Disparate affinities of antifolates for folylpolyglutamate synthetase from human leukemia cells. Blood 1997; 90(3):1241-1245.
61. Li WW, Lin JT, Tong WP et al. Mechanisms of natural resistance to antifolates in human soft tissue sarcomas. Cancer Res 1992; 52(6):1434-1438.
62. Longo GS, Gorlick R, Tong WP et al. gamma-Glutamyl hydrolase and folylpolyglutamate synthetase activities predict polyglutamylation of methotrexate in acute leukemias. Oncol Res 1997; 9(5):259-263.
63. Guo W, Healey JH, Meyers PA et al. Mechanisms of methotrexate resistance in osteosarcoma. Clin Cancer Res 1999; 5(3):621-627.
64. Zhao R, Assaraf YG, Goldman ID. A mutated murine reduced folate carrier (RFC1) with increased affinity for folic acid, decreased affinity for methotrexate, and an obligatory anion requirement for transport function. J Biol Chem 1998; 273(30):19065-19071.
65. Li W, Fan J, Hochhauser D et al. Lack of functional retinoblastoma protein mediates increased resistance to antimetabolites in human sarcoma cell lines. Proc Natl Acad Sci USA 1995; 92(22):10436-10440.
66. Li W, Fan J, Banerjee D et al. Overexpression of p21(waf1) decreases G2-M arrest and apoptosis induced by paclitaxel in human sarcoma cells lacking both p53 and functional Rb protein. Mol Pharmacol 1999; 55(6):1088-1093.
67. Fan J, Bertino JR. Functional roles of E2F in cell cycle regulation. Oncogene 1997; 14(10):1191-1200.
68. Alt FW, Kellems RE, Bertino JR et al. Selective multiplication of dihydrofolate reductase genes in methotrexate-resistant variants of cultured murine cells 1978. Biotechnology 1992; 24:397-410.
69. Carman MD, Schornagel JH, Rivest RS et al. Resistance to methotrexate due to gene amplification in a patient with acute leukemia. J Clin Oncol 1984; 2(1):16-20.
70. Cowan KH, Goldsmith ME, Levine RM et al. Dihydrofolate reductase gene amplification and possible rearrangement in estrogen-responsive methotrexate-resistant human breast cancer cells. J Biol Chem 1982; 257(24):15079-15086.
71. Curt GA, Carney DN, Cowan KH et al. Unstable methotrexate resistance in human small-cell carcinoma associated with double minute chromosomes. N Engl J Med 1983; 308(4):199-202.
72. Fischer GA. Increased levels of folic acid reductase as a mechanism of resistance to amethopterin in leukemic cells. Biochem Pharmacol 1961; 7:75-77.
73. Horns Jr RC, Dower WJ, Schimke RT. Gene amplification in a leukemic patient treated with methotrexate. J Clin Oncol 1984; 2(1):2-7.
74. Srimatkandada S, Medina WD, Cashmore AR et al. Amplification and organization of dihydrofolate reductase genes in a human leukemic cell line, K-562, resistant to methotrexate. Biochemistry 1983; 22(25):5774-5781.
75. Dedhar S, Hartley D, Fitz-Gibbons D et al. Heterogeneity in the specific activity and methotrexate sensitivity of dihydrofolate reductase from blast cells of acute myelogenous leukemia patients. J Clin Oncol 1985; 3(11):1545-1552.
76. Domin BA, Cheng YC, Hakala MT. Properties of dihydrofolate reductase from a methotrexate-resistant subline of human KB cells and comparison with enzyme from KB parent cells and mouse S180 AT/3000 cells. Mol Pharmacol 1982; 21(1):231-238.
77. Flintoff WF, Essani K. Methotrexate-resistant Chinese hamster ovary cells contain a dihydrofolate reductase with an altered affinity for methotrexate. Biochemistry 1980; 19(18):4321-4327.

78. Goldie JH, Dedhar S, Krystal G. Properties of a methotrexate-insensitive variant of dihydrofolate reductase derived from methotrexate-resistant L5178Y cells. J Biol Chem 1981; 256(22):11629-11635.
79. Melera PW, Davide JP, Oen H. Antifolate-resistant Chinese hamster cells. Molecular basis for the biochemical and structural heterogeneity among dihydrofolate reductases produced by drug-sensitive and drug-resistant cell lines. J Biol Chem 1988; 263(4):1978-1990.
80. Gorlick R, Goker E, Trippett T et al. Defective transport is a common mechanism of acquired methotrexate resistance in acute lymphocytic leukemia and is associated with decreased reduced folate carrier expression. Blood 1997; 89(3):1013-1018.
81. Rowinsky E. Microtubule-targeting natural products. In: Kufe DW, Pollock RE, Weichselbaum RR, Bast RC, Gansler TS, Holland JF, Frei E, eds. Cancer Medicine 6. Vol 1. Hamilton: BC Decker, 2003:799.
82. Jordan MA, Toso RJ, Thrower D et al. Mechanism of mitotic block and inhibition of cell proliferation by taxol at low concentrations. Proc Natl Acad Sci USA 1993; 90(20):9552-9556.
83. Dumontet C, Sikic BI. Mechanisms of action of and resistance to antitubulin agents: Microtubule dynamics, drug transport, and cell death. J Clin Oncol 1999; 17(3):1061-1070.
84. Torres K, Horwitz SB. Mechanisms of Taxol-induced cell death are concentration dependent. Cancer Res 1998; 58(16):3620-3626.
85. Wang LG, Liu XM, Kreis W et al. The effect of antimicrotubule agents on signal transduction pathways of apoptosis: A review. Cancer Chemother Pharmacol 1999; 44(5):355-361.
86. Huisman C, Ferreira CG, Broker LE et al. Paclitaxel triggers cell death primarily via caspase-independent routes in the nonsmall cell lung cancer cell line NCI-H460. Clin Cancer Res 2002; 8(2):596-606.
87. Horwitz SB, Cohen D, Rao S et al. Taxol: Mechanisms of action and resistance. J Natl Cancer Inst Monogr 1993; (15):55-61.
88. Cabral F, Wible L, Brenner S et al. Taxol-requiring mutant of Chinese hamster ovary cells with impaired mitotic spindle assembly. J Cell Biol 1983; 97(1):30-39.
89. Cabral FR. Isolation of Chinese hamster ovary cell mutants requiring the continuous presence of taxol for cell division. J Cell Biol 1983; 97(1):22-29.
90. Kavallaris M, Kuo DY, Burkhart CA et al. Taxol-resistant epithelial ovarian tumors are associated with altered expression of specific beta-tubulin isotypes. J Clin Invest 1997; 100(5):1282-1293.
91. Nicoletti MI, Valoti G, Giannakakou P et al. Expression of beta-tubulin isotypes in human ovarian carcinoma xenografts and in a sub-panel of human cancer cell lines from the NCI-Anticancer Drug Screen: Correlation with sensitivity to microtubule active agents. Clin Cancer Res 2001; 7(9):2912-2922.
92. Ranganathan S, Benetatos CA, Colarusso PJ et al. Altered beta-tubulin isotype expression in paclitaxel-resistant human prostate carcinoma cells. Br J Cancer 1998; 77(4):562-566.
93. Dubik D, Dembinski TC, Shiu RP. Stimulation of c-myc oncogene expression associated with estrogen-induced proliferation of human breast cancer cells. Cancer Res 1987; 47(24 Pt 1):6517-6521.
94. Dubik D, Shiu RP. Mechanism of estrogen activation of c-myc oncogene expression. Oncogene 1992; 7(8):1587-1594.
95. Murphy LJ, Murphy LC, Friesen HG. Estrogen induction of N-myc and c-myc proto-oncogene expression in the rat uterus. Endocrinology 1987; 120(5):1882-1888.
96. Prall OW, Sarcevic B, Musgrove EA et al. Estrogen-induced activation of Cdk4 and Cdk2 during G1-S phase progression is accompanied by increased cyclin D1 expression and decreased cyclin-dependent kinase inhibitor association with cyclin E-Cdk2. J Biol Chem 1997; 272(16):10882-10894.
97. Altucci L, Addeo R, Cicatiello L et al. 17beta-Estradiol induces cyclin D1 gene transcription, p36D1-p34cdk4 complex activation and p105Rb phosphorylation during mitogenic stimulation of G(1)-arrested human breast cancer cells. Oncogene 1996; 12(11):2315-2324.
98. Planas-Silva MD, Weinberg RA. Estrogen-dependent cyclin E-cdk2 activation through p21 redistribution. Mol Cell Biol 1997; 17(7):4059-4069.
99. Prall OW, Carroll JS, Sutherland RL. A low abundance pool of nascent p21WAF1/Cip1 is targeted by estrogen to activate cyclin E*Cdk2. J Biol Chem 2001; 276(48):45433-45442.
100. Osborne CK, Shou J, Massarweh S et al. Crosstalk between estrogen receptor and growth factor receptor pathways as a cause for endocrine therapy resistance in breast cancer. Clin Cancer Res 2005; 11(2 Pt 2):865s-870s.
101. Razandi M, Alton G, Pedram A et al. Identification of a structural determinant necessary for the localization and function of estrogen receptor alpha at the plasma membrane. Mol Cell Biol 2003; 23(5):1633-1646.
102. McKenna NJ, Lanz RB, O'Malley BW. Nuclear receptor coregulators: Cellular and molecular biology. Endocr Rev 1999; 20(3):321-344.

103. Osborne CK, Zhao H, Fuqua SA. Selective estrogen receptor modulators: Structure, function, and clinical use. J Clin Oncol 2000; 18(17):3172-3186.
104. Parker MG. Steroid and related receptors. Curr Opin Cell Biol 1993; 5(3):499-504.
105. Shou J, Massarweh S, Osborne CK et al. Mechanisms of tamoxifen resistance: Increased estrogen receptor-HER2/neu cross-talk in ER/HER2-positive breast cancer. J Natl Cancer Inst 2004; 96(12):926-935.
106. Kato S, Endoh H, Masuhiro Y et al. Activation of the estrogen receptor through phosphorylation by mitogen-activated protein kinase. Science 1995; 270(5241):1491-1494.
107. Font de Mora J, Brown M. AIB1 is a conduit for kinase-mediated growth factor signaling to the estrogen receptor. Mol Cell Biol 2000; 20(14):5041-5047.
108. Nicholson RI, McClelland RA, Robertson JF et al. Involvement of steroid hormone and growth factor cross-talk in endocrine response in breast cancer. Endocr Relat Cancer 1999; 6(3):373-387.
109. Osborne CK, Bardou V, Hopp TA et al. Role of the estrogen receptor coactivator AIB1 (SRC-3) and HER-2/neu in tamoxifen resistance in breast cancer. J Natl Cancer Inst 2003; 95(5):353-361.
110. Smith CL, Nawaz Z, O'Malley BW. Coactivator and corepressor regulation of the agonist/antagonist activity of the mixed antiestrogen, 4-hydroxytamoxifen. Mol Endocrinol 1997; 11(6):657-666.
111. Buzdar A, Harvey H. Aromatase Inhibitors. In: Kufe DW, Pollock RE, Weichselbaum RR, Bast RC, Gansler TS, Holland JF, Frei E, eds. Cancer Medicine 6. Vol 1. Hamilton: BC Decker, 2003:947.
112. Ellis MJ, Coop A, Singh B et al. Letrozole is more effective neoadjuvant endocrine therapy than tamoxifen for ErbB-1- and/or ErbB-2-positive, estrogen receptor-positive primary breast cancer: Evidence from a phase III randomized trial. J Clin Oncol 2001; 19(18):3808-3816.
113. Howell A, Cuzick J, Baum M et al. Results of the ATAC (Arimidex, Tamoxifen, Alone or in Combination) trial after completion of 5 years' adjuvant treatment for breast cancer. Lancet 2005; 365(9453):60-62.
114. Buzdar A, Harvey H. Aromatase Inhibitors. In: Kufe DW, Pollock RE, Weichselbaum RR, Bast RC, Gansler TS, Holland JF, Frei E, eds. Cancer Medicine 6. Vol 1. Hamilton: BC Decker, 2003:957.
115. Yu D, Hung MC. Overexpression of ErbB2 in cancer and ErbB2-targeting strategies. Oncogene 2000; 19(53):6115-6121.
116. Nagata Y, Lan KH, Zhou X et al. PTEN activation contributes to tumor inhibition by trastuzumab, and loss of PTEN predicts trastuzumab resistance in patients. Cancer Cell 2004; 6(2):117-127.
117. Dieras V, Fumoleau P, Romieu G et al. Randomized parallel study of doxorubicin plus paclitaxel and doxorubicin plus cyclophosphamide as neoadjuvant treatment of patients with breast cancer. J Clin Oncol 2004; 22(24):4958-4965.
118. Buzdar AU, Singletary SE, Theriault RL et al. Prospective evaluation of paclitaxel versus combination chemotherapy with fluorouracil, doxorubicin, and cyclophosphamide as neoadjuvant therapy in patients with operable breast cancer. J Clin Oncol 1999; 17(11):3412-3417.
119. Paridaens R, Dirix L, Lohrisch C et al. Mature results of a randomized phase II multicenter study of exemestane versus tamoxifen as first-line hormone therapy for postmenopausal women with metastatic breast cancer. Ann Oncol 2003; 14(9):1391-1398.
120. Nabholtz JM, Buzdar A, Pollak M et al. Anastrozole is superior to tamoxifen as first-line therapy for advanced breast cancer in postmenopausal women: Results of a North American multicenter randomized trial. Arimidex Study Group. J Clin Oncol 2000; 18(22):3758-3767.
121. Milla-Santos A, Milla L, Portella J et al. Anastrozole versus tamoxifen as first-line therapy in postmenopausal patients with hormone-dependent advanced breast cancer: A prospective, randomized, phase III study. Am J Clin Oncol 2003; 26(3):317-322.
122. Mouridsen H, Gershanovich M, Sun Y et al. Phase III study of letrozole versus tamoxifen as first-line therapy of advanced breast cancer in postmenopausal women: Analysis of survival and update of efficacy from the International Letrozole Breast Cancer Group. J Clin Oncol 2003; 21(11):2101-2109.
123. Eiermann W, Paepke S, Appfelstaedt J et al. Preoperative treatment of postmenopausal breast cancer patients with letrozole: A randomized double-blind multicenter study. Ann Oncol 2001; 12(11):1527-1532.
124. Bonneterre J, Thurlimann B, Robertson JF et al. Anastrozole versus tamoxifen as first-line therapy for advanced breast cancer in 668 postmenopausal women: Results of the Tamoxifen or Arimidex Randomized Group Efficacy and Tolerability study. J Clin Oncol 2000; 18(22):3748-3757.
125. Vogel CL, Cobleigh MA, Tripathy D et al. Efficacy and safety of trastuzumab as a single agent in first-line treatment of HER2-overexpressing metastatic breast cancer. J Clin Oncol 2002; 20(3):719-726.
126. Norbury C, Nurse P. Animal cell cycles and their control. Annu Rev Biochem 1992; 61:441-470.

127. Hartwell LH, Weinert TA. Checkpoints: Controls that ensure the order of cell cycle events. Science 1989; 246(4930):629-634.
128. Pardee AB. A restriction point for control of normal animal cell proliferation. Proc Natl Acad Sci USA 1974; 71(4):1286-1290.
129. Evans T, Rosenthal ET, Youngblom J et al. Cyclin: A protein specified by maternal mRNA in sea urchin eggs that is destroyed at each cleavage division. Cell 1983; 33(2):389-396.
130. Pines J. Cyclins: Wheels within wheels. Cell Growth Differ 1991; 2(6):305-310.
131. Ohtsubo M, Theodoras AM, Schumacher J et al. Human cyclin E, a nuclear protein essential for the G1-to-S phase transition. Mol Cell Biol 1995; 15(5):2612-2624.
132. Keyomarsi K, Herliczek TW. The role of cyclin E in cell proliferation, development and cancer. Prog Cell Cycle Res 1997; 3:171-191.
133. Zhang HS, Gavin M, Dahiya A et al. Exit from G1 and S phase of the cell cycle is regulated by repressor complexes containing HDAC-Rb-hSWI/SNF and Rb-hSWI/SNF. Cell 2000; 101(1):79-89.
134. Montagnoli A, Fiore F, Eytan E et al. Ubiquitination of p27 is regulated by Cdk-dependent phosphorylation and trimeric complex formation. Genes Dev 1999; 13(9):1181-1189.
135. Perez-Roger I, Kim SH, Griffiths B et al. Cyclins D1 and D2 mediate myc-induced proliferation via sequestration of p27(Kip1) and p21(Cip1). Embo J 1999; 18(19):5310-5320.
136. Alevizopoulos K, Vlach J, Hennecke S et al. Cyclin E and c-Myc promote cell proliferation in the presence of p16INK4a and of hypophosphorylated retinoblastoma family proteins. Embo J 1997; 16(17):5322-5333.
137. Cassia R, Moreno-Bueno G, Rodriguez-Perales S et al. Cyclin E gene (CCNE) amplification and hCDC4 mutations in endometrial carcinoma. J Pathol 2003; 201(4):589-595.
138. Schraml P, Bucher C, Bissig H et al. Cyclin E overexpression and amplification in human tumours. J Pathol 2003; 200(3):375-382.
139. Strohmaier H, Spruck CH, Kaiser P et al. Human F-box protein hCdc4 targets cyclin E for proteolysis and is mutated in a breast cancer cell line. Nature 2001; 413(6853):316-322.
140. Rajagopalan H, Jallepalli PV, Rago C et al. Inactivation of hCDC4 can cause chromosomal instability. Nature 2004; 428(6978):77-81.
141. Bloom J, Pagano M. Deregulated degradation of the cdk inhibitor p27 and malignant transformation. Semin Cancer Biol 2003; 13(1):41-47.
142. Koepp DM, Schaefer LK, Ye X et al. Phosphorylation-dependent ubiquitination of cyclin E by the SCFFbw7 ubiquitin ligase. Science 2001; 294(5540):173-177.
143. Harwell RM, Porter DC, Danes C et al. Processing of cyclin E differs between normal and tumor breast cells. Cancer Res 2000; 60(2):481-489.
144. Akli S, Zheng PJ, Multani AS et al. Tumor-specific low molecular weight forms of cyclin E induce genomic instability and resistance to p21, p27, and antiestrogens in breast cancer. Cancer Res 2004; 64(9):3198-3208.
145. Furstenthal L, Kaiser BK, Swanson C et al. Cyclin E uses Cdc6 as a chromatin-associated receptor required for DNA replication. J Cell Biol 2001; 152(6):1267-1278.
146. Ma T, Van Tine BA, Wei Y et al. Cell cycle-regulated phosphorylation of p220(NPAT) by cyclin E/Cdk2 in Cajal bodies promotes histone gene transcription. Genes Dev 2000; 14(18):2298-2313.
147. Zhao J, Kennedy BK, Lawrence BD et al. NPAT links cyclin E-Cdk2 to the regulation of replication-dependent histone gene transcription. Genes Dev 2000; 14(18):2283-2297.
148. Fisk HA, Winey M. The mouse Mps1p-like kinase regulates centrosome duplication. Cell 2001; 106(1):95-104.
149. Okuda M, Horn HF, Tarapore P et al. Nucleophosmin/B23 is a target of CDK2/cyclin E in centrosome duplication. Cell 2000; 103(1):127-140.
150. Shanahan F, Seghezzi W, Parry D et al. Cyclin E associates with BAF155 and BRG1, components of the mammalian SWI-SNF complex, and alters the ability of BRG1 to induce growth arrest. Mol Cell Biol 1999; 19(2):1460-1469.
151. Wang C, Chua K, Seghezzi W et al. Phosphorylation of spliceosomal protein SAP 155 coupled with splicing catalysis. Genes Dev 1998; 12(10):1409-1414.
152. Seghezzi W, Chua K, Shanahan F et al. Cyclin E associates with components of the premRNA splicing machinery in mammalian cells. Mol Cell Biol 1998; 18(8):4526-4536.
153. Hara E, Hall M, Peters G. Cdk2-dependent phosphorylation of Id2 modulates activity of E2A-related transcription factors. Embo J 1997; 16(2):332-342.
154. Deed RW, Hara E, Atherton GT et al. Regulation of Id3 cell cycle function by Cdk-2-dependent phosphorylation. Mol Cell Biol 1997; 17(12):6815-6821.
155. Ohtsubo M, Roberts JM. Cyclin-dependent regulation of G1 in mammalian fibroblasts. Science 1993; 259(5103):1908-1912.

156. Akli S, Keyomarsi K. Cyclin E and its low molecular weight forms in human cancer and as targets for cancer therapy. Cancer Biol Ther 2003; 2(4 Suppl 1):S38-47.
157. Park JW. Liposome-based drug delivery in breast cancer treatment. Breast Cancer Res 2002; 4(3):95-99.
158. Carlson BA, Dubay MM, Sausville EA et al. Flavopiridol induces G1 arrest with inhibition of cyclin-dependent kinase (CDK) 2 and CDK4 in human breast carcinoma cells. Cancer Res 1996; 56(13):2973-2978.
159. Losiewicz MD, Carlson BA, Kaur G et al. Potent inhibition of CDC2 kinase activity by the flavonoid L86-8275. Biochem Biophys Res Commun 1994; 201(2):589-595.
160. Gray NS, Wodicka L, Thunnissen AM et al. Exploiting chemical libraries, structure, and genomics in the search for kinase inhibitors. Science 1998; 281(5376):533-538.
161. Senderowicz AM, Headlee D, Stinson SF et al. Phase I trial of continuous infusion flavopiridol, a novel cyclin-dependent kinase inhibitor, in patients with refractory neoplasms. J Clin Oncol 1998; 16(9):2986-2999.
162. Thomas JP, Tutsch KD, Cleary JF et al. Phase I clinical and pharmacokinetic trial of the cyclin-dependent kinase inhibitor flavopiridol. Cancer Chemother Pharmacol 2002; 50(6):465-472.
163. Schiller JH, Harrington D, Belani CP et al. Comparison of four chemotherapy regimens for advanced nonsmall-cell lung cancer. N Engl J Med 2002; 346(2):92-98.
164. Shapiro GI, Supko JG, Patterson A et al. A phase II trial of the cyclin-dependent kinase inhibitor flavopiridol in patients with previously untreated stage IV nonsmall cell lung cancer. Clin Cancer Res 2001; 7(6):1590-1599.
165. Kris MG, Natale RB, Herbst RS et al. Efficacy of gefitinib, an inhibitor of the epidermal growth factor receptor tyrosine kinase, in symptomatic patients with nonsmall cell lung cancer: A randomized trial. Jama 2003; 290(16):2149-2158.
166. Meijer L, Borgne A, Mulner O et al. Biochemical and cellular effects of roscovitine, a potent and selective inhibitor of the cyclin-dependent kinases cdc2, cdk2 and cdk5. Eur J Biochem 1997; 243(1-2):527-536.
167. McClue SJ, Blake D, Clarke R et al. In vitro and in vivo antitumor properties of the cyclin dependent kinase inhibitor CYC202 (R-roscovitine). Int J Cancer 2002; 102(5):463-468.
168. Benson C, White J, Twelves C et al. A phase I trial of the oral cyclin dependent kinase inhibitor CYC202 in patients with advanced malignancy. Chicago, Il: Paper presented at: Annual Meeting of the American Society of Clinical Oncology, 2003.
169. Pierga J, Faiver S, Vera K. A phase I and pharmacokinetic (PK) trial of CYC202, a novel oral cyclin-dependent kinase (CDK) inhibitor, in patients (pts) with advanced solid tumors. Chicago, Il: Paper presented at: Annual Meeting of the American Society of Clinical Oncology, 2003.
170. Kim KS, Kimball SD, Misra RN et al. Discovery of aminothiazole inhibitors of cyclin-dependent kinase 2: Synthesis, X-ray crystallographic analysis, and biological activities. J Med Chem 2002; 45(18):3905-3927.
171. Shapiro G, Lewis N, Bai S et al. A phase I study to determine the safety and pharmacokinetics (PK) of BMS-387032. Chicago, Il: Paper presented at: Annual Meeting of the American Society of Clinical Oncology, 2003.
172. McCormick J, Gadgeel M, Helmke W et al. Phase I study of BMS-387032, a cyclin dependent kinase (CDK) 2 inhibitor. Chicago, Il: Paper presented at: Annual Meeting of the American Society of Clinical Oncology, 2003.
173. Jones S, Burris H, Kies M et al. A phase I study to determine the safety and pharmacokinetics (PK) of BMS-387032 given intravenously every three weeks in patients with metastatic refractory solid tumors. Chicago, Il: Paper presented at: Annual Meeting of the American Society of Clinical Oncology, 2003.
174. Akinaga S, Nomura K, Gomi K et al. Effect of UCN-01, a selective inhibitor of protein kinase C, on the cell-cycle distribution of human epidermoid carcinoma, A431 cells. Cancer Chemother Pharmacol 1994; 33(4):273-280.
175. Akiyama T, Yoshida T, Tsujita T et al. G1 phase accumulation induced by UCN-01 is associated with dephosphorylation of Rb and CDK2 proteins as well as induction of CDK inhibitor p21/Cip1/WAF1/Sdi1 in p53-mutated human epidermoid carcinoma A431 cells. Cancer Res 1997; 57(8):1495-1501.
176. Akiyama T, Shimizu M, Okabe M et al. Differential effects of UCN-01, staurosporine and CGP 41 251 on cell cycle progression and CDC2/cyclin B1 regulation in A431 cells synchronized at M phase by nocodazole. Anticancer Drugs 1999; 10(1):67-78.
177. Chen X, Lowe M, Keyomarsi K. UCN-01-mediated G1 arrest in normal but not tumor breast cells is pRb-dependent and p53-independent. Oncogene 1999; 18(41):5691-5702.

178. Kawakami K, Futami H, Takahara J et al. UCN-01, 7-hydroxyl-staurosporine, inhibits kinase activity of cyclin-dependent kinases and reduces the phosphorylation of the retinoblastoma susceptibility gene product in A549 human lung cancer cell line. Biochem Biophys Res Commun 1996; 219(3):778-783.
179. Seynaeve CM, Stetler-Stevenson M, Sebers S et al. Cell cycle arrest and growth inhibition by the protein kinase antagonist UCN-01 in human breast carcinoma cells. Cancer Res 1993; 53(9):2081-2086.
180. Shimizu E, Zhao MR, Nakanishi H et al. Differing effects of staurosporine and UCN-01 on RB protein phosphorylation and expression of lung cancer cell lines. Oncology 1996; 53(6):494-504.
181. Usuda J, Saijo N, Fukuoka K et al. Molecular determinants of UCN-01-induced growth inhibition in human lung cancer cells. Int J Cancer 2000; 85(2):275-280.
182. Wang Q, Fan S, Eastman A et al. UCN-01: A potent abrogator of G2 checkpoint function in cancer cells with disrupted p53. J Natl Cancer Inst 1996; 88(14):956-965.
183. Busby EC, Leistritz DF, Abraham RT et al. The radiosensitizing agent 7-hydroxystaurosporine (UCN-01) inhibits the DNA damage checkpoint kinase hChk1. Cancer Res 2000; 60(8):2108-2112.
184. Graves PR, Yu L, Schwarz JK et al. The Chk1 protein kinase and the Cdc25C regulatory pathways are targets of the anticancer agent UCN-01. J Biol Chem 2000; 275(8):5600-5605.
185. Sarkaria JN, Busby EC, Tibbetts RS et al. Inhibition of ATM and ATR kinase activities by the radiosensitizing agent, caffeine. Cancer Res 1999; 59(17):4375-4382.
186. Sato S, Fujita N, Tsuruo T. Interference with PDK1-Akt survival signaling pathway by UCN-01 (7-hydroxystaurosporine). Oncogene 2002; 21(11):1727-1738.
187. Testa JR, Bellacosa A. AKT plays a central role in tumorigenesis. Proc Natl Acad Sci USA 2001; 98(20):10983-10985.
188. Sausville EA, Arbuck SG, Messmann R et al. Phase I trial of 72-hour continuous infusion UCN-01 in patients with refractory neoplasms. J Clin Oncol 2001; 19(8):2319-2333.
189. Senderowicz A, Headlee D, Lush R et al. Phase I trial of infusional UCN-01, a novel protein kinase inhibitor, in patients with refractory neoplasms. Amsterdam, Holland: Paper presented at: 10th National Cancer Institute-European Organization fro Research on Treatment of Cancer Symposium, 1998.
190. Shao RG, Cao CX, Shimizu T et al. Abrogation of an S-phase checkpoint and potentiation of camptothecin cytotoxicity by 7-hydroxystaurosporine (UCN-01) in human cancer cell lines, possibly influenced by p53 function. Cancer Res 1997; 57(18):4029-4035.
191. Jones CB, Clements MK, Wasi S et al. Enhancement of camptothecin-induced cytotoxicity with UCN-01 in breast cancer cells: Abrogation of S/G(2) arrest. Cancer Chemother Pharmacol 2000; 45(3):252-258.
192. Hsueh CT, Kelsen D, Schwartz GK. UCN-01 suppresses thymidylate synthase gene expression and enhances 5-fluorouracil-induced apoptosis in a sequence-dependent manner. Clin Cancer Res 1998; 4(9):2201-2206.
193. Bunch RT, Eastman A. Enhancement of cisplatin-induced cytotoxicity by 7-hydroxystaurosporine (UCN-01), a new G2-checkpoint inhibitor. Clin Cancer Res 1996; 2(5):791-797.
194. Garcia HH, Brar GA, Nguyen DH et al. Indole-3-carbinol (I3C) inhibits cyclin-dependent kinase-2 function in human breast cancer cells by regulating the size distribution, associated cyclin E forms, and subcellular localization of the CDK2 protein complex. J Biol Chem 2005; 280(10):8756-8764.
195. Chen X, Danes C, Lowe M et al. Activation of the estrogen-signaling pathway by p21(WAF1/CIP1) in estrogen receptor-negative breast cancer cells. J Natl Cancer Inst 2000; 92(17):1403-1413.
196. Barboule N, Chadebech P, Baldin V et al. Involvement of p21 in mitotic exit after paclitaxel treatment in MCF-7 breast adenocarcinoma cell line. Oncogene 1997; 15(23):2867-2875.
197. Yu D, Jing T, Liu B et al. Overexpression of ErbB2 blocks Taxol-induced apoptosis by upregulation of p21Cip1, which inhibits p34Cdc2 kinase. Mol Cell 1998; 2(5):581-591.
198. Bacus SS, Yarden Y, Oren M et al. Neu differentiation factor (Heregulin) activates a p53-dependent pathway in cancer cells. Oncogene 1996; 12(12):2535-2547.
199. Sugiyama K, Shimizu M, Akiyama T et al. Combined effect of navelbine with medroxyprogesterone acetate against human breast carcinoma MCF-7 cells in vitro. Br J Cancer 1998; 77(11):1737-1743.
200. Rao S, Porter DC, Chen X et al. Lovastatin-mediated G1 arrest is through inhibition of the proteasome, independent of hydroxymethyl glutaryl-CoA reductase. Proc Natl Acad Sci USA 1999; 96(14):7797-7802.

CHAPTER 5

p53, BRCA1 and Breast Cancer Chemoresistance

Kimberly A. Scata and Wafik S. El-Deiry*

Abstract

The tumor suppressor genes p53 and BRCA1 are involved in hereditary as well as sporadic breast cancer development and therapeutic responses. While p53 mutations contribute to resistance to chemo- and radiotherapy, BRCA1 dysfunction leads to enhanced sensitivity to DNA damaging therapeutic agents. The biochemical pathways used by p53 and BRCA1 for signaling tumor suppression involve some cross-talk including repression of BRCA1 transcription by p53 and altered selectivity of p53-dependent gene activation by BRCA1. In this chapter we review clinical and preclinical data implicating p53 and BRCA1 in breast cancer chemosensitivity. We discuss the known signaling pathways downstream of p53 or BRCA1 that contribute to their modulation of therapeutic responses, and we discuss the implications of p53 or BRCA1 mutation in therapeutic design.

Introduction

A woman's chances of being diagnosed with breast cancer increase with age resulting in a lifetime risk of 1 in 7 (ACS, 2005). Breast cancer accounts for roughly 32% of the cancers diagnosed in women in the U.S. and accounts for 15% of the 272,00 predicted cancers deaths in 2004 (ACS). Currently, only a fraction of patients respond to postsurgical chemo- and/or radiotherapy. It is therefore critical to identify markers that will (1) help define which women benefit from chemotherapy and (2) identify the most appropriate therapeutic regimen for the patient. While the search to identify such prognostic indicators has been going on for years (i.e., factors that will predict the patients' outcome), only recently have efforts begun to look for predictive factors, (i.e., factors that will predict response to a given treatment) as well.

Early studies were limited by the available technologies and it was only possible to measure several "candidate" genes at a time to try to determine whether or not a candidate could be used as a prognostic or a predictive indicator. Microarrays and other "high throughput" technologies now allow evaluation of global changes in gene expression, histology or chromosome number without any preconceived notion as to what might be important. Microarrays, for example, can be used to screen through thousands of genes in order to identify prognostic signatures as well as predictive signatures. Such signatures should enable a more accurate diagnosis and hopefully direct the choice of the most appropriate therapy for the disease. This technology is already playing a role in determining which patients with early stage breast cancer may benefit from chemotherapy. Oncotype DX uses a signature of 21 genes to determine the likelihood of recurrence at ten years from patient samples.[1] This assay is used for a very defined population,

*Corresponding Author: Wafik S. El-Deiry—University of Pennsylvania, 415 Curie Boulevard, CRB 437A, Philadelphia, Pennsylvania 19104, U.S.A. Email: wafik@mail.med.upenn.edu

Breast Cancer Chemosensitivity, edited by Dihua Yu and Mien-Chie Hung.

specifically, newly diagnosed, Stage I or II, ER positive patients (Genomic Health website: www.genomichealth.com). Clearly there is a need for predictive factors for patients who do not fall into these categories. Of the vast number of genetic alterations in cancers, it is important to determine the significance of an alteration with respect to treatment outcome. Specifically, two genes that are important in the development of breast cancer are the tumor suppressors BRCA1 and p53. Both of these genes have been intensively studied in the laboratory and the clinic.

To understand the importance of BRCA1 and p53 or, any tumor suppressor gene, one must first understand its function in the cell. At the simplest level, a dividing cell has two functions: duplicate its genome exactly and segregate one copy of the genome to each daughter cell. To distill the cell's life to such simplicity belies the complexity of the events that must be coordinated in order for these two "simple" processes to occur. Many disparate events must be coordinated in an ordered fashion such that the cell moves forward at the proper time and only in the forward direction. Acting as the cell's policemen, tumor suppressor genes ensure that the cell acts according to the rules for an orderly life. When tumor suppressors are incapacitated, the cell begins a life of chaos that can result in the uncontrolled growth that could lead to cancer. In familial syndromes, such as Li-Fraumeni or hereditary breast cancer, one copy of a tumor suppressor gene has been inactivated in the germline. Patients with such mutations have an increased risk of cancer. Knudson postulated this increased risk of cancer when he stated that the rate-limiting step for tumor formation is the inactivation of the tumor suppressor's second allele.[2] Originally interpreted to mean an inactivating mutation in the second allele or loss of heterozygosity (LOH), it is now clear that epigenetic events such as methylation of CpG islands can also contribute to the loss of function of a tumor suppressor.[3]

Among tumor suppressors, p53, arguably the best characterized, is still enigmatic. Frequently referred to as the "guardian of the genome", p53 is mutated in roughly 50% of all cancers.[4] p53 seems to control genome stability by monitoring and regulating key events in a cell's life. Given p53's importance in maintaining genomic stability, it is believed that other components of the p53 pathway are inactivated in the remaining cancers.[5]

Activated by intrinsic and extrinsic stresses, such as DNA damage, oncogene activation, hypoxia and reactive oxygen species, p53 directs the cell to inhibit cell cycle progression, steer the cell to senescence, or induce apoptosis. p53 can act as a transcriptional activator and repressor but may induce apoptosis in a transcription-independent manner as well (reviewed in refs. 6, 7). Consistent with its role in regulating the response to cellular stress, some patients with Li-Fraumeni syndrome, who are susceptible to a wide variety of cancers, carry p53 mutations.[8] It has been of great interest to determine whether or not p53 is a prognostic indicator for the chemo- and radio-responsiveness of tumors. The increased incidence of breast cancers in Li-Fraumeni patients as well as the observation that p53 is mutated in roughly 20% of breast cancers suggested that p53 should be assessed for its prognostic/predictive significance in breast cancer.[9]

After genetic linkage studies suggested a candidate for a breast cancer suppressor gene on 17q21, positional cloning techniques were used to identify BRCA1.[10,11] The impact of the BRCA1 mutation is not immediately obvious if one considers that familial breast cancer accounts for only 10% of all breast cancer cases.[12] The impact of mutations in this gene becomes obvious, however, when one realizes that the BRCA1 carrier has an 80% lifetime risk of breast cancer and a 40% risk of ovarian cancer.[13] These high levels of susceptibility, as well as the early belief that this tumor suppressor would also be mutated in sporadic cancers (and therefore lead to insights into the mechanisms of sporadic breast cancer) sparked a flurry of work to decipher the function of the BRCA1 gene.[14] Despite this activity, BRCA1 remains enigmatic. A role for BRCA1 has been found in the regulation of transcription (both positive and negative), centrosome number, protein stability, cell cycle checkpoints as well as chromatin remodeling, DNA damage repair and X-inactivation.[15-20] Unfortunately, the crucial function of BRCA1 is still unclear. In addition, the key question "How can a ubiquitously expressed protein such as BRCA1 give rise to a very limited tumor spectrum?" remains unanswered. In the case of some proteins, tissue-specific tumor types can be attributed to the expression of homologous proteins in overlapping tissues. This is clearly not the case for BRCA1. Furthermore, it is still unclear which function or functions might be the most important for tumor progression.

Despite the many questions remaining about the functions of p53 and BRCA1, several pieces of evidence suggest that it is important to look at whether or not these proteins might be used as prognostic or predictive factors for treating breast cancer. In the case of BRCA1, carriers of a germline mutation are frequently younger at the time of disease onset than patients with sporadic tumors.[10] It is therefore important to determine whether this early onset is a prognostic indicator. Furthermore, given its function in DNA repair, it is important to understand if the BRCA1 mutation carriers have a better or worse prognosis after chemo and/or radiation therapy, or if there is an increased risk of secondary cancers due to adjuvant therapy. In the case of p53, its status as guardian of the genome and its role in promoting apoptosis make it a prime target for a prognostic and/or a predictive factor. This review will discuss and compare data from basic research including tissue culture and animal models with data from clinical studies.

p53

Initially identified as an SV40 large T antigen interacting protein, p53 was believed to be an oncogene.[5] Further studies demonstrated that p53, mutated in roughly 50% of all cancers, is a tumor suppressor gene.[5] Although p53 has many functions ascribed to it, most, if not all of its tumor suppressor activity can be ascribed to its function as a transcriptional activator.[21] p53 can also repress transcription, regulate the G1 checkpoint, induce apoptosis via direct interactions at the mitochondria, bind to damaged DNA and regulate recombination (reviewed in refs. 7, 22). Consisting of 393 amino acids, p53 can be divided into several functional domains. The transcriptional activation domain resides at the amino-terminus followed by a proline rich region important for p53-mediated repression. DNA binding activity resides in the middle of the protein. To activate transcription, p53 must form tetramers, which is mediated by the carboxy-terminal oligomerization domain. The carboxy-terminus also functions as a regulatory domain (see ref. 23).

As a transcription factor, p53 has been shown to activate numerous genes including cell cycle arrest genes, DNA repair genes, apoptotic genes and anti-angiogenic genes.[7] Because of the potential to activate genes that can kill the cell, p53 activity must be tightly regulated. In unstressed cells, p53 is rapidly degraded via ubiquitin mediate proteolysis. Hdm2, the ubiquitin E3 ligase primarily responsible for targeting p53 for degradation, is itself transcriptionally activated by p53. This feedback loop serves to keep p53 (and hdm2) levels under control. Numerous cellular stresses such as DNA damage, hypoxia, nucleotide depletion or oncogene activation, induce kinases that phosphorylate p53 and activate a cascade of p53 post-translational modifications, which stabilize and activate the p53 protein. Among the post-translational modifications identified on p53 are phosphorylation, ubiquitylation, sumoylation, neddylation and acetylation (reviewed in refs. 24, 25). Methylation has also recently been described.[26] In addition to inhibiting the ability of hdm2 to interact with p53, phosphorylation also allows the carboxy-terminus to be acetylated (reviewed in refs. 24, 25). Although widely believed to be important for stabilization and activation of p53, recently a knock-in mouse in which the seven carboxy terminal lysines were mutated to arginines shows a remarkably normal response to adriamycin treatment.[27]

Regardless of the significance of p53 acetylation, phosphorylation inhibits the p53-mdm2 interaction and permits p53 to accumulate. Activated p53 is able to bind to p53 consensus sequences in target genes and activate their transcription. The best characterized target gene is $p21^{waf1}$, a cyclin dependent kinase (cdk) inhibitor.[28-30] While p21 seems to be the primary p53 target responsible for inhibiting cell cycle progression in G1 phase, there are a plethora of p53 targets involved in both intrinsic and extrinsic apoptotic pathways. These pro-apoptotic genes fall into three categories: the first consists of genes such as death receptor 5 (DR5), which act at the cell surface to mediate death signals through the TNF family of ligands. Genes such as Puma, Noxa, Bid, Bax and Bak, which act at the mitochondrial membrane, activate apoptosis in response to intrinsic death signals. The final group of pro-apoptotic p53 target genes includes the PIDD and the PIG genes, which seem to have some role in generating or reacting to reactive oxygen species (reviewed in refs. 31,32). Although different apoptotic targets have been implicated as "key" in different studies, it seems clear that the "key" apoptosis inducer(s) is (are) tissue and perhaps stressor specific.[33-35]

In addition to its role in inducing pro-apoptotic genes, there is evidence that p53 is also able to act directly at the mitochondrial membrane to induce the release of cytochrome c.[31,36] p53 also has the ability to repress gene transcription of some anti-apoptotic genes, thereby tipping the balance to favor apoptosis.[31,36]

As mentioned previously, the prevalence of p53 mutations in cancers along with its key role in guarding genomic stability, makes it an obvious candidate factor with prognostic and/or predictive significance for various cancers. Although recent meta-analysis has demonstrated the mutant p53 correlates with more aggressive breast cancer and poorer survival,[37] early studies were contradictory. Recent observations have explained these contradictory results and have shed light on important considerations for future studies. Studies in which immunohistochemistry has been used to assess p53 status have mistakenly designated cells positive for p53 as "p53 mutant" while those not staining (or with low staining) as p53 wild-type. This system of scoring for p53 mutants derived from the belief that p53 mutations stabilized the protein. While this is certainly true for some p53 mutations, it is not true for all. In fact, closer examination of p53 mutants via DNA sequencing has shown that roughly 30% of p53 mutations do not stabilize the p53 protein.[38,39] Additionally, mutations in exons 4, 9 and 10 are more likely to be frameshift mutations and therefore negative in IHC studies.[39] Nonsense mutations are also likely to be undetected by IHC. Furthermore, Soussi and Bernoud suggest that splice site mutations that impact p53 structure have been underestimated. Importantly, in the stressful tumor environment, it is likely that wild-type p53 will be activated therefore giving a false positive result.

Studies in which DNA sequencing has been used have also been flawed in some cases. Roughly 40% of studies have analyzed exons 5-8 of p53 while nearly 14% of mutations reside outside of this region.[39] These recent sequence-based studies suggest many reasons for the controversial findings regarding p53 mutations and patient diagnosis (reviewed in ref. 39). When first published, however, such data were perplexing because in vitro data demonstrate a convincing role for p53 in cell cycle arrest and apoptosis (reviewed in ref. 7). These studies suggested that the inability to induce apoptosis should result in disease progression in vivo. Further evidence using mouse models supported this hypothesis. For example, p53-null mice showed a tumor-prone phenotype and p53-dependent apoptosis contributes to this tumor phenotype (reviewed in ref. 40). Examination of p53 mutations using the entire gene has demonstrated a relationship between p53 mutation and resistance to chemo- and radiotherapies.[39] More specifically, several groups have noted that mutations in the Loop2/Loop 3 (L2/L3) domains of p53 correlate with poor prognosis although one study contradicted these findings (reviewed in ref. 39).

In studies that examined p53 status and response to therapy, several groups showed that mutations in the L2/L3 domains of p53 correlate with progressive breast cancer growth in response to doxorubicin as well as to a combined 5'FU/MMC regimen.[38,41-43] Another study demonstrated that patients with breast cancers containing mutations in L3 or in the DNA contact residues of p53 showed a poor response to tamoxifen treatment.[44] Interestingly, Bertheau and colleagues found that there was no correlation between neoadjuvant treatment and clinical response but found that 8/14 patients with mutant p53 had a complete histological response to a cyclophosphamide/epirubicin regimen. In contrast, none of the 36 patients with wild-type p53 showed a complete histological response.[45] It is unclear why there should be such a dramatic difference in results between the studies. In the Bertheau study, it is not clear why tumors expressing wild-type p53 did not undergo apoptosis in response to therapy. It is possible that these cancers were dysfunctional for some other part of the p53 pathway. Finally, it is possible that the tumors expressing wild-type p53 may have upregulated a proliferative or pro-survival pathway that was able to compensate for p53-mediated apoptosis.

These studies suggest that future studies should be designed to identify the specific p53 mutation in each patient as well as loss of wild-type p53 function. The importance of identifying the exact mutation is supported by a paper from the Borresen-Dale laboratory, which showed that a common polymorphism at codon 72 may affect the function of p53 mutations in patients with breast cancer.[46] Patients with the Pro72 polymorphism were more likely to have wild-type p53 while patients homozygous for the Arg72 polymorphism (or heterozygous) were likely to have a mutant p53. Furthermore, this mutation was likely to be on the Arg allele.[46] In head and neck cancers, the mutations in the Arg72 polymorphism correlated with decreased responsiveness to cisplatin therapy and Arg72 mutants that inhibited p73, a p53 family member with pro-apoptotic ability, had a particularly poor prognosis.[47] A follow-up study demonstrated that the Arg allele induces apoptosis in response to various chemotherapeutic agents while the Pro allele is more likely to induce G1 arrest.[48] The increased apoptosis noted with the Arg allele correlated with an increased ability to induce Puma and Noxa, but not bax, mdm2 or p21. The authors also demonstrated that patients with advanced head and neck cancers were more likely to respond to chemo if they retained a wild-type p53 Arg allele. A recent study demonstrated that roughly 40% of women carrying Arg/Pro or Arg/Arg had a good pathological response to chemotherapy (either CTF, CAF or FEC regimen) while those women with Pro/Pro alleles had a much lower response (13%) to therapy.[49] Furthermore, the women with Pro/Pro alleles were more likely to have positive axillary lymph nodes. Although small, this study suggests that this common polymorphism may play an important role in determining the appropriate therapeutic regimen for women with breast cancer.

As stated above, the clinical data demonstrated that specific mutations, i.e., the loop mutations, correlated with poor outcome. Mouse models and in vitro studies demonstrate that different p53 mutations have very different effects on cellular biology. When one compares the phenotypes seen in the various p53 knock-in mice to p53-null mice it becomes very clear that all mutations are not equivalent nor are they all null.[40] NMR studies have demonstrated that the disruptiveness of the L2/L3 mutations are dependent upon the amino acid change as well as the amino acid mutated.[50,51] Other mutations shown to have predictive value (i.e., resistance to chemotherapy) disrupted the structure of p53 to the extent that they were rendered null mutations.[41] Cell culture studies have demonstrated that there are true "gain-of-function" alleles of p53[52-55] as well as alleles that exert dominant-negative effects on the remaining wild-type p53 allele (reviewed in ref. 39). There is also evidence that some p53 mutants can interact with and impair the ability of p73, a p53 family member, to induce apoptosis.[47,56] A recent study of 251 primary breast tumors demonstrated that dysregulated expression of genes repressed or activated by p53 correlated with the development of distant metastases within 5 years and poorer survival.[57] These data suggest that at a minimum, identifying the specific p53 mutation will be important for determining the best therapeutic regimen for a patient. Ideally, one will also screen for activation of downstream p53 targets as this will indicate whether or not the p53 pathway is intact.

BRCA1

BRCA1 is an equally enigmatic protein. Despite being cloned more than 10 years ago and having multiple functions attributed to it, the key question surrounding BRCA1 remains unanswered: "Why do BRCA1 mutations result in relatively cell-type specific cancers when BRCA1 is ubiquitously expressed and has important roles in transcription, DNA repair, cell cycle checkpoints, and centrosome regulation?" For years models have been proposed to explain this problem (reviewed in ref. 58). It is possible that the mammary and ovarian epithelium receive greater exposure to damage that requires a BRCA1 response. For example, the mammary gland is subject to high levels of estrogen metabolites that can form adducts with DNA.[58] It is also possible that there is a functional homolog of BRCA1 in other tissues that is not present in the breast and ovary. Another model suggests that the mammary and ovarian epithelium are more likely to have a delayed apoptotic response to the loss of BRCA1, thereby allowing damaged cells to survive long enough to gain another mutation that allows them to become cancerous.[59] Monteiro suggests that it is possible that the breast and ovary have different (increased) rates of mitotic recombination, therefore allowing a second hit to occur preferentially in these tissues.[58]

After extensive mapping, BRCA1 was cloned in 1994.[10,11] The initial enthusiasm about cloning BRCA1 has given way to an appreciation for the protein's complexity. Part of the problem is the size: the 1863 amino acid nuclear phospho-protein had few recognizable motifs. The amino-terminus RING domain, initially believed to mediate protein-protein or protein-nucleic acid interactions is now known to function as an E3 ubiquitin ligase. Although BRCA1 homodimers have some ligase activity, in vitro assays show that activity increases dramatically when the protein heterodimerizes with BARD1.[60-63] The best-characterized function of ubiquitylation is to target a protein for ubiquitin-mediated proteolysis via the proteasome pathway.[64] Recent work has shown that ubiquitylation may also serve as a protein-targeting mechanism.[64] The unusual lysine 6 linkage mediated by the BRCA1-BARD1 heterodimer may in fact be a targeting signal.[65] There are several known in vitro targets for the ubiquitin ligase activity of BRCA1 including p53, nucleophosmin, the C-terminal domain of the large subunit of RNA polymerase II, H2AX, topoisomerase II and γ-tubulin.[60,66-71] With the exception of Pol II, Topoisomerase II and γ-tubulin, the substrates have been identified in in vitro reactions and the significance of the ubiquitylation in vivo is unclear. Pol II is ubiquitylated in response to UV irradiation thereby transiently inhibiting 3'end processing of mRNA transcripts.[70] Topoisomerase II is ubiquitylated in a BRCA1-dependent manner and loss of either BRCA1 or topoisomerase II expression results in defective DNA decatenation.[68] The inability to properly decatenate sister chromatids results in lagging chromosomes and missegregation of chromosomes. Neither of these functions, however, are not specific to breast and ovarian cells.

Interestingly, the only evidence for a breast/ovarian-specific function for BRCA1 is related to its ligase activity. In breast cancer cells but not other cell lines, BRCA1/BARD1 regulates centrosome number via the ubiquitylation of γ-tubulin, a centrosome component.[67] Inhibition of ubiquitylation of γ-tubulin results in an aberrant increase in centrosome number thereby resulting in genomic instability, a hallmark of BRCA1 tumors.[67] Apparently, in nonmammary cell types, there is another protein able to regulate γ-tubulin but in mammary epithelial cells, this "backup" mechanism is not present. At the present time, it is unknown if ovarian cells also lack this "backup" regulator of centrosome number. It is also unknown how this relates to development of breast cancer in vivo.

The carboxy-terminus of BRCA1 has two tandem BRCT domains. The BRCT domain is present in many DNA damage responsive genes and mediates protein-protein interactions.[72] BRCA1 has been shown to play a role in DNA repair via homologous recombination, nonhomologous end-joining, transcription-coupled repair and global genomic repair.[17,18,20,73,74] BRCA1 interacts with numerous proteins and has been suggested to play a scaffolding role in response to damage.[75] Given the capacity to participate in numerous types of repair, it is of great interest to determine how BRCA1 selects which pathway to use.

In addition to the role in DNA damage signaling, the BRCT domains are important in the transcriptional role of BRCA1.[76] One of the two interactions between p53 and BRCA1 occurs via the BRCT domains.[76] This domain also seems to be important for the transcriptional activity of BRCA1.[76] The acidic nature of BRCA1, particularly at its C-terminus suggested that BRCA1 might be a transcription factor.[77] BRCA1 has been shown to interact with many transcription factors including p53.[78] This interaction increases the expression of p53 reporter constructs and increases p21 expression.[78] BRCA1 is also able to activate p21 expression independently of p53.[79] BRCA1 interacts with the transcriptional repressor, CtIP via the BRCT domains, thereby repressing expression of p21.[80] Upon DNA damage, the interaction is released, allowing BRCA1 to activate transcription.[80]

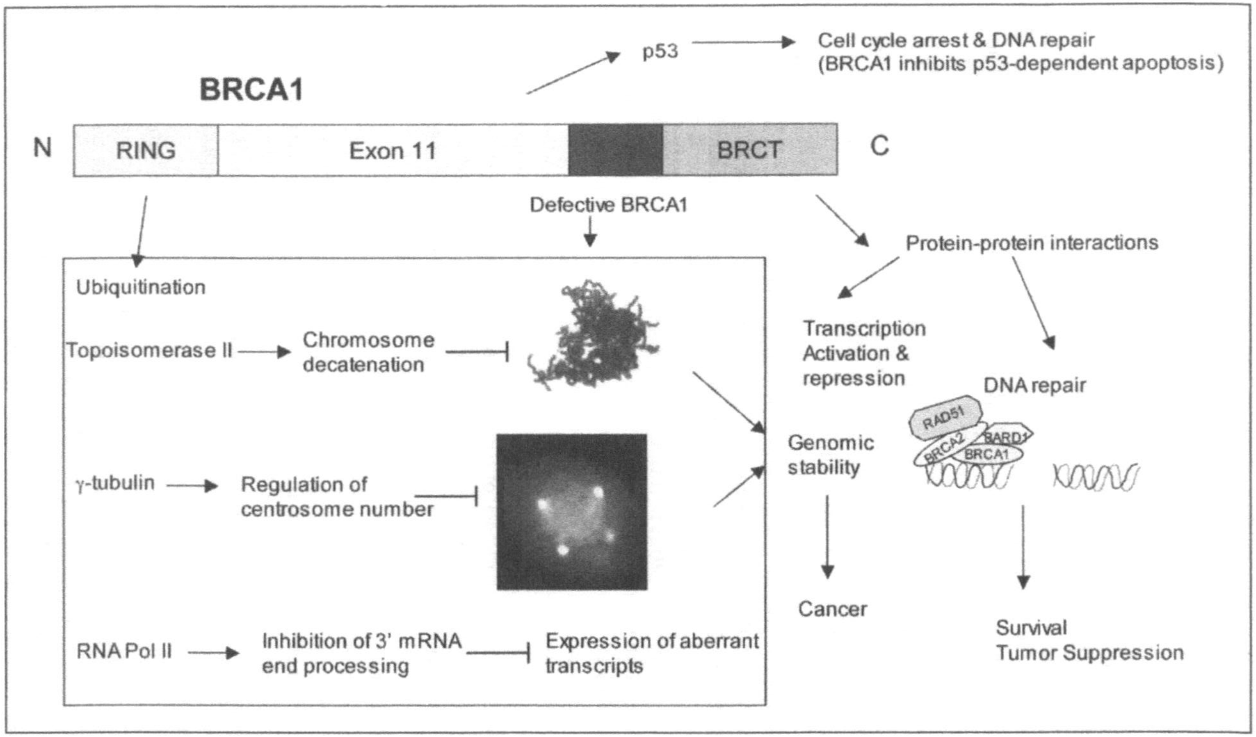

Figure 1. Many functions have been attributed to BRCA1. The amino terminus contains a RING domain, which interacts with BARD1, thereby forming a ubiquitin ligase. The ligase activity is responsible for ubiquitylation of several substrates including: Topoisomerase II, which results in proper chromosome condensation and decatenation. Ubiquitylation of γ-tubulin results in the regulation of centrosome number, thereby inhibiting genomic instability. In the presence of UV damage, the heterodimer ubiquitylates the large subunit of RNA Polymerase II, thereby preventing the 3' end processing of mRNA. The carboxy-terminus is important for the protein-protein interactions required for transcription regulation and repair of DNA damage. The p53-BRCA1 interaction results in expression of genes associated with cell cycle arrest and DNA repair preferentially over apoptotic genes. Reproduced with permission from: Lou Z et al. Nat Struct Mol Biol 2005; 12(7):589-93; ©2005, Nature Publishing Group.[68]

Additional research has shown that BRCA1 is able to regulate several checkpoints, including the G2/M checkpoint, the inter-S-phase checkpoint and the DNA decatenation checkpoint.[68,73,81] Intriguingly, BRCA1 seems to be playing a role in X-inactivation in mammary epithelial cells.[16] Despite the flurry of activity, the regulation of centrosome number and the role in X-inactivation are the only known breast specific functions of BRCA1.

BRCA1 in Sporadic Breast Cancer

By virtue of its tumor suppressor function and the germline mutations seen in familial breast cancer, it was reasonable to hypothesize that BRCA1 would be mutated in sporadic breast cancers as well. BRCA1 shows frequent loss of heterozygosity (LOH) in sporadic breast tumors. Using samples from 72 cases of sporadic breast cancer, 36 showed LOH at loci in and around BRCA1.[14] Another study looked at 120 sporadic breast tumors ranging from Stage I to Stage III saw LOH at the BRCA1 locus in 50% of the cases.[82] Interestingly, there was also loss of heterozygosity at BRCA2 and p53 in 28% of the BRCA1 LOH cases.[82] If there is LOH at a disease-associated locus, the gene of interest usually has a mutation in the remaining allele.[82] Unfortunately, the search for mutations in BRCA1 in sporadic cancers, however, has not yielded many. The study by Futreal and colleagues identified only 3 BRCA1 mutations in the 36 breast cancer cases they screened.[14] Garcia-Patino and colleagues screened 105 patients with sporadic breast cancers and identified only six frameshift mutations (leading to putative truncation mutants) and three missense mutations of unknown consequences.[83] One report demonstrated a germline mutation in the 5'UTR of BRCA1, which decreases translation efficiency.[84] Clearly, mutation of the coding regions of the remaining BRCA1 allele is not playing an important role in sporadic breast tumors.

Other data, however, clearly support a role for loss of BRCA1 in sporadic tumors. For example, Thompson and colleagues demonstrated that BRCA1 mRNA levels are lower in invasive breast cancers than in normal mammary epithelial cells and DCIS[85] and a study by Wilson and colleagues demonstrated that normal breast tissue or invasive lobular cancers or lower grade ductal carcinomas all had higher BRCA1 protein levels than high grade tumors[86] suggesting that BRCA1 expression is lost during cancer progression. While these studies investigated relatively small numbers of samples, all of the invasive tumors had decreased levels of BRCA1. Another mechanism for silencing the BRCA1 locus seems to be through epigenetic means. Several groups have investigated BRCA1 methylation status and demonstrated that BRCA1 is silenced by methylation in 11 to 30% of sporadic tumors.[87,88] High levels of BRCA1 promoter methylation were seen in medullary breast carcinomas, which share morphological features with the BRCA1 familial tumors.[89] Evidence for abnormal expression of other DNA repair proteins, including hRAD51, BARD1, and components of the mismatch repair pathway, suggests that perhaps there is an important role for DNA repair associated proteins in breast cancers.[90]

Many groups have generated BRCA1 mice. A recent review extensively summarizes the mouse phenotypes.[91] Therefore, only a few pertinent points will be commented on here. First, the BRCA1 targeting vector was important to the observed phenotype. For example, the targeting vector that allowed expression of a BRCA1 splice product survived longer than those mice with insertions in exons 2, 5 or 11 (see Moynahan). This suggests that identifying a patient's mutation may be important for predicting her response to therapy. Furthermore, the genetic background was shown to be important by Ludwig and colleagues who showed that an exon 11 insertion was tolerated in the 129/Sv and the 129/Sv/MF1 strains. Although mammary development was normal in these mice, 85% showed spontaneous tumors, including mammary tumors.[91] In order to better address the function of BRCA1 in breast cancer, conditional mutants of BRCA1 have been generated.[92] Using either Cre driven by the WAP promoter or by the MMTV promoter, 25% of the conditional knock-out mice showed spontaneous mammary tumor formation by two years of age.[93] The number of mice developing tumors increased when the conditional BRCA1 mice were crossed with p53+/- mice.[93] The interaction between BRCA1 and p53 will be discussed extensively later in the chapter.

BRCA1 and Response to Therapy

Drugs that impair microtubule dynamics are important chemotherapeutic agents in the treatment of breast cancer.[94] In tissue culture models of breast cancer, exogenous expression of BRCA1 increases the sensitivity to taxol, vincristine and vinorelbine, suggesting that these would not be good chemotherapeutic agents for use in patients with a germ-line mutation of BRCA1 or who show low/no expression of BRCA1.[95-99] Perhaps due to the resistance to microtubule disrupting agents in vitro, there are no studies published in which these agents are used to treat patients with BRCA1 mutations.

Early studies of BRCA1 demonstrated that BRCA1 colocalized with the homologous recombination protein Rad51 during S phase.[95] It was subsequently demonstrated that BRCA1 plays a role in double strand break repair.[91] BRCA1 has also been implicated in nonhomologous end joining, a more error prone pathway for the repair of DSBs, global genomic repair and transcription coupled repair.[94] Based on these studies, one might predict that patients with a mutant form of BRCA1 should be more responsive to chemotherapeutic agents that induce DNA damage. The results, in fact, are mixed (see Table 1). Exogenous expression of BRCA1 in breast cancer cells does not affect the cells' response to 5FU, cyclophosphamide or hydroxyurea

Table 1. Phenotype when WT BRCA1 is expressed

Drug	Bhattaycharyya JBC (2000)	Thangaraju JBC (2000)	Lafarge Oncogene (2001)	Mullan Oncogene (2001)	Quinn Cancer Res (2003)	Tassone (2003) BJCancer
5FU	NT	NT	NT	NT	No change	NT
Etoposide	NT	NT	Increases resistance	NT	Increases resistance	NT
Cisplatin	Increases resistance	NT	Increases resistance	Increases sensitivity*	Increases resistance	Increases resistance
Bleomycin	NT	NT	NT	NS	Increases resistance	NT
Doxorubicin	NT	NT	NT	No difference	NT	Decreases resistance
Vinorelbine	NT	NT	NT	NT	Increases sensitivity	NT
Paclitaxel	NT	Increases sensitivity	Increases sensitivity	Increases sensitivity	Increases sensitivity	Increases sensitivity
IR	Increases resistance	Decreases resistance	NT	NT	NT	NT
Vincristine	NT	NT	Increases sensitivity	Increases sensitivity	NT	NT
	ES cells[1]	Breast cancer cells[4]	Breast cancer[5]	MBR-62 bcl cells [3]	Breast cancer[2]	Breast cancer[2]

NT = not tested; NS = data not shown in the reference

1. WT vs. Δ11/Δ11
2. Hcc1937
3. Authors say they had no change in sensitivity for 5FU, cyclophosphamide, bleomycin, HU (data not shown)
4. T47D, Mcf7, Hcc1937 and ovarian line OV177

* Note: Authors use a concentration of cisplatin that kills both sets of cells.

(HU), suggesting that women with germline mutations in BRCA1 should respond in a manner similar to other breast cancer patients. One study using etoposide showed that exogenous BRCA1 expression increased resistance to the drug, suggesting that this might be a good choice for BRCA1 carriers.[96] Two studies using different models obtained conflicting results with respect to doxorubicin. The human breast cancer cell line Hcc1937, with a 5382 ins C mutation in its only BRCA1 allele, has been used extensively to study the role of BRCA1.[97] Reconstituting expression of wild-type BRCA1 in Hcc1937 resulted in an increased sensitivity to doxorubicin.[98] In a breast cancer cell line expressing a tetracycline regulated BRCA1 derived from the MDA-MB231 breast cancer line, doxorubicin treatment did not differentially affect the cells when BRCA1 was induced.[99] Studies using cisplatin have also shown conflicting results. Several studies have shown that exogenous wild-type BRCA1 increases resistance to cisplatin while one study reported no difference (see Table 1). The clinical results are suggestive of a positive response to chemotherapy for BRCA1 patients. In response to anthracyclines, BRCA1 carriers showed a better pathological response than noncarriers.[100] Another study showed that BRCA1 carriers, although having many characteristics that correlate with poor survival, were more likely to receive adjuvant chemotherapy and/or radiotherapy. Although the follow up time was rather short, these patients didn't show worse overall survival than nonmutation carriers.[101]

As mentioned above, the tumor type-specificity of BRCA1 suggested that estrogen might play a role in breast and ovarian tumor formation. Subsequent studies demonstrated that BRCA1 is able to inhibit ERα function in a ligand-independent manner.[102] This itself is rather puzzling given the fact that most BRCA1-related breast cancers are ER negative.[101] It is possible that loss of BRCA1 allows ERα to signal excessively and that after some early tumor-promoting activity, ERα mediated signaling is no longer required. This is entirely possible, as it is well known that sporadic breast cancers frequently lose ERα expression.[103] A recent study looking at expression of ER and PR in the benign epithelium close to BRCA1 tumors found that there is a trend toward higher ER expression in the normal epithelium of BRCA1 carriers and a significant increase in PR expression in the benign epithelium near the tumor margins.[104] Because this benign tissue retained the WT BRCA1 allele it seems that haploinsufficiency of BRCA1 is responsible for some changes in the "normal" epithelium. It is also possible that these changes are non-cell autonomous and are being induced by the nearby tumor tissue. The hypothesis that BRCA1 is inducing non-cell autonomous changes is supported by a conditional knock-out mouse in which BRCA1 is only inhibited in the granulosa cells.[105] Of the 59 homozygous knock-out mice sacrificed, 40 of them had visible cysts in the ovary or uterine horn. None of the age matched control mice had tumors. Furthermore, the tumor tissue had characteristics of epithelial cells while lacking characteristics of granulosa cells. It seems that BRCA1 regulates a substance produced by the granulosa cells that inhibits ovarian and uterine tumor formation. Recent studies have shown that there is a pattern of promoter methylation seen in the ER negative cancers that is not seen in the ER positive tumor cell lines or in normal mammary epithelial cells.[106] Furthermore, loss of ER expression results downregulation and methylation of ER signaling targets as well as an increase in the expression of proteins involved in gene silencing such as HDAC1 and DNMT1 and 3b.[106,107] How loss of BRCA1 might affect these downstream events requires further experimentation.

Understanding the interaction between these two proteins has important implications for BRCA1 mutation carriers as well as in sporadic breast cancers without BRCA1 (or with low expression of BRCA1). After adjuvant tamoxifen treatment was shown to reduce the risk of contralateral breast cancer, it was of interest to determine whether tamoxifen might be a useful prophylactic agent for women with a high risk for breast cancer.[108] Studies clearly demonstrated a reduction in the number of ER positive breast cancers. The initial belief was that, because the majority of BRCA1 breast cancers are ER negative, tamoxifen would not be useful for these women. Analysis of data from the National Adjuvant Breast and Bowel Project (NSABP-P1) Breast Cancer Prevention Trial showed that tamoxifen did not reduce the risk of breast cancer in carriers of BRCA1 mutations.[108] Unfortunately, there were only 8 patients with BRCA1 mutations that were diagnosed with breast cancer: 5 received tamoxifen and 3 received the placebo. The 95% confidence interval was quite large: 0.32-10.70. In a later case-control study of 848 patients with BRCA1 mutations, tamoxifen treatment reduced the risk of contralateral breast cancer by 50%.[108a] Perhaps surprisingly, tamoxifen induced mammary gland proliferation and induced earlier adenocarcinoma formation than placebo treated BRCA1 co/co MMTV-cre/p53+/- mice.[103] The reason for this difference is not clear but may have to do with the fact that the mice started with a loss of both BRCA1 alleles, which is not the case for carriers of a germline mutation of BRCA1. Perhaps the haploinsufficiency of BRCA1 in humans is important mediating the tamoxifen effect.

p53 and BRCA1

The relationship between p53 and BRCA1 is complex. p53, activated in response to cellular stresses is believed to activate pro-apoptotic targets when levels are high enough to activate the promoters of these genes, which usually have greater divergence from the p53 consensus site.[109] When BRCA1 is overexpressed, however, p53 is stabilized but is preferentially directed to cell cycle arrest targets and DNA repair targets.[110-112] The result of this preferential activation is that cells are more resistant to chemotherapeutics.[110,111] The severity of a p53 response (i.e., apoptosis) mandates that p53 be tightly regulated. In order to keep p53 under control, it is regulated by a number of feedback loops. For example, mdm2, the ubiquitin ligase primarily responsible for degrading p53, is also a direct transcriptional target of p53.[113] There is evidence that BRCA1 and p53 are also able to regulate each other. While BRCA1 stabilizes and increases p53's activity towards cell cycle arrest and repair targets, at some point in the DNA damage response, p53 is able to downregulate the expression of BRCA1.[114,115] The downregulation of BRCA1 by p53 could be the tipping point in the decision between life and death.

In addition to these studies, mouse studies demonstrate a genetic interaction between the two proteins. Early attempts to make a BRCA1 knockout mouse showed that homozygous inactivation of BRCA1 resulted in embryonic lethality.[91] Crossing the BRCA1 heterozygotes with $p53^{+/-}$ mice allowed the mice to survive a few more days. In these mice, inhibition of BRCA1 results in genomic instability, which in turn activates p53. When p53 is eliminated, the cells exhibiting genomic instability are no longer eliminated, thereby allowing the mice to survive a few extra days. Recent studies using MEFs demonstrate that MEFs null for p53 and BRCA1 are more sensitive to topoisomerase I and II inhibitors and platinum compounds than $p53^{-/-}$; $BRCA1^{+/-}$ MEFs. The double-null MEFs showed no change in sensitivity to anti-metabolites or taxanes. It remains to be tested whether or not this holds true in mammary epithelial cells (specifically breast cancer cells).[116] These results are also important to consider in the clinics because p53 is frequently mutated in the BRCA1-associated breast cancers.[116]

Interestingly, the spectrum of p53 mutations observed in patients with familial breast cancer (specifically BRCA1-associated breast cancer) is quite different from the spectrum of mutations seen in any other type of cancer and includes several mutations never seen in any other cancer.[117] Even more surprising is the fact that there is often more than one p53 mutation, which is not due to a more generalized mutator phenotype.[118] Recent analysis of some of the p53 mutations found in BRCA1 or BRCA2 related breast cancers showed surprising results: many of the mutants retained transcriptional activation activity and the ability to induce apoptosis but were unable to suppress colony formation.[119] This study was very interesting but there are several important considerations. First and foremost, the authors addressed the significance of these mutants in nonbreast cell lines. Second, the authors dissected out the various functions of p53 via specific assays, their phenotypes were sometimes subtle and it is unclear how these various differences might affect tumorigenicity in vivo. Finally, the cells were also positive for BRCA1 (and BRCA2) expression. Given the fact that BRCA1 interacts with p53 to activate transcription, it would be interesting to know how these p53 mutants act in BRCA1-null lines (or lines expressing BRCA1 with clinically relevant mutations), a question that is now more amenable to study via the RNAi technology. It will be important to identify the functions of these mutants in the context of breast cancer models, specifically those in which BRCA1 function is inhibited.

Screening Mutations via Yeast Experiments

For several years now, different groups have tried to use activation of genes in yeast as an assay for differentiating between disease causing mutants and benign polymorphisms.[120] The use of a yeast reporter assay to identify p53 mutations in tumor samples has been used in several studies.[53] The authors of one study maintain that the assay is sensitive enough to identify p53 mutations in as few as 15% of the tumor cells. This is a "quick and dirty" way to identify which patients have a p53 mutation but in light of the arguments given above, it seems important that once identified, these samples should be sequenced in order to identify the exact mutation present in the tumor. A recent paper by Furuta and colleagues suggest that yeast or mammalian transcriptional activation assays must be interpreted with care.[121] The authors used a three dimensional Mcf10A model system, which differentiates into acinar structures in culture. Inhibition of wild-type BRCA1 with an adenovirus expressing BRCA1 shRNA gave rise to cells unable to form acinar structures. The group further went on to address the effects of two disease-associated mutations, Q356R and M1775R. Although Q356R had growth suppressing abilities almost equal to that of wild-type BRCA1, the M1775R mutant was severely compromised in its ability to suppress cell number. Because both are disease-associated mutations, the question arises as to whether BRCA1-mutated tumors may be further stratified with regard to responsiveness to chemo- and/or radiation therapy and suggests that the mutation seen in BRCA1 patients should be included in any clinical studies focused on chemo- and radiation responsiveness. Although an important question, it is very likely to be difficult to address experimentally. Currently, there are more than 1500 known BRCA1 mutations (BIC website: http://research.nhgri.nih.gov/projects/bic). The size of the BRCA1 locus was the original impetus for using the yeast model to identify mutations. Another caveat for using the yeast system affects both p53 and BRCA1-both proteins have functions beyond transcriptional control, as seen in the Furuta paper, so merely identifying mutants with problems in transcriptional regulation may not be sufficient to identify how they will respond to therapy. It may be useful to score all mutants in homologous recombination, ubiquitylation or decatenation assays or other assays that evaluate tumorigenicity or protection from DNA damage.

Summary and Future Questions

Ultimately, an important question to ask is "Is BRCA1 a druggable target?" Recently, two groups reported that poly-(ADP-ribose) polymerase (PARP) inhibitors could be used as single agent treatments in cells with DNA repair defects.[122,123] Their hypothesis was based on the role of PARP in single strand break repair. If PARP is inhibited, DNA replication proceeds in the presence of the single strand break, thereby turning a minor damage event into a double strand break. Double strand break repair can occur via the nonmutagenic homologous recombination pathway or via one of two more mutagenic pathways: single strand annealing or nonhomologous end-joining (reviewed in ref. 124). Bryant, Farmer and their colleagues decided to test whether PARP inhibition might be more lethal if the homologous recombination pathway was inactivated. Bryant and colleagues showed that siRNA-mediated knock-down of BRCA2 in either Mcf7 or MDA-MB-231 breast cancer cell lines resulted in increased death when treated with PARP inhibitors. Both groups demonstrated that PARP inhibitors decreased tumor burden in nude mice injected with cells lacking BRCA2. There are several caveats to these experiments. First, the Bryant group had a relatively small number of animals and the three animals that responded showed responses from minor to complete remission. The Farmer group had an impressive response to the treatment in mice containing BRCA2-null tumors, however, they used ES cells as a model and looked at tumor formation, i.e., began their experiment prior to the presence of discernable tumors. While both studies are exciting and intriguing, they bring up some questions. First: are all mutations that inhibit homologous recombination equal? There was an early controversy about whether the PARP inhibitors were effective against a naturally occurring BRCA2 mutation seen in the CAPAN-1 pancreatic cell line. One follow-up study demonstrated that PARP-inhibitors had no effect on cell survival in vitro.[125] However, when

CAPAN-1 cells were treated with the more potent PARP inhibitors used by the Ashworth laboratory, these cells were very sensitive to PARP inhibition.[126] A second caveat is that neither group used a breast cell line for their in vivo experiments. Would their results be as good in a more relevant mouse model of breast cancer? What happens in mammary tumors induced by loss of BRCA2? Are there differences between mammary tumors induced by BRCA1 compared to BRCA2? Farmer and colleagues looked at BRCA1-deficient and BRCA2-deficient cell lines and found similar results in vitro, making it tempting to speculate that BRCA1 mutant cancers would behave in a manner similar to the BRCA2 mutant cancers. While BRCA2 seems to be primarily involved in homologous recombination, BRCA1 has also been implicated in transcriptional activation and repression, chromatin remodeling, centrosome regulation, and regulation of the G1/S and G2/M checkpoints. Might any or all of these functions of BRCA1 not related to its role in homologous recombination affect the response of these cells to PARP inhibitors? Clinical studies show that there is a difference in BRCA1 tumors compared to BRCA2 tumors with respect to gene expression and histology (see above). But, as mentioned previously, BRCA2 is frequently lost in sporadic tumors (as is BRCA1). The tumors in which both tumor suppressors have been lost seem to have features of BRCA1-mutated familial breast cancers.[82] This is consistent with the double BRCA1/BRCA2 knock-out mice, which clearly show that BRCA1 is epistatic over BRCA2.[127] Without a clearer understanding of how and why these differences arise, it seems premature to extrapolate data from the BRCA2-deficient tumors. Caveats aside, PARP inhibition is an exciting new chemotherapeutic avenue waiting to be explored.

A diagnosis of breast cancer, as for any cancer, is a frightening experience. For women who carry a BRCA1 mutation, the probability of such a diagnosis increases dramatically. While a cure for all of these women is still a long way off, recent data suggests that we should be optimistic. Clinical, mouse and tissue culture data support the use of anthracyclines while suggesting that agents such as paclitaxel are unlikely to be useful when treating the BRCA1 mutation carriers. Exciting data suggests that even for BRCA1 carriers who have ER negative tumors, tamoxifen may be beneficial for some of these women. Finally, preclinical studies on the PARP-inhibitors suggest that we may soon have tumor specific agents that have minimal effects on normal tissues, thereby minimizing side effects. The challenge is to determine who benefits and why.

References

1. Hedenfalk I, Duggan D, Chen Y et al. Gene-expression profiles in hereditary breast cancer. N Engl J Med 2001; 344(8):539-48.
2. Knudson AG. Cancer genetics. Am J Med Genet 2002; 111(1):96-102.
3. Jones PA, Baylin SB. The fundamental role of epigenetic events in cancer. Nat Rev Genet 2002; 3(6):415-28.
4. Lane DP, Lain S. Therapeutic exploitation of the p53 pathway. Trends Mol Med 2002; 8(4 Suppl):S38-42.
5. Levine AJ. p53, the cellular gatekeeper for growth and division. Cell 1997; 88(3):323-31.
6. Bode AM, Dong Z. Post-translational modification of p53 in tumorigenesis. Nat Rev Cancer 2004; 4(10):793-805.
7. El-Deiry WS. The role of p53 in chemosensitivity and radiosensitivity. Oncogene 2003; 22(47):7486-95.
8. Kleihues P, Schauble B, zur Hausen A et al. Tumors associated with p53 germline mutations: A synopsis of 91 families. Am J Pathol 1997; 150(1):1-13.
9. Gasco M, Yulug IG, Crook T. TP53 mutations in familial breast cancer: Functional aspects. Hum Mutat 2003; 21(3):301-6.
10. Hall JM, Lee MK, Newman B et al. Linkage of early-onset familial breast cancer to chromosome 17q21. Science 1990; 250(4988):1684-9.
11. Miki Y, Swensen J, Shattuck-Eidens D et al. A strong candidate for the breast and ovarian cancer susceptibility gene BRCA1. Science 1994; 266(5182):66-71.
12. Wooster R, Weber BL. Breast and ovarian cancer. N Engl J Med 2003; 348(23):2339-47.
13. Ford D, Easton DF, Stratton M et al. Genetic heterogeneity and penetrance analysis of the BRCA1 and BRCA2 genes in breast cancer families. The Breast Cancer Linkage Consortium. Am J Hum Genet 1998; 62(3):676-89.

14. Futreal PA, Liu Q, Shattuck-Eidens D et al. BRCA1 mutations in primary breast and ovarian carcinomas. Science 1994; 266(5182):120-2.
15. Easton D, Ford D, Peto J. Inherited susceptibility to breast cancer. Cancer Surv 1993; 18:95-113.
16. Ganesan S, Silver DP, Greenberg RA et al. BRCA1 supports XIST RNA concentration on the inactive X chromosome. Cell 2002; 111(3):393-405.
17. Powell SN, Kachnic LA. Roles of BRCA1 and BRCA2 in homologous recombination, DNA replication fidelity and the cellular response to ionizing radiation. Oncogene 2003; 22(37):5784-91.
18. Deng CX, Wang RH. Roles of BRCA1 in DNA damage repair: A link between development and cancer. Hum Mol Genet 2003; 12(Suppl 1):R113-23.
19. D'Andrea AD, Grompe M. The Fanconi anaemia/BRCA pathway. Nat Rev Cancer 2003; 3(1):23-34.
20. Kerr P, Ashworth A. New complexities for BRCA1 and BRCA2. Curr Biol 2001; 11(16):R668-76.
21. Nister M, Tang M, Zhang XQ et al. p53 must be competent for transcriptional regulation to suppress tumor formation. Oncogene 2005; 24(22):3563-73.
22. Oren M. Decision making by p53: Life, death and cancer. Cell Death Differ 2003; 10(4):431-42.
23. Prives C, Hall PA. The p53 pathway. J Pathol 1999; 187(1):112-26.
24. Anderson CW, Appella E, Sakaguchi K. Posttranslational modifications involved in the DNA damage response. J Protein Chem 1998; 17(6):527.
25. Xu Y. Regulation of p53 responses by post-translational modifications. Cell Death Differ 2003; 10(4):400-3.
26. Chuikov S, Kurash JK, Wilson JR et al. Regulation of p53 activity through lysine methylation. Nature 2004; 432(7015):353-60.
27. Krummel KA, Lee CJ, Toledo F et al. The C-terminal lysines fine-tune p53 stress responses in a mouse model but are not required for stability control or transactivation. Proc Natl Acad Sci USA 2005; 102(29):10188-93.
28. El-Deiry WS, Harper JW, O'Connor PM et al. WAF1/CIP1 is induced in p53-mediated G1 arrest and apoptosis. Cancer Res 1994; 54(5):1169-74.
29. El-Deiry WS, Tokino T, Velculescu VE et al. WAF1, a potential mediator of p53 tumor suppression. Cell 1993; 75(4):817-25.
30. Harper JW, Adami GR, Wei N et al. The p21 Cdk-interacting protein Cip1 is a potent inhibitor of G1 cyclin-dependent kinases. Cell 1993; 75(4):805-16.
31. Haupt S, Berger M, Goldberg Z et al. Apoptosis - the p53 network. J Cell Sci 2003; 116(Pt 20):4077-85.
32. Sax JK, El-Deiry WS. p53 downstream targets and chemosensitivity. Cell Death Differ 2003; 10(4):413-7.
33. Burns TF, Bernhard EJ, El-Deiry WS. Tissue specific expression of p53 target genes suggests a key role for KILLER/DR5 in p53-dependent apoptosis in vivo. Oncogene 2001; 20(34):4601-12.
34. Fei P, El-Deiry WS. p53 and radiation responses. Oncogene 2003; 22(37):5774-83.
35. Fei P, Bernhard EJ, El-Deiry WS. Tissue-specific induction of p53 targets in vivo. Cancer Res 2002; 62(24):7316-27.
36. Fridman JS, Lowe SW. Control of apoptosis by p53. Oncogene 2003; 22(56):9030-40.
37. Pharoah PD, Day NE, Caldas C. Somatic mutations in the p53 gene and prognosis in breast cancer: A meta-analysis. Br J Cancer 1999; 80(12):1968-73.
38. Geisler S, Lonning PE, Aas T et al. Influence of TP53 gene alterations and c-erbB-2 expression on the response to treatment with doxorubicin in locally advanced breast cancer. Cancer Res 2001; 61(6):2505-12.
39. Soussi T, Beroud C. Assessing TP53 status in human tumours to evaluate clinical outcome. Nat Rev Cancer 2001; 1(3):233-40.
40. Lozano G, Liu G. Mouse models dissect the role of p53 in cancer and development. Semin Cancer Biol 1998; 8(5):337-44.
41. Geisler S, Borresen-Dale AL, Johnsen H et al. TP53 gene mutations predict the response to neoadjuvant treatment with 5-fluorouracil and mitomycin in locally advanced breast cancer. Clin Cancer Res 2003; 9(15):5582-8.
42. Aas T, Geisler S, Eide GE et al. Predictive value of tumour cell proliferation in locally advanced breast cancer treated with neoadjuvant chemotherapy. Eur J Cancer 2003; 39(4):438-46.
43. Aas T, Borresen AL, Geisler S et al. Specific p53 mutations are associated with de novo resistance to doxorubicin in breast cancer patients. Nat Med 1996; 2(7):811-4.
44. Berns EM, Foekens JA, Vossen R et al. Complete sequencing of TP53 predicts poor response to systemic therapy of advanced breast cancer. Cancer Res 2000; 60(8):2155-62.
45. Bertheau P, Plassa F, Espie M et al. Effect of mutated TP53 on response of advanced breast cancers to high-dose chemotherapy. Lancet 2002; 360(9336):852-4.

46. Langerod A, Bukholm IR, Bregard A et al. The TP53 codon 72 polymorphism may affect the function of TP53 mutations in breast carcinomas but not in colorectal carcinomas. Cancer Epidemiol Biomarkers Prev 2002; 11(12):1684-8.
47. Bergamaschi D, Gasco M, Hiller L et al. p53 polymorphism influences response in cancer chemotherapy via modulation of p73-dependent apoptosis. Cancer Cell 2003; 3(4):387-402.
48. Sullivan A, Syed N, Gasco M et al. Polymorphism in wild-type p53 modulates response to chemotherapy in vitro and in vivo. Oncogene 2004; 23(19):3328-37.
49. Xu Y, Yao L, Ouyang T et al. p53 Codon 72 polymorphism predicts the pathologic response to neoadjuvant chemotherapy in patients with breast cancer. Clin Cancer Res 2005; 11(20):7328-33.
50. Wong KB, DeDecker BS, Freund SM et al. Hot-spot mutants of p53 core domain evince characteristic local structural changes. Proc Natl Acad Sci USA 1999; 96(15):8438-42.
51. Bullock AN, Henckel J, Fersht AR. Quantitative analysis of residual folding and DNA binding in mutant p53 core domain: Definition of mutant states for rescue in cancer therapy. Oncogene 2000; 19(10):1245-56.
52. Blandino G, Levine AJ, Oren M. Mutant p53 gain of function: Differential effects of different p53 mutants on resistance of cultured cells to chemotherapy. Oncogene 1999; 18(2):477-85.
53. Campomenosi P, Monti P, Aprile A et al. p53 mutants can often transactivate promoters containing a p21 but not Bax or PIG3 responsive elements. Oncogene 2001; 20(27):3573-9.
54. Monti P, Campomenosi P, Ciribilli Y et al. Tumour p53 mutations exhibit promoter selective dominance over wild type p53. Oncogene 2002; 21(11):1641-8.
55. Weisz L, Zalcenstein A, Stambolsky P et al. Transactivation of the EGR1 gene contributes to mutant p53 gain of function. Cancer Res 2004; 64(22):8318-27.
56. Strano S, Blandino G. p73-mediated chemosensitivity: A preferential target of oncogenic mutant p53. Cell Cycle 2003; 2(4):348-9.
57. Wei CL, Wu Q, Vega VB et al. A global map of p53 transcription-factor binding sites in the human genome. Cell 2006; 124(1):207-19.
58. Monteiro AN. BRCA1: The enigma of tissue-specific tumor development. Trends Genet 2003; 19(6):312-5.
59. Elledge SJ, Amon A. The BRCA1 suppressor hypothesis: An explanation for the tissue-specific tumor development in BRCA1 patients. Cancer Cell 2002; 1(2):129-32.
60. Mallery DL, Vandenberg CJ, Hiom K. Activation of the E3 ligase function of the BRCA1/BARD1 complex by polyubiquitin chains. Embo J 2002; 21(24):6755-62.
61. Chen A, Kleiman FE, Manley JL et al. Autoubiquitination of the BRCA1*BARD1 RING ubiquitin ligase. J Biol Chem 2002; 277(24):22085-92.
62. Baer R, Ludwig T. The BRCA1/BARD1 heterodimer, a tumor suppressor complex with ubiquitin E3 ligase activity. Curr Opin Genet Dev 2002; 12(1):86-91.
63. Lorick KL, Jensen JP, Fang S et al. RING fingers mediate ubiquitin-conjugating enzyme (E2)-dependent ubiquitination. Proc Natl Acad Sci USA 1999; 96(20):11364-9.
64. Pickart CM, Fushman D. Polyubiquitin chains: Polymeric protein signals. Curr Opin Chem Biol 2004; 8(6):610-6.
65. Wu-Baer F, Lagrazon K, Yuan W et al. The BRCA1/BARD1 heterodimer assembles polyubiquitin chains through an unconventional linkage involving lysine residue K6 of ubiquitin. J Biol Chem 2003; 278(37):34743-6.
66. Starita LM, Horwitz AA, Keogh MC et al. BRCA1/BARD1 ubiquitinate phosphorylated RNA polymerase II. J Biol Chem 2005.
67. Starita LM, Machida Y, Sankaran S et al. BRCA1-dependent ubiquitination of gamma-tubulin regulates centrosome number. Mol Cell Biol 2004; 24(19):8457-66.
68. Lou Z, Minter-Dykhouse K, Chen J. BRCA1 participates in DNA decatenation. Nat Struct Mol Biol 2005; 12(7):589-93.
69. Dong Y, Hakimi MA, Chen X et al. Regulation of BRCC, a holoenzyme complex containing BRCA1 and BRCA2, by a signalosome-like subunit and its role in DNA repair. Mol Cell 2003; 12(5):1087-99.
70. Kleiman FE, Wu-Baer F, Fonseca D et al. BRCA1/BARD1 inhibition of mRNA 3' processing involves targeted degradation of RNA polymerase II. Genes Dev 2005; 19(10):1227-37.
71. Sato K, Hayami R, Wu W et al. Nucleophosmin/B23 is a candidate substrate for the BRCA1-BARD1 ubiquitin ligase. J Biol Chem 2004; 279(30):30919-22.
72. Huyton T, Bates PA, Zhang X et al. The BRCA1 C-terminal domain: Structure and function. Mutat Res 2000; 460(3-4):319-32.
73. Lou Z, Chen J. BRCA proteins and DNA damage checkpoints. Front Biosci 2003; 8:s718-21.
74. Lane TF. BRCA1 and transcription. Cancer Biol Ther 2004; 3(6):528-33.

75. Hohenstein P, Giles RH. BRCA1: A scaffold for p53 response? Trends Genet 2003; 19(9):489-94.
76. Wang Q, Zhang H, Fishel R et al. BRCA1 and cell signaling. Oncogene 2000; 19(53):6152-8.
77. Monteiro AN, August A, Hanafusa H. Evidence for a transcriptional activation function of BRCA1 C-terminal region. Proc Natl Acad Sci USA 1996; 93(24):13595-9.
78. Zhang H, Somasundaram K, Peng Y et al. BRCA1 physically associates with p53 and stimulates its transcriptional activity. Oncogene 1998; 16(13):1713-21.
79. Somasundaram K, MacLachlan TK, Burns TF et al. BRCA1 signals ARF-dependent stabilization and coactivation of p53. Oncogene 1999; 18(47):6605-14.
80. Li S, Chen PL, Subramanian T et al. Binding of CtIP to the BRCT repeats of BRCA1 involved in the transcription regulation of p21 is disrupted upon DNA damage. J Biol Chem 1999; 274(16):11334-8.
81. Deming PB, Cistulli CA, Zhao H et al. The human decatenation checkpoint. Proc Natl Acad Sci USA 2001; 98(21):12044-9.
82. Hanby AM, Kelsell DP, Potts HW et al. Association between loss of heterozygosity of BRCA1 and BRCA2 and morphological attributes of sporadic breast cancer. Int J Cancer 2000; 88(2):204-8.
83. Garcia-Patino E, Gomendio B, Lleonart M et al. Loss of heterozygosity in the region including the BRCA1 gene on 17q in colon cancer. Cancer Genet Cytogenet 1998; 104(2):119-23.
84. Signori E, Bagni C, Papa S et al. A somatic mutation in the 5'UTR of BRCA1 gene in sporadic breast cancer causes downmodulation of translation efficiency. Oncogene 2001; 20(33):4596-600.
85. Thompson ME, Jensen RA, Obermiller PS et al. Decreased expression of BRCA1 accelerates growth and is often present during sporadic breast cancer progression. Nat Genet 1995; 9(4):444-50.
86. Wilson CA, Ramos L, Villasenor MR et al. Localization of human BRCA1 and its loss in high-grade, noninherited breast carcinomas. Nat Genet 1999; 21(2):236-40.
87. Staff S, Isola J, Tanner M. Haplo-insufficiency of BRCA1 in sporadic breast cancer. Cancer Res 2003; 63(16):4978-83.
88. Chen CM, Chen HL, Hsiau TH et al. Methylation target array for rapid analysis of CpG island hypermethylation in multiple tissue genomes. Am J Pathol 2003; 163(1):37-45.
89. Osin P, Lu YJ, Stone J et al. Distinct genetic and epigenetic changes in medullary breast cancer. Int J Surg Pathol 2003; 11(3):153-8.
90. Yoshikawa K, Ogawa T, Baer R et al. Abnormal expression of BRCA1 and BRCA1-interactive DNA-repair proteins in breast carcinomas. Int J Cancer 2000; 88(1):28-36.
91. Moynahan ME. The cancer connection: BRCA1 and BRCA2 tumor suppression in mice and humans. Oncogene 2002; 21(58):8994-9007.
92. Xu X, Wagner KU, Larson D et al. Conditional mutation of Brca1 in mammary epithelial cells results in blunted ductal morphogenesis and tumour formation. Nat Genet 1999; 22(1):37-43.
93. Brodie SG, Xu X, Qiao W et al. Multiple genetic changes are associated with mammary tumorigenesis in Brca1 conditional knockout mice. Oncogene 2001; 20(51):7514-23.
94. Kennedy RD, Quinn JE, Mullan PB et al. The role of BRCA1 in the cellular response to chemotherapy. J Natl Cancer Inst 2004; 96(22):1659-68.
95. Scully R, Chen J, Ochs RL et al. Dynamic changes of BRCA1 subnuclear location and phosphorylation state are initiated by DNA damage. Cell 1997; 90(3):425-35.
96. Quinn JE, Kennedy RD, Mullan PB et al. BRCA1 functions as a differential modulator of chemotherapy-induced apoptosis. Cancer Res 2003; 63(19):6221-8.
96a. Lafarge S, Sylvain V, Ferrara M, Bignon YJ. Inhibition of BRCA1 leads to increased chemoresistance to microtubule-interfering agents, an effect that involves the JNK pathway. Oncogene 2001; 20(45):6597-606.
96b. Thangaraju M, Kaufmann SH, Couch FJ. BRCA1 facilitates stress-induced apoptosis in breast and ovarian cancer cell lines. J Biol Chem 2000; 275(43):33487-96.
97. Tomlinson GE, Chen TT, Stastny VA et al. Characterization of a breast cancer cell line derived from a germ-line BRCA1 mutation carrier. Cancer Res 1998; 58(15):3237-42.
98. Tassone P, Tagliaferri P, Perricelli A et al. BRCA1 expression modulates chemosensitivity of BRCA1-defective HCC1937 human breast cancer cells. Br J Cancer 2003; 88(8):1285-91.
99. Mullan PB, Quinn JE, Gilmore PM et al. BRCA1 and GADD45 mediated G2/M cell cycle arrest in response to antimicrotubule agents. Oncogene 2001; 20(43):6123-31.
100. Chappuis PO, Goffin J, Wong N et al. A significant response to neoadjuvant chemotherapy in BRCA1/2 related breast cancer. J Med Genet 2002; 39(8):608-10.
101. El-Tamer M, Russo D, Troxel A et al. Survival and recurrence after breast cancer in BRCA1/2 mutation carriers. Ann Surg Oncol 2004; 11(2):157-64.
102. Fan S, Wang J, Yuan R et al. BRCA1 inhibition of estrogen receptor signaling in transfected cells. Science 1999; 284(5418):1354-6.

103. Jones LP, Li M, Halama ED et al. Promotion of mammary cancer development by tamoxifen in a mouse model of Brca1-mutation-related breast cancer. Oncogene 2005; 24(22):3554-62.
104. King TA, Gemignani ML, Li W et al. Increased progesterone receptor expression in benign epithelium of BRCA1-related breast cancers. Cancer Res 2004; 64(15):5051-3.
105. Chodankar R, Kwang S, Sangiorgi F et al. Cell-nonautonomous induction of ovarian and uterine serous cystadenomas in mice lacking a functional Brca1 in ovarian granulosa cells. Curr Biol 2005; 15(6):561-5.
106. Macaluso M, Cinti C, Russo G et al. pRb2/p130-E2F4/5-HDAC1-SUV39H1-p300 and pRb2/p130-E2F4/5-HDAC1-SUV39H1-DNMT1 multimolecular complexes mediate the transcription of estrogen receptor-alpha in breast cancer. Oncogene 2003; 22(23):3511-7.
107. Leu YW, Yan PS, Fan M et al. Loss of estrogen receptor signaling triggers epigenetic silencing of downstream targets in breast cancer. Cancer Res 2004; 64(22):8184-92.
108. King MC, Wieand S, Hale K et al. Tamoxifen and breast cancer incidence among women with inherited mutations in BRCA1 and BRCA2: National Surgical Adjuvant Breast and Bowel Project (NSABP-P1) Breast Cancer Prevention Trial. Jama 2001; 286(18):2251-6.
108a. Gronwald J, Tung N, Foulkes WD et al. Tamoxifen and contralateral breast cancer in BRCA1 and BRCA2 carriers: An update. Int J Cancer 2006; 118(9):2281-4.
109. Weber JD, Zambetti GP. Renewing the debate over the p53 apoptotic response. Cell Death Differ 2003; 10(4):409-12.
110. MacLachlan TK, Somasundaram K, Sgagias M et al. BRCA1 effects on the cell cycle and the DNA damage response are linked to altered gene expression. J Biol Chem 2000; 275(4):2777-85.
111. MacLachlan TK, Takimoto R, El-Deiry WS. BRCA1 directs a selective p53-dependent transcriptional response towards growth arrest and DNA repair targets. Mol Cell Biol 2002; 22(12):4280-92.
112. Ongusaha PP, Ouchi T, Kim KT et al. BRCA1 shifts p53-mediated cellular outcomes towards irreversible growth arrest. Oncogene 2003; 22(24):3749-58.
113. Brooks CL, Gu W. p53 Ubiquitination: Mdm2 and beyond. Mol Cell 2006; 21(3):307-15.
114. MacLachlan TK, Dash BC, Dicker DT, El-Deiry WS. Repression of BRCA1 through a feedback loop involving p53. J Biol Chem 2000; 275(41):31869-75.
115. Arizti P, Fang L, Park I et al. Tumor suppressor p53 is required to modulate BRCA1 expression. Mol Cell Biol 2000; 20(20):7450-9.
116. Fedier A, Steiner RA, Schwarz VA et al. The effect of loss of Brca1 on the sensitivity to anticancer agents in p53-deficient cells. Int J Oncol 2003; 22(5):1169-73.
117. Crook T, Crossland S, Crompton MR et al. p53 mutations in BRCA1-associated familial breast cancer. Lancet 1997; 350(9078):638-9.
118. Crook T, Brooks LA, Crossland S et al. p53 mutation with frequent novel condons but not a mutator phenotype in BRCA1- and BRCA2-associated breast tumours. Oncogene 1998; 17(13):1681-9.
119. Smith PD, Crossland S, Parker G et al. Novel p53 mutants selected in BRCA-associated tumours which dissociate transformation suppression from other wild-type p53 functions. Oncogene 1999; 18(15):2451-9.
120. Billack B, Monteiro AN. Methods to classify BRCA1 variants of uncertain clinical significance: The more the merrier. Cancer Biol Ther 2004; 3(5):458-9.
121. Furuta S, Jiang X, Gu B et al. Depletion of BRCA1 impairs differentiation but enhances proliferation of mammary epithelial cells. Proc Natl Acad Sci USA 2005.
122. Farmer H, McCabe N, Lord CJ et al. Targeting the DNA repair defect in BRCA mutant cells as a therapeutic strategy. Nature 2005; 434(7035):917-21.
123. Bryant HE, Schultz N, Thomas HD et al. Specific killing of BRCA2-deficient tumours with inhibitors of poly(ADP-ribose) polymerase. Nature 2005; 434(7035):913-7.
124. Hoeijmakers JH. Genome maintenance mechanisms for preventing cancer. Nature 2001; 411(6835):366-74.
125. Gallmeier E, Kern SE. Absence of specific cell killing of the BRCA2-deficient human cancer cell line CAPAN1 by Poly(ADP-ribose) polymerase inhibition. Cancer Biol Ther 2005; 4(7):703-6.
126. McCabe N, Lord CJ, Tutt AN et al. BRCA2-deficient CAPAN-1 cells are extremely sensitive to the inhibition of poly (ADP-Ribose) polymerase: An issue of potency. Cancer Biol Ther 2005; 4(9):934-6.
127. Ludwig T, Chapman DL, Papaioannou VE et al. Targeted mutations of breast cancer susceptibility gene homologs in mice: Lethal phenotypes of Brca1, Brca2, Brca1/Brca2, Brca1/p53, and Brca2/p53 nullizygous embryos. Genes Dev 1997; 11(10):1226-41.

Chapter 6

Integrin-Mediated Adhesion:
Tipping the Balance between Chemosensitivity and Chemoresistance

Mary M. Zutter*

Abstract

The integrin family of extracellular matrix receptors plays an important role in normal development, epithelial morphogenesis, angiogenesis, and in tumor progression and metastasis. Integrins cooperate with growth factor receptors to control many cellular functions including proliferation and cell survival. Integrin-mediated adhesion regulates many of the cell cycle checkpoints including activation of cyclin D/cdk4/6 complexes, expression of cyclin D genes, and regulation of levels of cyclin-dependent kinase inhibitors. In addition, integrin-mediated cell adhesion regulates apoptosis by modulating the activity of both the mitochondrial pathway and the death receptor pathways. Therefore, integrin-mediated adhesion modulates the decision of life or death. A role for tumor-matrix interactions in the acquisition of drug resistance has been reported for many cancers including breast cancer. Recent evidence suggests that integrin-mediated adhesion to the ECM may undermine the response of tumors to chemotherapeutic agents. Integrins have been shown to be readily accessible drug targets and are therefore attractive potential targets for combined modality chemotherapy.

The Integrin Family of Cell Adhesion Receptors

The integrin family of extracellular matrix (ECM) receptors plays an important role in normal development, epithelial morphogenesis, angiogenesis, and in tumor progression and metastasis.[1-5] Recent evidence suggests that integrin-mediated adhesion to the ECM may undermine the response of tumors to chemotherapeutic agents. Integrins have been shown to be readily accessible drug targets and are therefore attractive potential targets for combined modality chemotherapy.[6-9] This chapter reviews the molecular mechanisms by which integrins regulate cell survival, cell proliferation and the response to chemotherapeutic agents with a focus on breast cancer.

The integrins are a family of noncovalently associated heterodimeric adhesive receptors that play roles in mediating the cell-substrate and cell-cell interactions. Several general reviews are available.[10-12] At least 18 different α subunits and 8 different β subunits combine to form the 24 distinct heterodimers identified in humans.[11,12] These $\alpha\beta$ heterodimers act as receptors for a variety of ECM proteins including collagens, fibronectin and laminins. Some integrins act as receptors for soluble ligands, such as fibrinogen, while other integrins bind to counterreceptors on other cells, such as intracellular adhesion molecules.[13] Ligand specificity is a function of particular alpha/beta combination although great deal of redundancy is inherent in the system.

*Mary M. Zutter—Vanderbilt University Medical Center, Departments of Pathology and Cancer Biology, 4800B The Vanderbilt Clinic, Nashville, Tennessee 37232-5310, U.S.A. Email: mary.zutter@vanderbilt.edu

Breast Cancer Chemosensitivity, edited by Dihua Yu and Mien-Chie Hung.

The significant redundancy in adhesive receptors for small numbers of ECM ligands suggests that each receptor may mediate distinct sets of post-receptor ligand occupancy events. The degree of integrin activation may also contribute to ligand binding specificity.[14]

Integrins cooperate with growth factor receptors to control many cellular functions such as migration, proliferation and cell survival.[15] To accomplish this function, integrins link the ECM through the cell membrane to the cytoskeleton and many cytoplasmic signaling and structural proteins such as focal adhesion kinase (FAK), integrin-linked kinase (ILK), Src-family kinases, α-actinin, paxillin, as well as many others.[16-24] FAK interacts with cytoplasmic domains of the integrin β subunit and cytoplasmic adaptor proteins that also bind to the β cytoplasmic domain such as talin, vinculin, and paxillin. FAK is activated by autophosphorylation which results in the recruitment of additional focal adhesion adaptors including tensin and p130cas.[25-30] This activation leads to downstream activation of Crk and c-Jun N-terminal kinase (JNK).[31] The Src-family kinases also phosphorylate FAK to create a binding site for the adaptor protein Grb2. Grb2 then recruits mSOS and activates Ras and the extracellular signal-regulated kinase (ERK) mitogen-activated protein kinase (MAPK) cascade.[26] Activated FAK also binds and activates the phosphoinositide 3-OH kinase (PI 3-kinase).[32] Activation of the Ras/MAPK, PI3K, and JunK cascades stimulate cell proliferation and inhibit cell death.

Following ligand binding, the α1β1, α5β1, and αvβ3 integrins have been shown to interact with the membrane protein caveolin-1 through the transmembrane portion of the integrin α subunit.[16,17] This interaction has been shown to recruit the Src-family kinases Fyn and Yes to the integrin signaling complex and to activate the Shc-pathway which leads to activation of the Ras-ERK MAPK pathway and promotes cell proliferation and cell survival.[16,17] Other integrin α subunits use their extracellular or transmembrane domains to interact with other proteins. For example, the α3β1 and the α6β1 integrins bind to members of a family of four transmembrane domain containing proteins that activate phosphatidylinositol-4-OH kinase.[33] The integrin associated protein has been shown to bind to the extracellular portion of the α subunit of the αIIbβ3, αvβ3, and α2β1 in a complex with thrombospondin.[34,35] Other molecules that have been shown to interact with integrin α subunits include calreticulin, F-actin, paxillin, Nisharin, Mss4, and BIN1.[36-40]

ILK is a serine-threonine protein kinase that also interacts with β integrin subunits to mediate downstream signals via interactions with a diverse set of signaling molecules, in a manner similar to but not identical to FAK. ILK interacts with paxilllin,[41-43] but in addition, ILK also interacts with a family of novel adaptors called the double zinc finger domain (LIM)-only proteins that include PINCH-1 and PINCH-2, and the F-actin binding proteins, parvins to activate PI3K/Akt and stimulate cell survival.[43-46] Overexpression of ILK in epithelial cells stimulates anchorage-independent cell cycle progression and suppresses anoikis.[47,48]

Integrins and Cell Proliferation

Integrins modulate cell proliferation and cell cycle progression at many steps.[49] As cells progress through the cell cycle from a quiescent state in G0 towards duplication of their DNA in S-phase, the first gap-phase, G1, serves as a checkpoint to ensure that the cell is in the proper environment to proceed with replication. Integrins regulate many of the cell cycle checkpoints. Integrin-mediated adhesion is required to activate the cyclin D/cdk4/6 complexes and in some cell types is required for cyclin D expression. Integrin ligation synergizes with signals from growth factor receptors to activate the JNK and ERK MAPK pathways that ultimately lead to increase transcription of cyclin D genes.[50] In some cells, robust activation of the growth factor-induced activation of Raf, MAPK, Rac or PI3K requires integrin-mediated cell adhesion.[51,52]

During the late G1 phase of the cell cycle, up-regulation of cyclin E and activation of the the cyclin E/cdk2 kinase is required for S-phase entry. In many cell types this checkpoint is also coordinately regulated by growth factor activation and integrin-mediated adhesion. Adhesion regulates the levels of the cyclin-dependent kinase inhibitors (CKI) of the p21 family (p21, p27, and p57). Cell adhesion to the ECM down-regulates the levels of p21 and p27, thereby decreasing levels available to inhibit the cdk2/cyclin E complex.[53-55] In addition, the adhesion-dependent up-regulation of cyclin D1, increased cyclin D/cdk4/6 complexes act to sequester p21 and p27.[56-58] Thus through changes in the inhibitor levels, redistribution of inhibitor, or a combination of both, integrin-dependent adhesion reduces the amount of free p21 and p27.

Integrins and Apoptosis

Epithelial cells and endothelial cells require attachment to matrix for cell survival through the inhibition of apoptosis.[59,60] Programmed cell death (apoptosis) is the active process of cell suicide induced by the withdrawal of survival factors. Although many diverse stimulae can induce apoptosis, apoptosis that is induced by the loss of adhesion or inadequate or inappropriate cell-matrix interaction is called "anoikis", based on the Greek word for homelessness. Anoikis is a normal biologic process used by both epithelial cells and endothelial cells to maintain appropriate cell numbers and tissue organization.[59,60] In epithelial and endothelial cells detachment from matrix leads to increased expression of p53, a stimulus of apoptosis. In other cells, including malignant cells such as melanoma and sarcoma, and transformed fibroblasts, matrix detachment leads to decreased levels of p53 and changes in p14/p19 Arf after DNA damage that result in cell survival and genomic instability.[61] Therefore, in these cell types, loss of adhesion-mediated regulation of apoptosis may contribute to both malignancy and therapy resistance.

The expression of specific oncogenes including overexpression of bcl-2 or transformation of epithelial cells with v-Ha-ras, v-src, or treatment with phorbol ester renders certain cells anchorage-independent. In human cancers, loss of control of apoptosis contributes to the development of mammary, gastrointestinal and lung cancers.[62-65]

To understand how integrin-mediated adhesion controls apoptosis we need a brief overview of apoptosis. Apoptosis occurs when the delicate balance of pro-survival and pro-apoptotic stimuli shifts toward apoptosis.[66] Our current understanding is that apoptosis occurs via two separate pathways, the death receptor-dependent (extrinsic) pathway and the independent (intrinsic or mitochondrial) pathway. As extensively reviewed by Martin and Vuori, integrin-mediated cell adhesion has been shown to modulate the activity of both the mitochondrial pathway and the death receptor pathway.[67] The mitochondrial pathway is mediated by a balance of anti-apoptotic and pro-apoptotic proteins of the Bcl-2 family. The Bcl-2 family regulates the decision between life and death through interaction of pro- and anti-apoptotic family members. Integrin-mediated adhesion influences apoptosis via numerous mechanisms including enhanced transcription of the Bcl-2 gene, erk/MAPK-induced phosphorylation of Bad, preventing its proapoptotic function, and erk/MAPK-induced phosphorylation of Bcl-2, preventing its unbiquitin-mediated degradation.[68-74] Stromblad et al demonstrated that disruption of integrin-mediated adhesion leads to apoptosis or cell cycle arrest through activation of p53.[76] p53 regulates the transcription of proapoptotic Bcl-2 family members, including Bax. Increased expression of Bax then alters Bcl-2/Bax ratio in favor of cell death.[77,78]

In mammary epithelial cells, Gilmore et al showed that detachment of normal primary mammary epithelial cells induced a rapid translocation of Bax to the mitochondria. Bax translocation occurred prior to activation of the caspase cascade.[74] In mammary epithelial cells, Bid also translocates to the mitochondria with identical kinetics to Bad following cell detachment. Bid translocation is required for anoikis.[79] In contrast to the role of Bax and Bid is mediating anoikis, the role of Bim is controversial. Reginato et al demonstrated that Bim was a critical mediator of anoikis using MCF-10A cells. In fact, after detachment of MCF-10A cells from the matrix only levels of Bim changed and not levels of Bad, Bim, Bmf or Bid.[80] In contrast, Wang et al reported that apoptosis following deadhesion of a primary mouse mammary epithelial cells and a mouse mammary epithelial cell line did not utilize Bim.[81] Further studies are required to resolve these issues. Additional mechanisms by which cell adhesion regulates anoikis were reviewed by Hazlehurst et al.[82] In some cases proapoptotic Bcl-2 family members such as Bmf and Bim are sequestered to the cytoskeleton in normal healthy cells and released following matrix detachment or UV exposure.[83-86]

Integrin ligation activates many kinase/phosphatase signaling molecules implicated not only in the control of cell proliferation but also in anoikis.[87-91] Following integrin ligation FAK activates PI3K and the serine-threonine kinase Akt directly or indirectly via the p130cas pathway. FAK activation also leads to activation of c-Jun.[92-75] FAK-mediated activation of PKB/Akt inactivates the pro-apoptotic proteins, Bad and caspase 9, thereby inhibiting apoptosis.[96,97] In addition, FAK has been shown to regulate the expression of caspase inhibitors of the IAP (inhibitor of apoptosis) family by a proposed mechanism involving PI3K/Akt activation of the NF-κB pathway.[66,98]

ILK binding to the cytoplasmic domain of the β integrins results in transient phosphorylation of PI3K/Akt. Overexpression of ILK leads to prolonged activation of Akt and suppression of anoikis. Expression of dominant-negative ILK results in decreased phosphorylation of Akt and enhanced apoptosis.[99-101]

Integrins and Cancer

One of the earliest suggestions that integrins play a role in differentiation and malignancy came from studies of the malignant transformation of cells in culture. Plantefaber and Hynes demonstrated that oncogenic transformation of rodent fibroblasts with Rous sarcoma virus encoding the src oncogene or murine sarcoma virus encoding the ras oncogene led to reduced expression of the α5β1 integrin and two other unidentified integrins.[102] Expression of the α3β1 integrin was retained. Later, Dedhar and Saulier demonstrated that treatment of a human osteogenic sarcoma cell line (HOS) with N-methyl-N'-nitro-N-nitrosoguanidine (MMNG), a potent carcinogen, altered integrin expression. Increased expression of the α6β1, α2β1, α1β1 integrins contrasted with reduction in αvβ3 expression and no change in α5β1 or α3β1.[103] In both examples, transformation was associated with morphologic alterations and increased invasiveness, suggesting that changes in integrin expression might contribute to changes in cell phenotype associated with malignant transformation.

In the human breast, oncogenesis results from amplification and/or over-expression of members of the epidermal growth factor receptor (EGF-R) family. EGF-R is over-expressed in approximately 40% of ductal carcinomas.[104] In approximately 30% of other breast cancers, the *c-erbB2/c-neu* oncogene is either amplified or over-expressed.[103-107] Moreover, over-expression of *c-erbB2* is an independent risk factor for poor prognosis suggesting that this particular proto-oncogene may be involved in cancer progression. Over-expression of *c-erbB2* in an immortalized human mammary epithelial cell line derived from luminal epithelial cells disrupted morphogenesis in three-dimensional collagen gels when compared to parental cells in which the *c-erbB2* proto-oncogene was not over-expressed.[108] Over-expression of *c-erbB2* in this model system resulted in decreased expression of the α2 integrin subunit protein and mRNA.[109,110] These findings suggested an inverse correlation between *c-erbB2* expression and the expression of the $\alpha2\beta_1$ integrin.

Using biopsy samples of human cancers, many investigations including our own have described alterations in expression or cellular localization of many different integrin heterodimers in carcinoma of the breast, colon, prostate, lung, pancreas, and skin. The α2β1, α3β1, and the α6β4 integrins are expressed at high levels in most normal epithelial cells and associated with the differentiated epithelial phenotype. Loss of expression of any one of these receptors has been associated with loss of differentiation in malignancies. In contrast, expression of the α5β1 and the αvβ3 integrin is increased, decreased, or unchanged in different malignancies. Increased expression of the αvβ3 integrin has been associated with a poor clinical course in a number of malignancies, including epithelial malignancies and melanoma.[111-117]

Integrin-Mediated Drug Resistance

Cancer therapies including both ionizing irradiation and chemotherapy selectively induce apoptosis of rapidly dividing cells.[9] Cancer recurrence and metastasis is thought to result from incomplete killing of all tumor cells. Tumors that recur are at a selective advantage, will not respond to the same chemotherapeutic and/or radiation therapeutic regimens and ultimately become more aggressive. The ability of radiation and chemotherapy to damage DNA is a critical component of therapeutic intervention. DNA damage leads to apoptosis via a complex pathway regulated by p53 family members. P53, a tumor suppressor protein, serves as a molecular switch deciding between apoptosis and cell cycle arrest following DNA damage and other genotoxic stresses (reviewed in ref. 9). As discussed above, integrin-mediated adhesion modulates many pathways that lead to the decision of life or death.

Resistance to chemotherapy may be an intrinsic property of the tumor cells or tumor cells may develop resistance to chemotherapy during treatment. The mechanisms for drug resistance include drug efflux, drug inactivation, alterations in drug targets, processing of drug-induced damage and evasion of apoptosis, modulation of cell cycle checkpoint mediators, and changes in the downstream mediators of the apoptotic pathways.[9] Many of these mechanisms such as changing patterns of drug efflux, the ability to inactivate the drug or alter the drug target depend on the tumor cell alone. We now appreciate that cell adhesion may play an important role in controlling drug responsiveness.[8]

The earliest works demonstrating a contribution of integrin-mediated adhesion to drug resistance was in small cell lung cancer (SCLC). Fridman et al demonstrated that the majority of SCLC cell lines adhered to laminin and that adhesion to laminin resulted in resistance to a number chemotherapeutic agents including etoposide, cisplatinum, doxorubicin, and nitrogen mustard. The cells that failed to adhere to laminin were not resistant to these chemotherapeutic agents.[118] Although a role for a specific integrin was not directly addressed, these data suggested that integrin-mediated adhesion was at least in part responsible for the chemoresistance. Kraus et al later showed that chemoresistance of SCLC cells resulted from the induction of apoptosis-resistant variants that up-regulated the expression of the integrin subunits α2, β3 and β4. The resistance to chemotherapy required matrix adhesion, activation of the Akt and MAP kinase pathways, increased levels of phosphorylated Bad protein and activation of NF-kappa B.[119]

In vivo, SCLC is surrounded by an ECM-rich environment composed of fibronectin, collagen IV, and tenascin both at primary sites and at sites of tumor metastasis. Cells from several small cell lung cancer cell lines expressed many of the β1 integrins including the α2β1, α3β1, α6β1 and αvβ1 integrins and adhered to fibronectin, laminin and type IV collagen in a β1 integrin-dependent manner. Integrin-mediated adhesion to all three matrix components protected the cells from apoptosis induced by doxorubicin, cyclophosphamide and cis-plantinum.[120] Protection from cell death was prevented by the tyrosine phosphorylation inhibitor Tyrphostin-25 or by an anti-β1 integrin antibody. The integrin-dependent, anti-apoptotic effect was a consequence of blocking the proteolytic cleavage of pro-caspase-3.[121] Hartmann et al also demonstrated that activation of integrins or the chemokine receptor CXCR4 resulted in increased chemotherapy resistance by SCLC cells.[122] Similar observations were made using the nonsmall cell carcinoma cell line A549.[123] Attachment of these cells to fibronectin resulted in chemoresistance to irradiation, cisplatin, paclitaxel or mitomycin C.

Integrin expression predicts chemoresistance in vivo. Okita et al investigated the relationship between expression of the β1 integrin family, p53 expression, and resistance to chemotherapy in a cohort of patients with SCLC. The overall survival of patients with elevated expression of either both β1 integrin and p53, or β1 integrin, or p53 was significantly worse than that of patients without high level expression of β1 integrin or p53. When the association between survival and prognostic factors, including gender, age, performance status, clinical stage were examined by the Cox proportional hazards model, expression of both β1 integrin and p53 were independent risk factors (hazard ratio = 0.394, $p = 0.0005$).[124]

Integrin-mediated adhesion also modulates the expression of proteins that block apoptosis. Increased expression of proteins that block apoptosis, including X-linked inhibitor of apoptosis (XIAP) and survivin were associated with decreased overall survival, increased recurrence and resistance to radiotherapy in nonsmall cell lung and colon cancer.[125-128] Cao et al demonstrated that inhibition of survivin or XIAP greatly increased irradiation-induced cell death by stimulating apoptosis and inhibiting cell survival. Using an in vivo mouse model of lung cancer, the combination of radiotherapy with inhibition of survivin or XIAP inhibited tumor growth better than either agent alone.[129] β1 integrin-mediated adhesion of the aggressive PC3 prostate adenocarcinoma cell line to fibronectin resulted in up-regulation of survivin and protection from tumor necrosis factor-α (TNF-α)-induced apoptosis.[130] Inhibition of survivin activity or decreased expression of survivin in PC3 cells using dominant-negative or antisense survivin constructs obliterated the adhesion-mediated protection from apoptosis. In addition, adhesion to fibronectin resulted in increased expression of survivin protein via a PI3K/ Akt dependent pathway.[130]

In contrast to the importance of the β1 integrin family described above, drug-induced apoptosis of glioma cell lines was inhibited by adhesion to vitronectin, but not to fibronectin. Either of the two classic vitronectin-binding integrins, αvβ3 or αvβ5, conferred chemoresistance to topoisomerase. The chemoresistance observed with vitronectin was associated with increased expression of two antiapoptotic proteins, Bcl-2 and Bcl-X(L), with a consequent increase in the ratios for Bcl-2:Bax and Bcl-X(L):Bax.[13] A role for tumor-matrix interactions in the acquisition of drug resistance had been reported for pancreatic cancer, colon cancer, and ovarian cancer.[131-135]

Integrins and Chemoresistance in Breast Cancer

MCF7 cells are a relatively well-differentiated and nonmetastatic human breast cancer cell line. Nista et al reported that adriamycin-resistant MCF-7 cells demonstrated increased expression of the α5β1 integrin in comparison to the MCF-7 parental cells. Augmented adhesion to fibronectin correlated with inhibition of apoptosis in response to serum starvation.[136] Narita et al demonstrated that the α6β1 but not the α2β1 integrin was also upregulated in an adriamycin-resistant MCF-7 cell line.[137] The highly metastatic MDA-MB-435 breast carcinoma cell line with targeted deletion of the α6β1 integrin exhibited growth suppression and increased apoptosis in comparison to the parental tumors in vivo. These data suggested that expression of the α6β1 integrin in breast carcinoma cell lines facilitated tumorigenesis and promoted tumor cell survival in distant organs.[138]

To directly address the impact of integrin-mediated cell adhesion on drug-induced apoptosis, Aoudjit and Vuori evaluated whether cell adhesion to different matrices altered the response of MDA-MB-231 and MDA-MB-435, both highly metastatic breast cancer cell lines that are sensitive to the microtubule-targeting chemotherapeutic agents paclitaxel and vincristine.[139] MDA-MB-231 cells adhered well and survived to an equal extent on type I collagen, fibronectin, laminin-1, and vitronectin under baseline conditions. However, when MDA-MB-231 cells were treated with increasing concentration of paclitaxel, cells plated on uncoated dishes or on nonmatrix coated (poly-L lysine) died by apoptosis, as demonstrated by increased DNA fragmentation. Cell attachment to type I collagen or fibronectin inhibited apoptosis and DNA fragmentation. In contrast, adhesion to vitronectin and laminin-1 failed to protect from drug-induced apoptosis. The ability of type I collagen and fibronectin to protect MDA-MB-231 cells from apoptosis was dependent specifically on two integrins, the α2β1 and α5β1. In contrast, MDA-MB-435 cells demonstrated different chemoresistant properties. Adhesion to neither type I collagen nor fibronectin protected MDA-MB-435 cells from paclitaxel-induced apoptosis. However, adhesion of MDA-MB-435 cells to laminin-1 via the α6β1 integrin provided protection from apoptosis. These studies demonstrated that protection from drug-induced apoptosis is exquisitely matrix protein-, integrin- and cell type-dependent. In these studies, integrin-mediated regulation of drug-induced cell death was dependent on PI3K/Akt activation, inhibition of cytochrome c release from the mitochondria and sustained high level expression of Bcl-2.[139]

Menendez et al utilized the MCF7 cell model to evaluate the effect of overexpression of the CYR61 (CCN1; the human homolog of a mouse immediate early response gene, *Cyr61*) gene.[140] CYR61 was identified as a gene that was differentially expressed in invasive and metastatic human breast cancer cells.[141] CYR61, a secreted protein that associates with the cell surface and extracellular matrix, is a member of the Cysteine-rich 61/Connective tissue growth factor/nephroblastoma-overexpressed (CCN) gene family of angiogenic and growth regulators.[142-148] In breast cancer CYR61 is a downstream effector of heregulin (HRG). Overexpression of CYR61 without HRG or Her-2/neu (erbB-2) oncogene overexpression promoted hormone independence and drug resistance, and enhanced the metastatic phenotype.[148-152] CYR61 functioned in an autocrine-paracrine manner through interaction with multiple integrins (αvβ3, αvβ5, α6β1 and αIIbβ3). Overexpression of CYR61 in HRG-negative MCF-7 cells resulted in markedly increased expression (greater than 200 fold) of the αvβ3 integrin and rendered the cells resistant to paclitaxel-induced cytotoxicity. In fact the resistance of breast cancer cells to Taxol-induced cytotoxicity was blocked by functional inhibition of the integrin αvβ3. These results suggested that CYR61 modulates chemosensitivity or chemoresistance to therapeutic agents such as Taxol via interaction with the αvβ3 integrin. In addition, Menendez et al demonstrated that functional antagonism of the αvβ3 integrin enhanced Taxol-induced apoptosis, although antagonism of the integrin alone reduced cell proliferation but failed to increase apoptosis, suggesting a synergism between integrin antagonism and chemotherapy. Additional data utilizing this system suggested a model in which CYR61 overexpression in breast cancer cells led to increased expression of the αvβ3 integrin and subsequent αvβ3 integrin activation via CYR61 of the Raf/MEK1/MEK2-ERK1/ERK2 cascade. These data suggest the potential clinical use of peptidomimetic integrin-antagonists in combination with chemotherapy for breast cancer.[140]

Kayaselcuk et al compared the expression of survivin in biopsies of malignant breast cancer to nonmalignant, intraductal epithelial neoplasia. Breast cancer cells exhibited increased survivin expression. Increased expression of survivin in this series of breast cancer was associated with increased expression of Bcl-2 and p53.[153] In other studies of breast cancer, survivin expression was associated with clinical features including high histologic grade, p53 gene mutation and loss of heterozygosity at chromosome 17p13.1.[154-155] A direct correlation between survivin expression and integrin-mediated drug resistance in breast cancer has not been reported, however the data from prostate cancer suggests an important association.

The Tumor Microenvironment and Breast Cancer

The important roles that three-dimensional (3-D) tissue architecture and the microenvironment play on normal mammary gland development, tumor progression and chemotherapy resistance have emerged from studies by numerous investigators (review by Zahir and Weaver).[156] The mammary gland and breast cancers develop within a 3-D environment where cells are influenced by the extracellular matrix and other cells. Teicher et al (1990) demonstrated using EMT-6 mammary tumor cells grown in mice that the tumors rapidly developed drug resistance when treated with a number of different drugs including cis-diamminedichloroplatinum (II) (CDDP), carboplatin, cyclophosphamide (CTX), or thiotepa.[157] Cells treated in 2D cell culture with the same agents failed to develop drug resistance. These in vivo selected drug-resistant variants expressed their resistance when cultured in vitro as spheroids. The term "multicellular drug resistance" was coined to refer to therapeutic resistance of cancer cells in 3-D tumor spheroids in vitro or solid tumors in vivo to chemotherapy, ionizing radiation, and F_c-dependent host defense mechanisms.[158,159] The resistance behavior of cancer cells in 3-D spheroids mimicked that of solid tumors in vivo. The resistance was in part mediated by changes in tissue architecture, cell-cell adhesion, integrin expression, and ECM organization.[160] Green et al demonstrated that much of the drug resistant phenotype resulting from tumor spheroid formation was dependent on E-cadherin mediated cell-cell adhesion.[161]

Additional roles that integrins play in the 3-D architecture of the mammary gland and the influence of integrins on breast cancer survival have been described. Boudreau demonstrated that tissue architecture was crucial for homeostasis, suppression of apoptosis, and maintenance of differentiated mammary epithelial cell phenotype.[162] Weaver et al subsequently demonstrated that polarized 3-D glandular structures surrounded by basement membrane conferred protection from apoptosis in both nonmalignant and malignant mammary epithelial cells. The resistance to apoptosis required ligation of the β4 integrin and NF-κB activation.[163] Apoptosis resistance was acquired in breast cancers via autocrine synthesis of laminin-5 and a positive feedback loop from α6β4 integrin-laminin-5 ligation and anchorage-independent survival.[164]

Conclusions

In summary, integrin ligation plays an important role in modulating the decision of a cancer cell to live or to die. In modulating cell survival, the integrin family determines the impact of chemotherapy and can determine the acquisition of the drug resistant phenotype. Since integrins have been shown to be readily accessible drug targets, future directions should include the evaluation of integrin antagonists in combined modality chemotherapy.[6-9]

Reference

1. Loftus JC, Smith JW, Ginsberg MH. Integrin-mediated cell adhesion: The extracellular face. J Biol Chem 1994; 269(41):25235-25238.
2. Hood JD, Cheresh DA. Role of integrins in cell invasion and migration. Nat Rev Cancer 2002; 2(2):91-100.
3. Guo W, Giancotti FG. Integrin signalling during tumour progression. Nat Rev Mol Cell Biol 200; 5(10):816-826.
4. Hynes RO. Cell adhesion: Old and new questions. Trends Cell Biol 1999; 9(12):M33-M37.
5. Gustafsson E, Fassler R. Insights into extracellular matrix functions from mutant mouse models. Exp Cell Res 2000; 261(1):52-68.
6. Hynes RO. A reevaluation of integrins as regulators of angiogenesis. Nat Med 2002; 8(9):918-921.
7. Tucker GC. Inhibitors of integrins. Current Opinion in Pharmacology 2002; 2:394-401.
8. Damiano JS. Integrins as novel drug targets for overcoming innate drug resistance. Curr Cancer Drug Targets 2002; 2(1):37-43.
9. Longley DB, Johnston PG. Molecular mechanisms of drug resistance. J Pathology 2005; 205:275-292.
10. Zutter MM, Sun H, Santoro S. Altered integrin expression and the malignant phenotype: The contribution of multiple integrated integrin receptors. J Mammary Gland Biol Neoplasia 1998; 3(2):191-200.
11. Hynes RO. Integrins: Bidirectional, allosteric signaling machines. Cell 2002; 110(6):673-687.
12. Hynes RO, Zhao Q. The evolution of cell adhesion. J Cell Biol 2000; 150(2):F89-F96.
13. Clark EA, Brugge JS. Integrins and signal transduction pathways: The road taken. Science 1995; 268(5208):233-239.
14. Liddington RC, Ginsberg MH. Integrin activation takes shape. J Cell Biol 2002; 158(5):833-839.
15. Giancotti FG, Rouslahti E. Integrin signaling. Science 1999; 285:1028-1032.
16. Wary KK, Mainiero F, Isakoff SJ et al. The adaptor protein Shc couples a class of integrins to the control of cell cycle progression. Cell 1996; 87(4):733-743.
17. Wary KK, Mariotti A, Zurzolo C et al. A requirement for caveolin-1 and associated kinase Fyn in integrin signaling and anchorage-dependent cell growth. Cell 1998; 94(5):625-634.
18. Guan JL, Shalloway D. Regulation of focal adhesion-associated protein tyrosine kinase by both cellular adhesion and oncogenic transformation. Nature 199; 358(6388):690-692.
19. Schaller MD, Borgman CA, Cobb BS et al. pp125FAK a structurally distinctive protein-tyrosine kinase associated with focal adhesions. Proc Natl Acad Sci USA 1992; 89(11):5192-5196.
20. Schaller MD, Parsons JT. Focal adhesion kinase and associated proteins. Curr Opin Cell Biol 1994; 6(5):705-710.
21. Hannigan GE, Leung-Hagesteijn C, Fitz-Gibbon L et al. Regulation of cell adhesion and anchorage-dependent growth by a new beta 1-integrin-linked protein kinase. Nature 1996; 379(6560):91-96.
22. Dedhar S, Hannigan GE. Integrin cytoplasmic interactions and bidirectional transmembrane signalling. Curr Opin Cell Biol 1996; 8(5):657-669.

23. Lewis JM, Baskaran R, Taagepera S et al. Integrin regulation of c-Abl tyrosine kinase activity and cytoplasmic-nuclear transport. Proc Natl Acad Sci USA 1996; 93(26):15174-15179.
24. Delcommenne M, Tan C, Gray V et al. Phosphoinositide-3-OH kinase-dependent regulation of glycogen synthase kinase 3 and protein kinase B/AKT by the integrin-linked kinase. Proc Natl Acad Sci USA 1998; 95(19):11211-11216.
25. Schaller MD, Hildebrand JD, Parsons JT. Complex formation with focal adhesion kinase: A mechanism to regulate activity and subcellular localization of Src kinases. Mol Biol Cell 1999; 10(10):3489-3505.
26. Schlaepfer DD, Hanks SK, Hunter T et al. Integrin-mediated signal transduction linked to Ras pathway by GRB2 binding to focal adhesion kinase. Nature 1994; 372(6508):786-791.
27. Richardson A, Parsons T. A mechanism for regulation of the adhesion-associated proteintyrosine kinase pp125FAK. Nature 1996; 380(6574):538-450, (Erratum in Nature 1996; 381(6585):810).
28. Vuori K, Hirai H, Aizawa S et al. Introduction of p130cas signaling complex formation upon integrin-mediated cell adhesion: A role for Src family kinases. Mol Cell Biol 1996; 16(6):2606-2613.
29. Frisch SM, Vuori K, Ruoslahti E et al. Control of adhesion-dependent cell survival by focal adhesion kinase. J Cell Biol 1996; 134(3):793-799.
30. Schlaepfer DD, Broome MA, Hunter T. Fibronectin-stimulated signaling from a focal adhesion kinase-c-Src complex: Involvement of the Grb2, p130cas, and Nck adaptor proteins. Mol Cell Biol 1997; 17(3):1702-1713.
31. Oktay M, Wary KK, Dans M et al. Integrin-mediated activation of focal adhesion kinase is required for signaling to Jun NH2-terminal kinase and progression through the G1 phase of the cell cycle. J Cell Biol 1999; 145(7):1461-1469.
32. Chen HC, Appeddu PA, Parsons JT et al. Interaction of focal adhesion kinase with cytoskeletal protein talin. J Biol Chem 1995; 270(28):16995-16999.
33. Berditchevski F, Chang S, Bodorova J et al. A novel link between integrins, transmembrane-4 superfamily proteins (CD63 and CD81), and phosphatidylinositol 4-kinase. J Biol Chem 1997; 272(5):2595-2598.
34. Lindberg FP, Gresham HD, Schwarz E et al. Molecular cloning of integrin-associated protein: An immunoglobulin family member with multiple membrane-spanning domains implicated in alpha v beta 3-dependent ligand binding. J Cell Biol 1993; 123(2):485-496.
35. Chung J, Gao A, Frazier WA. Thrombspondin acts via integrin-associated protein to activate the platelet integrin alphaIIbbeta3. J Biol Chem 1997; 272(23):14740-14746.
36. Rojiani MV, Finlay BB, Gray V et al. In vitro interaction of a polypeptide homologous to human Ro/SS-A antigen (calreticulin) with a highly conserved amino acid sequence in the cytoplasmic domain of integrin alpha subunits. Biochemistry 1991; 30(41):9859-9866.
37. Kieffer JD, Plopper G, Ingber DE et al. Direct binding of F actin to the cytoplasmic domain of the alpha 2 integrin chain in vitro. Biochem Biophys Res Commun 1995; 217(2):466-474.
38. Liu S, Thomas SM, Woodside DG et al. Binding of paxillin to alpha4 integrins modifies integrin-dependent biological responses. Nature 1999; 402(6762):676-681.
39. Alahari SK, Lee JW, Juliano RL. Nischarin, a novel protein that interacts with the integrin alpha5 subunit and inhibits cell migration. J Cell Biol 2000; 151(6):1141-1154.
40. Wixler V, Laplantine E, Geerts D et al. Identification of novel interaction partners for the conserved membrane proximal region of alpha-integrin cytoplasmic domains. FEBS Lett 1999; 445(2-3):351-355.
41. Hannigan GE, Leung-Hagesteijn C, Fitz-Gibbon L et al. Regulation of cell adhesion and anchorage-dependent growth by a new beta 1-integrin-linked protein kinase. Nature 1996; 379(6560):91-96.
42. Nikolopoulos SN, Turner CE. Molecular dissection of actopaxin-integrin-linked kinase-paxillin interactions and their role in subcellular localization. J Biol Chem 2002; 277:1568-1575.
43. Nikolopoulos SN, Turner CE. Integrin-linked kinase (ILK) binding to paxillin LD1 motif regulates ILK localization to focal adhesions. J Biol Chem 2001; 276(26):23499-23505.
44. Olski TM, Noegel AA, Korenbaum E. Parvin: A 42-kDa focal adhesion protein, related to the alpha-actinin superfamily. J Cell Sci 2001; 114:525-538.
45. Tu Y, Huang Y, Zhang Z et al. A new focal adhesion protein that interacts with integrin-linked kinase and regulates cell adhesion and spreading. J Cell Biol 2001; 153:585-598.
46. Dedhar S. Cell-substrate interactions and signaling through ILK. Curr Opin Cell Biol 2000; 12:250-256.
47. Radeva G, Petrocelli T, Behrend E et al. Overexpression of the integrin-linked kinase promotes anchorage-independent cell cycle progression. J Biol Chem 1997; 272:13937-13944.
48. Attwell S, Roskelley C, Dedhar S. The integrin-linked kinase (ILK) suppresses anoikis. Oncogene 2000; 19:3811-3815.

49. Bottazzi MC, Assoian RK. The extracellular matrix and mitogenic growth factors control G1 phase cyclins and cyclin-dependent kinsae inhibitors. Trends Cell Biol 1997; 7:348-352.
50. Albanese C, Johnson J, Watanabe G et al. Transforming p21ras mutants and c-Ets-2 activate the cyclin D1 promoter through distinguishable regions. J Biol Chem 1995; 270(40):23589-23597.
51. Aplin AE, Howe AK, Juliano RL. Cell adhesion molecules, signal transduction and cell growth. Curr Opin Cell Biol 1999; 11(6):737-744.
52. Howe AK, Aplin AE, Juliano RL. Anchorage-dependent ERK signaling-mechanisms and consequences. Curr Opin Genet Dev 2000; 12(1):30-35.
53. Koyama H, Raines EW, Bornfeldt KE et al. Fibrillar collagen inhibits arterial smooth muscle proliferation through regulation of Cdk2 inhibitors. Cell 1996; 87(6):1069-1078.
54. Fang F, Orend G, Watanabe N et al. Dependence of cyclin E-CDK2 kinase activity on cell anchorage. Science 1996; 271(5248):499-502.
55. Zhu X, Ohtsubo M, Bohmer RM et al. Adhesion-dependent cell cycle progression linked to the expression of cyclin D1, activation of cyclin E-cdk2, and phosphorylation of the retinoblastoma protein. J Cell Biol 1996; 133(2):391-403.
56. Assoian RK. Anchorage-dependent cell cycle progression. J Cell Biol 1997; 136(1):1-4.
57. Assoian RK. Common sense signalling. Nat Cell Biol 2002; 4(8):E187-E188.
58. Sherr CJ, Roberts JM. Inhibitors of mammalian G1 cyclin-dependent kinases. Genes Dev 1995; 9(10):1149-1163.
59. Frisch SM, Francis H. Disruption of epithelial cell-matrix interactions induces apoptosis. J Cell Biol 1994; 124(4):619-626.
60. Meredith Jr JE, Fazeli B, Schwartz MA. The extracellular matrix as a cell survival factor. Mol Biol Cell 1993; 4(9):953-961.
61. Lewis JM, Truong TN, Schwartz MA. Integrins regulate the apoptotic response to DNA damage through modulation of p53. Proc Natl Acad Sci USA 2002; 99(6):3627-3632.
62. Yawata A, Adachi M, Okuda H et al. Prolonged cell survival enhances peritoneal dissemination of gastric cancer cells. Oncogene 1999; 16:2681-2686.
63. Streuli CH, Gilmore AP. Adhesion-mediated signaling in the regulation of mammary epithelial cell survival. J Mammary Gland Biol Neoplasia 1999; 4:183-191.
64. Shanmugathasan M, Jothy S. Apoptosis, anoikis and their relevance to the pathobiology of colon cancer. Pathol Int 2000; 50:273-279.
65. Wei L, Yang Y, Yu Q. Tyrosine kinase-dependent P13-kinase and mitogen-activated protein-kinase-independent signaling pathways prevent lung adenocarcinoma anoikis. Cancer Res 2001; 61:2439-2444.
66. Frisch SM, Screaton RA. Anoikis mechanisms. Curr Opin Cell Biol 2001; 13(5):555-562.
67. Martin SS, Vuori K. Regulation of Bcl-2 proteins during anoikis and amorphosis. Biochim Biophys Acta 2004; 1692(2-3):145-157.
68. Matter ML, Ruoslahti E. A signaling pathway from the alpha5beta1 and alpha(v)beta3 integrins that elevates bcl-2 transcription. J Biol Chem 2001; 276:27757-27763.
69. Zhang Z, Vuori K, Reed JC et al. The alpha 5 beta 1 integrin supports survival of cells on fibronectin and up-regulates Bcl-2 expression. Proc Natl Acad Sci USA 1995; 92:6161-6165.
70. Scheid MP, Duronio V. Dissociation of cytokine-induced phosphorylation of Bad and activation of PKB/akt: Involvement of MEK upstream of Bad phosphorylation. Proc Natl Acad Sci USA 1998; 95:7439-7444.
71. Scheid MP, Schubert KM, Duronio V. Regulation of bad phosphorylation and association with Bcl-x(L) by the MAPK/Erk kinase. J Biol Chem 1999; 274:31108-31113.
72. Breitschopf K, Haendeler J, Malchow P et al. Posttranslational modification of Bcl-2 facilitates its proteasome-dependent degradation: Molecular characterization of the involved signaling pathway. Mol Cell Biol 2000; 20:1886-1896.
73. Rytomaa M, Martins LM, Downward J. Involvement of FADD and caspase-8 signalling in detachment-induced apoptosis. Curr Biol 1999; 9:1043-1046.
74. Coll ML, Rosen K, Ladeda V et al. Increased Bcl-xL expression mediates v-Src-induced resistance to anoikis in intestinal epithelial cells. Oncogene 2002; 21:2908-2913.
75. Tiberio R, Marconi A, Fila C et al. Keratinocytes enriched for stem cells are protected from anoikis via an integrin signaling pathway in a Bcl-2 dependent manner. FEBS Lett 2002; 524:139-144.
76. Stromblad S, Becker JC, Yebra M et al. Suppression of p53 activity and p21WAF1/CIP1 expression by vascular cell integrin alphaVbeta3 during angiogenesis. J Clin Invest 1996; 98:426-433.
77. Miyashita T, Krajewski S, Krajewska et al. Tumor suppressor p53 is a regulator of bcl-2 and bax gene expression in vitro and in vivo. Oncogene 1994; 9:1799-1805.
78. Miyashita T, Reed JC. Tumor suppressor p53 is a direct transcriptional activator of the human bax gene. Cell 1995; 80:293-299.
79. Valentijn AJ, Gilmore AP. Translocation of full-length Bid to mitochondria during anoikis. J Biol Chem 2004; 279(31):32848-32857.

80. Reginato MJ, Mills KR, Paulus JK et al. Integrins and EGFR coordinately regulate the pro-apoptotic protein Bim to prevent anoikis. Nat Cell Biol 2003; 5(8):733-740.
81. Wang P, Gilmore AP, Streuli CH. Bim is an apoptosis sensor that responds to loss of survival signals delivered by epidermal growth factor but not those provided by integrins. J Biol Chem 2004; 279(40):41280-41285.
82. Hazlehurst LA, Landowski TH, Dalton WS. Role of the tumor microenvironment in mediating de novo resistance to drugs and physiological mediators of cell death. Oncogene 2003; 22(47):7396-7402.
83. Puthalakath H, Huang DC, O'Reilly LA et al. The proapoptotic activity of the Bcl-2 family member Bim is regulated by interaction with the dynein motor complex. Mol Cell 1999; 3:287-296.
84. Puthalakath H, Villunger A, O'Reilly LA et al. Bmf: A proapoptotic BH3-only protein regulated by interaction with the myosin V actin motor complex, activated by anoikis. Science 2001; 293:1829-1832.
85. Bouillet P, Strasser A. BH3-only proteins - evolutionarily conserved proapoptotic Bcl-2 family members essential for initiating programmed cell death. J Cell Sci 2002; 115:1567-1574.
86. Bouillet P, Purton JF, Godfrey DI et al. BH3-only Bcl-2 family member Bim is required for apoptosis of autoreactive thymocytes. Nature 2002; 415:922-926.
87. Frisch SM, Vuori K, Ruoslahti E et al. Control of adhesion-dependent cell survival by focal adhesion kinase. J Cell Biol 1996; 134(3):793-799.
88. Hungerford JE, Compton MT, Matter ML et al. Inhibition of pp125FAK in cultured fibroblasts results in apoptosis. J Cell Biol 1996; 135(5):1383-1390.
89. Ilic D, Almeida EA, Schlaepfer DD et al. Extracellular matrix survival signals transduced by focal adhesion kinase suppress p53-mediated apoptosis. J Cell Biol 1998; 143:547-560.
90. Almeida EA, Ilic D, Han Q et al. Matrix survival signaling: From fibronectin via focal adhesion kinase to c-Jun NH(2)-terminal kinase. J Cell Biol 2000; 149:741-754.
91. Khwaja A, Downward J. Lack of correlation between activation of Jun-NH2-terminal kinase and induction of apoptosis after detachment of epithelial cells. J Cell Biol 1997; 139(4):1017-1023.
92. Kiyokawa E, Hashimoto Y, Kurata T et al. Evidence that DOCK180 up-regulates signals from the CrkII-p130(Cas) complex. J Biol Chem 1998; 273:24479-24484.
93. Aoudjit F, Vuori K. Engagement of the alpha2beta1 integrin inhibits Fas ligand expression and activation-induced cell death in T cells in a focal adhesion kinase-dependent manner. Blood 2000; 95:2044-2051.
94. Gilmore AP, Metcalfe AD, Romer LH et al. Integrin-mediated survival signals regulate the apoptotic function of Bax through its conformation and subcellular localization. J Cell Biol 2000; 149:431-446.
95. Khwaja A, Rodriguez-Viciana P, Wennstrom S et al. Matrix adhesion and Ras transformation both activate a phosphoinositide 3-OH kinase and protein kinase B/Akt cellular survival pathway. EMBO J 1997; 16(10):2783-2793.
96. Datta SR, Dudek H, Tao X et al. Akt phosphorylation of BAD couples survival signals to the cell-intrinsic death machinery. Cell 1997; 91(2):231-241.
97. Cardone MH, Roy N, Stennicke HR et al. Regulation of cell death protease caspase-9 by phosphorylation. Science 1998; 282(5392):1318-1321.
98. Sonoda Y, Matsumoto Y, Funakoshi M et al. Anti-apoptotic role of FAK: Induction of inhibitor of apoptosis proteins and apoptosis suppression by the overexpression of FAK in a human leukemic cell line, HL-60. J Biol Chem 2000; 275:16309-16315.
99. Hannigan G, Troussard AA, Dedhar S. Integrin-linked kinase: A cancer therapeutic target unique among its ILK. Nature Reviews Cancer 2005; 5:51-63.
100. Persad S, Attwell S, Gray V et al. Inhibition of integrin-linked kinase (ILK) suppresses activation of protein kinase B/Akt and induces cell cycle arrest and apoptosis of PTEN-mutant prostate cancer cells. Proc Natl Acad Sci USA 2000; 97:3207-3212.
101. Attwell S, Roskelle C, Dedhar S. The integrin-linked kinase (ILK) suppresses anoikis. Oncogene 2000; 19:3811-3815.
102. Plantefaber LC, Hynes RO. Changes in integrin receptors on oncogenically transformed cells. Cell 1989; 56(2):281-290.
103. Dedhar S, Saulnier R. Alterations in integrin receptor expression on chemically transformed human cells: Specific enhancement of laminin and collagen receptor complexes. J Cell Biol 1990; 110(2):481-489.
104. Sainsbury JR, Nicholson S, Angus B et al. Epidermal growth factor receptor status of histological sub-types of breast cancer. Br J Cancer 1988; 58(4):458-460.
105. Slamon DJ, Clark GM, Wong SG et al. Human breast cancer: Correlation of relapse and survival with amplification of the HER-2/neu oncogene. Science 1987; 235(4785):177-182.
106. Slamon DJ, Godolphin W, Jones LA et al. Studies of the HER-2/neu proto-oncogene in human breast and ovarian cancer. Science 1989; 244(4905):707-712.

107. Liu E, Thor A, He M et al. The HER2 (c-erbB-2) oncogene is frequently amplified in in situ carcinomas of the breast. Oncogene 1992; 7(5):1027-1032.
108. Berdichevsky F, Gilbert C, Shearer M et al. Collagen-induced rapid morphogenesis of human mammary epithelial cells: The role of the alpha 2 beta 1 integrin. J Cell Sci 1992; 102(Pt 3):437-446.
109. D'Souza B, Berdichevsky F, Kyprianou N et al. Collagen-induced morphogenesis and expression of the alpha 2-integrin subunit is inhibited in c-erbB2-transfected human mammary epithelial cells. Oncogene 1993; 8(7):1797-1806.
110. Ye J, Xu RH, Taylor-Papadimitriou J et al. Sp1 binding plays a critical role in Erb-B2- and v-ras-mediated downregulation of alpha2-integrin expression in human mammary epithelial cells. Mol Cell Biol 1996; 16(11):6178-6189.
111. Zutter MM, Mazoujian G, Santoro SA. Decreased expression of integrin adhesive protein receptors in adenocarcinoma of the breast. Am J Pathol 1990; 137(4):863-870.
112. Koukoulis GK, Virtanen I, Korhonen M et al. Immunohistochemical localization of integrins in the normal, hyperplastic, and neoplastic breast. Correlations with their functions as receptors and cell adhesion molecules. Am J Pathol 1991; 139(4):787-799.
113. Pignatelli M, Cardillo MR, Hanby A et al. Integrins and their accessory adhesion molecules in mammary carcinomas: Loss of polarization in poorly-differentiated tumors. Hum Pathol 1992; 23(10):1159-1166.
114. Pignatelli M, Hanby AM, Stamp GW. Low expression of beta 1, alpha 2 and alpha 3 subunits of VLA integrins in malignant mammary tumours. J Pathol 1991; 165(1):25-32.
115. Pignatelli M, Smith MEF, Bodmer WF. Low expression of collagen receptors in moderate and poorly-differentiated colorectal adenocarcinomas. Brit J Cancer 1991; 61:636-638.
116. Stallmach A, von Lampe B, Matthes H et al. Diminished expression of integrin adhesion molecules on human colonic epithelial cells during the benign to malign tumour transformation. Gut 1992; 33(3):342-346.
117. Bonkhoff H, Stein U, Remberger K. Differential expression of alpha 6 and alpha 2 very late antigen integrins in the normal, hyperplastic, and neoplastic prostate: Simultaneous demonstration of cell surface receptors and their extracellular ligands. Hum Pathol 1993; 24(3):243-248.
118. Fridman R, Giaccone G, Kanemoto T et al. Reconstituted basement membrane (matrigel) and laminin can enhance the tumorigenicity and the drug resistance of small cell lung cancer cell lines. Proc Natl Acad Sci USA 1990; 87(17):6698-6702.
119. Kraus AC, Ferber I, Bachmann SO et al. In vitro chemo- and radio-resistance in small cell lung cancer correlates with cell adhesion and constitutive activation of AKT and MAP kinase pathways. Oncogene 2002; 21(57):8683-8695.
120. Sethi T, Rintoul RC, Moore SM et al. Extracellular matrix proteins protect small cell lung cancer cells against apoptosis: A mechanism for small cell lung cancer growth and drug resistance in vivo. Nat Med 1999; 5(6):662-668.
121. Buttery RC, Rintoul RC, Sethi T. Small cell lung cancer: The importance of the extracellular matrix. Int J Biochem Cell Biol 2004; 36(7):1154-1160.
122. Hartmann TN, Burger M, Burger JA. The role of adhesion molecules and chemokine receptor CXCR4 (CD184) in small cell lung cancer. J Biol Regul Homeost Agents 2004; 18:126-130.
123. Cordes N, Beinke C, Plasswilm L et al. Irradiation and various cytotoxic drugs enhance tyrosine phosphorylation and beta(1)-integrin clustering in human A549 lung cancer cells in a substratum-dependent manner in vitro. Strahlenther Onkol 2004; 180(3):157-164.
124. Oshita F, Kameda Y, Hamanaka N et al. High expression of integrin β1 and p53 is a greater poor prognostic factor than clinical stage in small-cell lung cancer. Am J Clin Oncol 200; 27:215-219.
125. Holcik M, Yeh C, Korneluk RG et al. Translational upregulation of X-linked inhibitor of apoptosis (XIAP) increases resistance to radiation induced cell death. Oncogene 2000; 19:4174-4177.
126. Rodel C, Haas J, Groth A et al. Spontaneous and radiation-induced apoptosis in colorectal carcinoma cells with different intrinsic radiosensitivities: Survivin as a radioresistance factor. Int J Radiat Oncol Biol Phys 2003; 55:1341-1347.
127. Monzo M, Rosell R, Felip E et al. A novel anti-apoptosis gene: Reexpression of survivin messenger RNA as a prognosis marker in nonsmall-cell lung cancers. J Clin Oncol 1999; 17:2100-2104.
128. Ikehara M, Oshita F, Kameda Y et al. Expression of survivin correlated with vessel invasion is a marker of poor prognosis in small adenocarcinoma of the lung. Oncol Rep 2002; 9:835-838.
129. Cao C, Mu Y, Hallahan DE et al. XIAP and survivin as therapeutic targets for radiation sensitization in preclinical models of lung cancer. Oncogene 2004; 23(42):7047-7052.
130. Fornaro M, Plescia J, Chheang S et al. Fibronectin protects prostate cancer cells from tumor necrosis factor-alpha-induced apoptosis via the AKT/survivin pathway. J Biol Chem 2003; 278(50):50402-50411.

131. Uhm JH, Dooley NP, Kyritsis AP et al. Vitronectin, a glioma-derived extracellular matrix protein, protects tumor cells from apoptotic death. Clin Cancer Res 1999; 5:1587-1594.
132. Miyamoto H, Murakami T, Tsuchida K et al. Tumor-stroma interaction of human pancreatic cancer: Acquired resistance to anticancer drugs and proliferation regulation is dependent on extracellular matrix proteins. Pancreas 2004; 28(1):38-44.
133. Burbridge MF, Venot V, Casara PJ et al. Decrease in survival threshold of quiescent colon carcinoma cells in the presence of a small molecule integrin antagonist. Mol Pharmacol 2003; 63(6):1281-1288.
134. Maubant S, Cruet-Hennequart S, Poulain L et al. Altered adhesion properties and alphav integrin expression in a cisplatin-resistant human ovarian carcinoma cell line. Int J Cancer 2002; 97(2):186-194.
135. Cruet-Hennequart S, Maubant S, Luis J et al. alpha(v) integrins regulate cell proliferation through integrin-linked kinase (ILK) in ovarian cancer cells. Oncogene 2003; 22(11):1688-1702.
136. Nista A, Leonetti C, Bernardini G et al. Functional role of alpha4beta1 and alpha5beta1 integrin fibronectin receptors expressed on adriamycin-resistant MCF-7 human mammary carcinoma cells. Int J Cancer 1997; 72:133-141.
137. Narita T, Kimura N, Sato M et al. Altered expression of integrins in adriamycin-resistant human breast cancer cells. Anticancer Res 1998; 18(1A):257-262.
138. Wewer UM, Shaw LM, Albrechtsen R et al. The integrin alpha 6 beta 1 promotes the survival of metastatic human breast carcinoma cells in mice. Am J Pathol 1997; 151:1191-1198.
139. Aoudjit F, Vuori K. Integrin signaling inhibits paclitaxel-induced apoptosis in breast cancer cells. Oncogene 2001; 20(36):4995-5004.
140. Menendez JA, Vellon L, Mehmi I et al. A novel CYR61-triggered 'CYR61-alphavbeta3 integrin loop' regulates breast cancer cell survival and chemosensitivity through activation of ERK1/ERK2 MAPK signaling pathway. Oncogene 2005; 24(5):761-779.
141. Tsai M-S, Hornby AE, Lakins J et al. Expression and function of CYR61, an angiogenic factor, in breast cancer cell lines and tumor biopsies. Cancer Res 2000; 60:5603-5607.
142. Brigstock DR. The connective tissue growth factor/cysteine-rich 61/nephroblastoma overexpressed (CCN) family. Endocr Rev 1999; 20:189-206.
143. Brigstock DR, Goldschmeding R, Katsube KI et al. Proposal for a unified CCN nomenclature. Mol Pathol 2003; 56:127-128.
144. Lester FL, Stephen C-TL. The CCN family of angiogenic regulators: The integrin connection. Exp Cell Res 1999; 248:44-57.
145. Perbal B. The CCN family of genes: A brief history. Mol Pathol 2001; 54:103-104.
146. Perbal B. CCN proteins: Multifunctional signalling regulators. Lancet 2004; 363:62-64.
147. Planque N, Perbal B. A structural approach to the role of CCN (CYR61/CTGF/NOV) proteins in tumourigenesis. Cancer Cell Int 2003; 3:15.
148. Menendez JA, Mehmi I, Griggs DW et al. The angiogenic factor CYR61 in breast cancer: Molecular pathology and therapeutic perspectives. Endocrine Rel Cancer 2003; 10:141-152.
149. Lupu R, Cardillo M, Cho C et al. The significance of heregulin in breast cancer tumor progression and drug resistance. Breast Cancer Res Treat 1996; 38:57-66.
150. Tsai M-S, Bogart DF, Castaneda JM et al. Cyr61 promotes breast tumorigenesis and cancer progression. Oncogene 2002; 21:8178-8185.
151. Tsai MS, Shamon-Taylor LA, Mehmi I et al. Blockage of heregulin expression inhibits tumorigenicity and metastasis of breast cancer. Oncogene 2003; 22:761-768.
152. Atlas E, Mehmi I, Cardilla M et al. Heregulin is sufficient for the promotion of tumorigenicity and metastasis of breast cancer cells in vivo. Mol Cancer Res 2003; 1:165-175.
153. Kayaselcuk F, Nursal TZ, Polat A et al. Expression of survivin, bcl-2, P53 and bax in breast carcinoma and ductal intraepithelial neoplasia (DIN 1a). J Exp Clin Cancer Res 2004; 23(1):105-112.
154. Chu JS, Shew JY, Huang CS. Immunohistochemical analysis of survivin expression in primary breast cancers. J Formos Med Assoc 2004; 103(12):925-931.
155. Tsuji N, Furuse K, Asanuma K et al. Mutations of the p53 gene and loss of heterozygosity at chromosome 17p13.1 are associated with increased survivin expression in breast cancer. Breast Cancer Res Treat 2004; 87(1):23-31.
156. Zahir N, Weaver VM. Death in the third dimension: Apoptosis regulation and tissue architecture. Curr Opin Genet Dev 2004; 14(1):71-80.
157. Teicher BA, Herman TS, Holden SA et al. Tumor resistance to alkylating agents conferred by mechanisms operative only in vivo. Science 1990; 247(4949 Pt 1):1457-1461.
158. Kerbel RS, Rak J, Kobayashi H et al. Multicellular resistance: A new paradigm to explain aspects of acquired drug resistance of solid tumors. Cold Spring Harbor Symp Quant Biol 1994; 59:661-672.

159. Kobayashi HS, Man SJ, Kapitain BA et al. Acquired multicellular-mediated resistance to alkylating agents in cancer. Proc Natl Acad Sci USA 1993; 90:3294-3298.
160. Santini MT, Rainaldi G, Indovina Pl. Apoptosis, cell adhesion and the extracellular matrix in the three-dimensional growth of multicellular tumor spheroids. Crit Rev Oncol Hematol 2000; 36(2-3):75-87.
161. Green SK, Francia G, Isidoro C et al. Antiadhesive antibodies targeting E-cadherin sensitize multicellular tumor spheroids to chemotherapy in vitro. Mol Cancer Ther 2004; 3(2):149-159.
162. Boudreau N, Werb Z, Bissell MJ. Suppression of apoptosis by basement membrane requires three-dimensional tissue organization and withdrawal from the cell cycle. Proc Natl Acad Sci USA 1996; 93(8):3509-3513.
163. Weaver VM, Lelievre S, Lakins JN et al. beta4 integrin-dependent formation of polarized three-dimensional architecture confers resistance to apoptosis in normal and malignant mammary epithelium. Cancer Cell 2002; 2(3):205-16.
164. Zahir N, Lakins JN, Russell A et al. Autocrine laminin-5 ligates alpha6beta4 integrin and activates RAC and NFkappaB to mediate anchorage-independent survival of mammary tumors. J Cell Biol 2003; 163(6):1397-1407.

CHAPTER 7

Insulin-Like Growth Factors and Breast Cancer Therapy

Xianke Zeng and Douglas Yee*

Abstract

Despite improvements in breast cancer therapy in recent years, additional therapies need to be developed. New therapies may have activity by themselves or may have utility in combination with other agents. Population, preclinical, and basic data suggest the insulin-like growth factor (IGF) system functions to maintain the malignant phenotype in breast cancer. Since the IGFs act via transmembrane tyrosine kinase receptors, targeting of the key receptors could provide a new pathway in breast cancer. In addition, IGF action enhances cell survival, so combination of anti-IGF therapy with conventional cytotoxic drugs could lead to synergistic effects. In this review, we will discuss the rationale for targeting the IGF system, potential methods to disrupt IGF signaling, and identify potential interactions between IGF inhibitors and other anti-tumor strategies. We will also identify important issues to consider when designing clinical trials.

Introduction

Breast cancer is the most common cancer in women responsible for over 40,000 deaths in the United States.[1] Since cancer death is almost always caused by growth of breast cancer in metastatic sites, systemic cytotoxic and endocrine therapies are commonly given. In operable breast cancer, systemic adjuvant therapy is employed to reduce the risk of recurrence and prolong overall survival. In women with metastatic disease, systemic therapy is given to control growth of breast cancer in distant organs. Both treatments essentially target metastatic disease; in the adjuvant setting, the goal is to eradicate any microscopic disease that is not clinically detectable while treatment of metastatic disease targets clinically obvious sites.

Though systemic treatment is clearly effective in both adjuvant and advanced settings, therapy is far from completely effective. In the adjuvant setting, administration of chemotherapy reduces the relative risk of recurrence by approximately 30%.[2] Identifying methods to enhance the cytotoxicity of chemotherapy are clearly needed. Since normal and cancer cells receive survival and proliferative signals from their extracellular environment, targeting of these signals could enhance the clinical benefit of chemotherapy. Indeed, trastuzumab, an antibody directed against the human epidermal growth factor receptor-2 (HER2), is commonly used in combination with chemotherapy for women with advanced cancer.[3] While the exact mechanism for the synergy between trastuzumab and chemotherapy is not completely understood, it is likely that trastuzumab renders cells more sensitive to the apoptotic effects of chemotherapy by attenuating cells survival pathways.

*Corresponding Author: Douglas Yee—Department of Pharmacology and Medicine, University of Minnesota Cancer Center, MMC 806, 420 Delaware Street SE, Minneapolis, MN 55455, U.S.A. Email: yeexx006@umn.edu

Breast Cancer Chemosensitivity, edited by Dihua Yu and Mien-Chie Hung.

Identifying additional survival pathways could also be used to enhance the benefits of chemotherapy. Insulin-like growth factors (IGFs) and the IGF signaling pathways play a role in development of the normal mammary gland. Numerous studies have now demonstrated that the IGF system regulates all of the key metastatic phenotypes in breast cancer cells: survival, proliferation, and metastasis.

Extensive data are available on the importance of IGF system in growth regulation of breast cancer cell lines.[4] The type I insulin-like growth factor receptor (IGF-IR) is significantly overexpressed[5] or hyperphosphorylated in tumor cells relative to normal breast epithelium and benign tumors.[5,6] In addition, several clinical studies also support a role for IGF-I in breast cancer risk.[7,8] In cell line model systems, IGF-IR conferred resistance to trastuzumab-induced growth inhibition[9] and a kinase inhibitor to IGF-IR increased radiosensitivity in some breast cancer cells lines.[10] Thus, data from population and laboratory studies provide a rationale for targeting IGF-IR in breast cancer. Moreover, disruption of IGF-IR may render cells more sensitive to apoptotic stimuli. Several reviews have already addressed the important role of the IGF system in breast cancer.[11-13] Here we will summarize the potential role of IGF action on breast cancer chemotherapy.

The IGF System

The IGF system involves the coordination of growth factors (IGF-I and IGF-II), cell surface receptors (IGF-IR, IGF-IIR, and the insulin receptor IR), six high affinity binding proteins (IGFBP-1 to 6), IGFBP proteases, and the downstream proteins involved in intracellular signaling distal to IGF-IR.

IGF-I and IGF-II are single-chain 7.5 kDa polypeptide growth factors with a high degree of homology to insulin. The main function of IGF-I is to act as an effector molecule of growth hormone (GH), which is fundamental to linear growth and development.[14] During puberty, pulsatile GH release from the pituitary stimulates expression of IGF-I in the liver. In addition to its endocrine role, it has been suggested that IGF-I may also have an important role in prenatal growth. Mice with a homozygous deletion of the IGF-I gene have a birth weights less than 60% of their wildtype littermates, these mice have a high post-natal mortality rate.[15] Thus, besides its endocrine role, IGF-I plays an important paracrine and autocrine role during normal development and growth of the organism. IGF-II expression is not regulated by GH. However, IGF-II has proliferative and antiapoptotic actions similar to IGF-I. In addition, IGF-II plays a fundamental role in embryonic and fetal growth, this was proven by IGF-II gene knockout mice, which survive but remain smaller than their wildtype littermates.[16] Interestingly, size at birth and height at age 14 have been linked to increase breast cancer risk suggesting an etiologic role for the IGFs and breast cancer development.[17]

The actions of IGFs can be modulated by interaction with a family of six insulin-like growth factor-binding proteins, IGFBP-1 to IGFBP-6, which share 40-60% amino acid identity. IGFBP3 is the largest and most abundant IGFBP, more than 75% IGF is confined to the vascular compartment as a ternary complex with IGFBP3 and the acid labile subunit. By binding IGF-I and IGF-II, IGFBPs regulated the bioavailability of IGFs in the circulation. IGFBP-3 has also been shown to competitively inhibit IGF action at the cellular level in the absence of IGF binding and exert IGF-independent proapoptotic and antiproliferative effects through the activation of caspases involved in a death receptor-mediated pathway.[18]

Although at least two receptors for IGFs exist, the primary signaling receptor through which both IGF-I and IGF-II exert their biological actions is the IGF-IR. Hybrid IGF-IR/insulin receptors also exist and could mediate IGF and insulin action in breast cancer cells.[19,20] Presence of the hybrid receptor adds an additional layer of complexity to IGF signaling.

Binding of ligand to the receptor induces autophosphorylation and activation of multiple downstream cell survival and proliferation signaling pathway via recruitment and tyrosine phosphorylation of specific adaptor/effector molecules.[21] While individual pathways have been linked to specific cancer phenotypes, it is clear that the intracellular signals initiated after IGF-IR activation constitute a network of interacting molecular events. It is likely oversimplistic to ascribe a specific behavior to a specific pathway. Nonetheless, activation of PI3 kinase downstream of the insulin receptor substrate-1 (IRS-1) adaptor protein has been linked to cell survival and regulation of several proteins involved in apoptosis.[22] Induction of cell growth and proliferation has been linked to both the PI3 kinase pathway and the MAP kinase pathway downstream of IGF-IR and its phosphorylated substrates, IRS-1 and Src-collagen homology (Shc) protein. The type II insulin-like growth factor receptor (IGF-IIR) lacks tyrosine kinase activity and appears to exert antiproliferative and proapoptotic activities by sequestration of IGF-II, reducing its availability for interaction with the IGF-IR.[23]

IGFs and Normal Mammary Tissue

IGFs play a key role in proliferation and survival in the mammary gland, particularly during puberty and pregnancy.[24,25] IGF-I is a potent mitogen for mammary epithelial cells and in combination with mammotrophic hormones, such as estrogen receptor (ER), it induces ductal growth in mammary gland explant cultures.[26] IGF-IR null mice have deficient mammary development with reductions in the number of terminal end buds, ducts and the per cent of the fat pad occupied by glandular elements. This phenotype is partially restored by administration of des(1-3) IGF-I.[27] In addition, IGF-I also plays a role in the maintenance of the adult mammary gland during lactation, lactating transgenic mice overexpressing the *Igf1* gene undergo ductal hypertrophy and fail to show normal mammary gland involution following weaning.[28] The same group also demonstrated that IGF-I slows the apoptotic loss of mammary epithelial cells during the declining phase of lactation.[29]

It is known that IGFs is one of the developmental/essential survival factors for the mammary gland, although other factors such as epidermal growth factor (EGF) and its homologues also deliver intracellular signals that suppress apoptosis. Direct evidence for IGFs as survival factors comes from culture studies.[24,30,31] IGF-I or IGF-II can suppress the apoptosis of mammary epithelial cells induced by serum withdrawal. It has been recently established that this is achieved through PI3K and MAPK signals that ultimately inhibit activity of a proapoptotic protein, BAD and enhance expression of another antiapoptotic protein Bcl-xL.[24,30] Indirect evidence came from the transgenic mice overexpressing IGFBP-5 in the mammary gland, these mice had reduced numbers of alveolar end buds, with decreased ductal branching and increased expression of the pro-apoptotic molecule caspase-3,and decreased expression of pro-survival molecules of the Bcl-2 family.[32]

IGF and Breast Cancer

The IGFs and IGF-IR function to promote proliferation, inhibit death and stimulate transformation in breast cancer cells.[11] High levels of serum IGF-I are associated with an increased risk of breast cancer in premenopausal women.[33] There is also substantial evidence that IGF expression occurs locally in breast cancer tissues. Although IGF-I is rarely expressed in primary breast cancer, IGF-II message is more frequently detectable in breast cancer cells compared to normal cells.[34] Moreover, the stroma is a rich source of both IGFs.

Studies in transgenic mice have revealed an important role of IGF-I in mammary tumorigenesis. Transgenic mice expressing human des(1-3) IGF-I (under the control of the rat whey acid protein) display an increased incidence of mammary tumors, with 53% of the mice developing mammary adenocarcinomas by 23 months of age.[35,36] Furthermore, data in a transgenic mouse system suggest that mice deficient in liver-expressed IGF-I have a reduced ability to develop mammary tumors.[37] In human studies, circulating IGF-I levels are higher in breast cancer patients than in controls. In addition, cohort studies have shown that higher levels of circulating IGF-I are associated with an increased risk of breast cancer in premenopausal women.[38]

The IGF-IR, the primary mediator of the biological actions of IGF-I, has been detected in a majority of primary breast tumor samples with overexpression in 30% to 40% of breast cancers.[39] Furthermore, IGF-IR autophosphorylation has been found to be elevated in human breast cancer suggesting that this is an active pathway in primary breast cancer.[6,40] Interestingly, a high level of IGF-IR in patients with breast cancer is associated with a greater recurrence risk of recurrence after local radiation therapy.[41]

Insulin receptor substrate-1 (IRS-1), the primary signaling molecule activated in response to IGF in MCF-7 human breast cancer cells, is reported to be overexpressed in some primary breast tumors and a high IRS-1 are associated with a decreased disease-free survival in a subset of patients with all tumors.[42,43] Activation of specific IRS species are associated with distinct biological effects.[44] Activation of IRS-1 signaling was associated with cell growth, whereas insulin receptor substrate-2 (IRS-2) signaling was associated with cell motility.[44,45] Nagle et al showed that mammary tumor cells obtained from IRS-2 knock-out mice were less invasive and more apoptotic in response to growth factor deprivation than their WT counterparts. In contrast, IRS-1(-/-) tumor cells, which express only IRS-2, were highly invasive and were resistant to apoptotic stimuli.[46] These data suggest that signaling pathways downstream of IGF-IR may ultimately be responsible for the malignant phenotype mediated by this growth factor signaling system.

Breast cancer cells also produce several IGFBPs that could modulate IGF action. In addition to the indirect modulation of IGF's mitogenic and antiapoptotic signals by ligand sequestration, IGFBPs also exert IGF-independent effects on cell survival. For example, IGFBP-4 and IGFBP-5 can rescue cells from ceramide or integrin-mediated apoptosis, which may account for the poor prognosis in breast cancers with high IGFBP-4 expression[13,47] whereas IGFBP-3 has been found to directly inhibit breast cancer cell growth without interacting with IGFs.[13,48]

Conventional Chemotherapy for Breast Cancer

It is well established that most chemotherapeutic agents eliminate cells by triggering apoptosis. Most currently approved agents target specific molecules required for a cell to traverse the cell cycle. Targets range from DNA itself, to enzymes (topoisomerases), or structural proteins (tubulin) required for cell division. There are three kinds of cells in a solid tumor: dividing cell that are continuously cycling, resting cell which may potentially enter the cell cycle, and those cells no longer able to divide. In breast cancer, Clarke et al have suggested that only a minority of cells contained within a tumor have the capacity to contribute to all these subpopulations of tumor cells leading to the idea that cancer stem cells exist.[49] Essentially only dividing cells are susceptible to the currently available chemotherapy drugs. It is the existence of resting or stem cells that makes it difficult to completely eradicate advanced tumors by chemotherapy; even after a substantial clinical response, a population of cells may still exist that have full capacity to enter the cell cycle.

The primary therapy of localized-early stage I and II- breast cancer is either breast-conserving surgery and radiation therapy or mastectomy with or without reconstruction.[50] Systemic adjuvant therapies designed to eradicate clinically undetectable microscopic deposits of cancer cells that may have spread from the primary tumor result in decreased recurrences and improved survival.[2,51] Adjuvant therapies include chemotherapy and hormonal therapy. In the adjuvant setting, chemotherapy is usually given in combination for 4-6 months. A wide variety of agents have been effective in breast cancer including DNA alkylators (cyclophosphamide), topoisomerase inhibitors/DNA intercalators (doxorubicin), anti-metabolites (5-fluorouracil, methotrexate), and tubulin interacting agents (paclitaxel).[2] Adjuvant chemotherapy effectively reduces the odds of recurrence and death by approximately 20% to 60% of patients. However, since this reduction of risk is not complete, substantial research effort is directed toward improving the benefits of chemotherapy. New target discovery and combination of new agents with "conventional" agents represent an active area of investigation.

IGF Signaling Confers Resistance to Chemotherapy

Besides inducing cell cycle progression, IGF-I also protects breast cancer cells from drug-induced apoptosis.[52-54] In fibroblasts, protection from apoptosis requires the tyrosine kinase activity of IGF-IR, as kinase defective receptors do not protect fibroblasts from apoptosis.[55] In addition to drug-induced apoptosis, IGF-IR activation appears to block other stimuli as well. For example, BNIP3 (Bcl-2/E1B 19 kDa interacting protein) is a proapoptotic member of the Bcl-2 family expressed in hypoxic regions of tumors. Treatment of the breast cancer cell line MCF-7 cells with IGF effectively protected these cells from BNIP3-induced cell death[56] likely via activation of PI3K and the Akt/PKB pathways.[55,57] Akt/PKB phosphorylates BAD, a member of the Bcl-2 family of proapoptotic proteins, phosphorylated BAD cannot heterodimerize with Bcl-2 or Bcl-Xl, remains in the cytosol and cell death is inhibited.[58] It has also been suggested that IGF can inhibit apoptosis by increasing the expression of Bcl-X_L at both the mRNA and protein level.[59] Thus, signaling from IGF-IR to multiple proteins involved in the intrinsic apoptotic pathways suggest a mechanism for protection from cell death signals.

IGFs provide resistance of breast cancer cells to chemotherapeutic agents.[53] IGF-I alters drug sensitivity of HBL100 human breast cancer cell by inhibition of apoptosis induced by diverse anticancer drugs, it increased cell survival of HBL100 cells treated with 5-FU, methotrexate, tamoxifen or camptothecin, but no changes were observed in Bcl-2 protein or Bax mRNA levels.[60] IGF-I signaling is also associated with resistance to the growth-inhibitory actions of trastuzumab by upregulation of ubiquitin-related $p27^{kip1}$ degradation and activation of the PI3K signaling pathway.[61] In breast cancer cells, IGF-I can activate JNK which is generally associated with a pro-apoptotic response. However, activation of Akt seems to override pro-apoptotic effects of JNK activation.[62] Thus targeting IGF-IR could enhance a pro-apoptotic response initiated by many different agents.

IGF-IR and DNA Repair

Several reports indicate that IGF-IR activation is also important in regulating DNA repair. Fibroblasts can be rescued from DNA damage by IGF-I via activation of the p38 MAP kinase signaling pathway.[63,64] Increasing IGF-IR expression increased radioresistance in breast tumor cells,[65] and delayed UVB-induced apoptosis by enhancing repair of DNA cyclobutane thymidine dimers in keratinocytes.[66,67] IGF-I stimulation supports homologous recombination-directed DNA repair (HRR) via an interaction between IRS-1 and Rad51, a key enzyme of HRR.[68] In contrast, IGF-I may actually inhibit the ability of A549 cells to repair potentially lethal DNA damage induced by radiation.[69] Though these observations are somewhat conflicting as they suggest IGF-IR may both enhance and inhibit DNA repair, it is possible that these differences relate to the varied experimental model systems and cell types studied. However, these experiments support a link between DNA repair and IGF-I action and the exact IGF effects may be context dependent.

Effects of Breast Cancer Therapy on the IGF System

On the other hand, breast cancer chemotherapy also affects IGFs. It has been reported that serum IGFBP-3 falls significantly following initiation of chemotherapy in breast cancer patients, those individuals with a decrease in IGFBP-3 greater than the median had significantly poorer survival (median survival 5.5 months vs 18 months).[70] Another clinical trial showed that plasma IGF-I concentration significantly decreased after the first cycle of cyclophosphamide, methotrexate and 5-fluorouracil adjuvant chemotherapy in breast cancer patients.[71] Retinoids such as fenretinide (4-HPR) inhibit breast cell growth while decreasing IGF-I and increase IGFBP-3.[72,73]

Proline analogues of melphalan can be effectively transported into the MDA-MB 231 cells, evoking higher cytotoxicity, with reduction in IGF-I receptor and MAP kinase expression.[74] Tamoxifen also affects the IGF system. IGFBP-3,4,6 levels are lower in breast cancer patients compared to normal controls and levels increased after tamoxifen treatment.[75,76] Raloxifene, a selective estrogen receptor modulator being tested in cancer prevention trials, significantly decreased IGF-I and IGF-I/IGFBP-3 ratios when compared to placebo.[77] Urokinase plasminogen activator(uPA) inhibitor -17 AAG inhibit MDA-MB-231 cell growth by inhibiting the IGF-IR and ultimately uPA, while expression of the IGF-IR and uPA in breast cancer is associated with poor survival.[78]

Thus, it is clear that anti-proliferative agents affect IGF system signaling. These associations do not prove a cause and effect relationship, however, given the role of IGF signaling in cell survival, the downregulation of this signaling pathway is consistent with the effects of many anti-cancer drugs.

Anti-IGF Strategies in Breast Cancer

Given the role for IGF signaling in many aspects of the malignant phenotype, it would be valuable to have reagents to disrupt IGF action. Several strategies to interrupt IGF signaling are currently under investigation, including endocrine maneuvers to suppress IGF production; antisense oligonucleotides to reduce functional IGF-IR levels; monoclonal antibodies, dominant negatives, and tyrosine kinase inhibitors to inhibit IGF-IR activation; and neutralization of IGF action using IGFBPs.[79,80]

Suppression of IGF Production

The majority of circulating IGF-I is produced by the liver in response to growth hormone (GH) released from the pituitary gland, acting through hepatic growth hormone receptors (GHR). GH-releasing hormone antagonists disrupt the pituitary production of GH and reduce circulating levels of GH and have been shown to inhibit the growth of a variety of cancers in animal model systems, including breast cancer.[81,82] Somatostatin and its analogues also inhibit the release of GH and thyroid-stimulating hormone from the pituitary gland. Preclinical studies on the anticancer activity of the somatostatin analog octreotide showed 50% reduction in tumor growth using two in vivo breast cancer models, ZR-75-1 breast xenografts and DMBA induced mammary tumors in rats. However, octreotide administered with tamoxifen did not improve response or survival in patients with metastatic breast cancer compared to tamoxifen alone.[83] While octreotide was able to reduce serum IGF-I levels, it was possible that this reduction was insufficient to block IGF-IR signaling.

More potent methods to disrupt endocrine IGF-I have been developed. GH-RH antagonists MZ-5-156 or JV-1-36 administered induced the growth-arrest of estrogen-independent MDA-MB-468 human breast cancers xenografted into nude mice.[84] Pegvisomant, a competitive antagonist of GHR, is the most potent therapy for reducing serum IGF-I levels in acromegalic patients and may have a role in cancer treatment.[84] However, these strategies to disrupt endocrine IGF-I do not address paracrine or autocrine sources of IGF-I. Furthermore, IGF-II is not under GH control, and merely suppressing serum IGF-I levels may be insufficient to block all IGF ligands.

Ligand Neutralization Using IGFBPs

Since IGF ligands are required to activate IGF-IR, disruption of ligand-receptor interactions is an attractive method to disrupt IGF signaling. Blockade of IGF-mediated cellular effects can be accomplished with overexpression of IGFBPs or by treatment with exogenous IGFBPs. IGFBP-1, either exogenously added or endogenously produced, has been observed to inhibit IGF-IR function resulting in inhibition of IGF-I mediated growth of MCF-7 breast cancer cells.[85,86] Silibinin has been shown to have anti-proliferative action against some malignant cell lines by increasing IGFBP3 mRNA and protein levels.[87] In vivo, treatment with polyethylene glycol-conjugated recombinant IGFBP-1, inhibited growth of MDA-MB-231 breast tumor xenografts and malignant ascites formation in the MDA-MB-435/LCC6 cells.[85] Similar neutralization of IGF ligands has been accomplished using the extracellular domain of

IGF-IR[88,89] and with neutralizing antibodies.[90] Thus, several methods to neutralize IGF ligand interaction with IGF-IR have been tested. Neutralization strategies have the advantage of targeting both IGF-I and IGF-II without the need to identify a specific receptor subtype.

Inhibition of IGF-IR Activation

Abundant evidence implicating IGF-IR is essential for the transformed phenotype and inhibition of apoptosis in breast cancer, targeting this receptor directly may be an effective cancer therapy. Antibody blockade of growth factor receptors is a proven strategy to inhibit receptor-mediated effects, with the effectiveness of trastuzumab against HER2 overexpressing breast cancers being a prime example. Several anti-IGF-IR antibodies have been developed and tested in preclinical model systems. α-IR3, the first monoclonal antibody directed against IGF-IR, inhibited clonal growth and blocked the mitogenic effects of exogenous IGF-I in breast cancer cells in vitro.[91] Interestingly, α-IR3 inhibited MDA-MB-231 tumor formation in athymic mice when administered at the time of tumor cell inoculation, but was ineffective against MCF-7 tumor xenografts. Since MCF-7 cells are sensitive to IGF-IR blockade in vitro, it was possible that the pharmacokinetic properties of the antibody are an important determinant of anti-tumor activity. A chimeric humanized single chain antibody(scFv-Fc), a partial agonist of IGF-IR, exhibited dose-dependent growth inhibition of IGF-IR-overexpressing NIH-3T3 cells, and significantly suppresses MCF-7 breast tumor growth in athymic mice.[92-94] EM164, a purely antagonistic anti-IGF-IR antibody, displays potent inhibitory activity against IGF-I and IGF-II, and serum-stimulated proliferation and survival of MCF-7 breast cancer cells.[95] A high-affinity fully human monoclonal antibody, A12, blocks IGF-I and IGF-II signaling and exhibits strong anti-tumor cell activity against MCF-7 xenograft tumors by enhancing apoptosis.[96]

An alternative strategy to inhibit IGF action is to target the tyrosine kinase activity of the receptor. Several members of the tyrphostin tyrosine kinase inhibitor family (e.g., AG10124, AG1024, AG538, and I-OMe AG538) were shown to competitively inhibit IGF-IR autophosphorylation and kinase activity in intact IGF-IR-overexpressing NIH-3T3 cells and to inhibit growth of MDA-MB468 and MCF-7 breast cancer cells in monolayer and colony formation in soft agar.[97,98] However, cross-reactivity of these compounds with the insulin receptor tyrosine kinase was reported due to the high degree of homology between the two receptors. Newer agents have been developed with apparently more selective IGF-IR activity.[99,100] However, it is uncertain if a highly selective IGF-IR tyrosine kinase inhibitor is desirable. Since insulin receptor may mediate some of the biological effects of the IGFs, it is possible that both IGF IR and insulin receptor will need to be inhibited in tumors. Using anti-sense oligonucleotides, Salatino et al have shown that specific targeting of the IGF-IR in mice inhibits tumor growth,[101] supporting the idea that specific inhibition of IGF-IR may block tumor growth. However, since mice have very low serum levels of IGF-II after birth,[16] it remains to be seen if targeted disruption of IGF-IR alone is sufficient to inhibit tumor growth in humans.

Combination of Anti-IGF Strategy with Chemotherapy

In theory, inhibition of survival pathways by blocking IGF-IR signaling while enhancing apoptotic stimuli has appeal. Combination of anti-IGFIR antibody αIR3 with doxorubicin resulted in increased cytotoxicity in IGF-I stimulated cells than with chemotherapy alone.[102] Similar enhancement of chemotherapy effects have been shown in Ewing's sarcoma cells.[103] Tyrphostin AG1024 (a tyrosine kinase inhibitor of IGF-IR) demonstrated a marked enhancement in radiosensitivity and amplification of radiation-induced apoptosis which was associated with increased expression of Bax, p53 and p21, and a decreased expression of Bcl-2.[10] Another study demonstrated that cotargeting IGF-IR and c-kit synergistic inhibit proliferation and induction of apoptosis in H209 small cell lung cancer cells.[104] There is also evidence that somatostatin analogues may enhance the effect of tamoxifen in animal models by suppressing plasma IGF-I and –II levels.[9]

In addition to conventional agents, it is also possible that anti-growth factor receptor strategies can be combined. Recent evidence shows that increased levels of IGF-IR signaling appears to interfere with the action of trastuzumab in breast cancer cell models that overexpress HER2/neu.[9] Thus, strategies that target IGF-IR signaling may prevent or delay development of resistance to trastuzumab.

Conclusion

The IGF-IR is a promising target in breast cancer therapy because it signals to multiple pathways required for maintenance of the malignant phenotype. Given the role for IGF-IR in cell survival, it is logical to combine anti-IGF therapies with conventional agents. Indeed, the preclinical data suggest that blockade of IGF-IR induces apoptosis and lowering a "survival threshold" with disruption of this signaling system should enhance chemotherapy efficacy.

However, there are several challenges that will need to be addressed before the idea that combination anti-IGF therapy and chemotherapy display synergy. First, there are many ways that IGF signaling could be targeted. As recently noted by Professor Baserga,[105] the potential anti-IGF strategies have gone from "rags to riches" in the course of a few short years. Clinical trials to test the most effective strategy will need to be completed before combination trials can begin. Second, the phenotypes regulated by IGF-IR are not restricted to survival alone. Since proliferation is also affected by IGF-IR, it will be important to consider scheduling and choice of chemotherapeutic agent when designing appropriate combinations. For example, it is possible that anti-metabolites would be less efficacious when combined with anti-IGF because of the requirement for cells in S-phase for anti-metabolites to function. Indeed, interference between hormonal therapy and chemotherapy has been noted in breast cancer[106] and it is possible that such interference could exist between anti-IGF therapy and certain drugs. On the other hand, agents that have a different mechanism of action, such as DNA alkylators or therapies that induce DNA strand breaks, may be enhanced by blocking IGF-IR due to the receptor's role in DNA repair. Careful preclinical studies will need to be performed before clinical trials should proceed with testing anti-IGF therapy with conventional cytotoxics. Lastly, the idea that multiple molecules are involved in growth factor signaling leads to the potential for "combination targeted therapy" trials. As mentioned, blockade of both HER2 and IGF-IR may have benefit in preclinical systems. It is also highly likely that blockade of IGF-IR and downstream signaling events (MAPK, Akt, etc), could be synergistic. Given the complexity of the cross-talk and feedback between these systems, preclinical studies should also be able to guide us with designing the optimal therapies.

However, it is clear that anti-IGF therapies will soon find their way into clinical trials. Hopefully, the vast experience with preclinical model systems will guide us in the optimal development of these agents.

Acknowledgements

This work was supported by Public Health Service grants CA74285, CA89652, and Cancer Center Support Grant P30 CA77398.

References

1. Jemal A, Tiwari RC, Murray T et al. Cancer statistics, 2004. CA Cancer J Clin 2004; 54(1):8-29.
2. Polychemotherapy for early breast cancer: An overview of the randomised trials. Early Breast Cancer Trialists' Collaborative Group. Lancet 1998; 352(9132):930-942.
3. Slamon DJ, Leyland-Jones B, Shak S et al. Use of chemotherapy plus a monoclonal antibody against HER2 for metastatic breast cancer that overexpresses HER2. N Engl J Med 2001; 344(11):783-792.
4. Yee D. The insulin-like growth factors and breast cancer—revisited. Breast Cancer Res Treat 1998; 47(3):197-199.
5. Papa V, Gliozzo B, Clark GM et al. Insulin-like growth factor-I receptors are overexpressed and predict a low risk in human breast cancer. Cancer Res 1993; 53(16):3736-3740.

6. Resnik JL, Reichart DB, Huey K et al. Elevated insulin-like growth factor I receptor autophosphorylation and kinase activity in human breast cancer. Cancer Res 1998; 58(6):1159-1164.
7. Li BD, Khosravi MJ, Berkel HJ et al. Free insulin-like growth factor-I and breast cancer risk. Int J Cancer 2001; 91(5):736-739.
8. Michels KB, Willett WC. Breast cancer—early life matters. N Engl J Med 2004; 351(16):1679-1681.
9. Lu Y, Zi X, Zhao Y et al. Insulin-like growth factor-I receptor signaling and resistance to trastuzumab (Herceptin). J Natl Cancer Inst 2001; 93(24):1852-1857.
10. Wen B, Deutsch E, Marangoni E et al. Tyrphostin AG 1024 modulates radiosensitivity in human breast cancer cells. Br J Cancer 2001; 85(12):2017-2021.
11. Pollak MN, Schernhammer ES, Hankinson SE. Insulin-like growth factors and neoplasia. Nat Rev Cancer 2004; 4(7):505-518.
12. Sachdev D, Yee D. The IGF system and breast cancer. Endocr Relat Cancer 2001; 8(3):197-209.
13. Moschos SJ, Mantzoros CS. The role of the IGF system in cancer: From basic to clinical studies and clinical applications. Oncology 2002; 63(4):317-332.
14. Laban C, Bustin SA, Jenkins PJ. The GH-IGF-I axis and breast cancer. Trends Endocrinol Metab 2003; 14(1):28-34.
15. Baker J, Liu JP, Robertson EJ et al. Role of insulin-like growth factors in embryonic and postnatal growth. Cell 1993; 75(1):73-82.
16. DeChiara TM, Efstratiadis A, Robertson EJ. A growth-deficiency phenotype in heterozygous mice carrying an insulin-like growth factor II gene disrupted by targeting. Natur 1990; 345(6270):78-80.
17. Ahlgren M, Melbye M, Wohlfahrt J et al. Growth patterns and the risk of breast cancer in women. N Engl J Med 2004; 351(16):1619-1626.
18. Kim HS, Ingermann AR, Tsubaki J et al. Insulin-like growth factor-binding protein 3 induces caspase-dependent apoptosis through a death receptor-mediated pathway in MCF-7 human breast cancer cells. Cancer Res 2004; 64(6):2229-2237.
19. Pandini G, Frasca F, Mineo R et al. Insulin/insulin-like growth factor I hybrid receptors have different biological characteristics depending on the insulin receptor isoform involved. J Biol Chem 2002; 277(42):39684-39695.
20. Pandini G, Vigneri R, Costantino A et al. Insulin and insulin-like growth factor-I (IGF-I) receptor overexpression in breast cancers leads to insulin/IGF-I hybrid receptor overexpression: Evidence for a second mechanism of IGF-I signaling. Clin Cancer Res 1999; 5(7):1935-1944.
21. Zhang X, Yee D. Tyrosine kinase signalling in breast cancer: Insulin-like growth factors and their receptors in breast cancer. Breast Cancer Res 2000; 2(3):170-175.
22. Ahmad S, Singh N, Glazer RI. Role of AKT1 in 17beta-estradiol- and insulin-like growth factor I (IGF-I)-dependent proliferation and prevention of apoptosis in MCF-7 breast carcinoma cells. Biochem Pharmacol 1999; 58(3):425-430.
23. Hankins GR, Desouza AT, Bentley RC et al. M6P/IGF2 receptor: A candidate breast tumor suppressor gene. Oncogene 1996; 12(9):2003-2009.
24. Marshman E, Green KA, Flint DJ et al. Insulin-like growth factor binding protein 5 and apoptosis in mammary epithelial cells. J Cell Sci 2003; 116(Pt 4):675-682.
25. Deeks S, Richards J, Nandi S. Maintenance of normal rat mammary epithelial cells by insulin and insulin-like growth factor 1. Exp Cell Res 1988; 174(2):448-460.
26. Richert MM, Wood TL. The insulin-like growth factors (IGF) and IGF type I receptor during postnatal growth of the murine mammary gland: Sites of messenger ribonucleic acid expression and potential functions. Endocrinology 1999; 140(1):454-461.
27. Ruan W, Kleinberg DL. Insulin-like growth factor I is essential for terminal end bud formation and ductal morphogenesis during mammary development. Endocrinology 1999; 140(11):5075-5081.
28. Hadsell DL, Greenberg NM, Fligger JM et al. Targeted expression of des(1-3) human insulin-like growth factor I in transgenic mice influences mammary gland development and IGF-binding protein expression. Endocrinology 1996; 137(1):321-330.
29. Hadsell DL, Bonnette SG, Lee AV. Genetic manipulation of the IGF-I axis to regulate mammary gland development and function. J Dairy Sci 2002; 85(2):365-377.
30. Gilmore AP, Valentijn AJ, Wang P et al. Activation of BAD by therapeutic inhibition of epidermal growth factor receptor and transactivation by insulin-like growth factor receptor. J Biol Chem 2002; 277(31):27643-27650.
31. Marshman E, Streuli CH. Insulin-like growth factors and insulin-like growth factor binding proteins in mammary gland function. Breast Cancer Res 2002; 4(6):231-239.
32. Tonner E, Barber MC, Allan GJ et al. Insulin-like growth factor binding protein-5 (IGFBP-5) induces premature cell death in the mammary glands of transgenic mice. Development 2002; 129(19):4547-4557.
33. Toniolo P, Bruning PF, Akhmedkhanov A et al. Serum insulin-like growth factor-I and breast cancer. Int J Cancer 2000; 88(5):828-832.

34. Paik S. Expression of IGF-I and IGF-II mRNA in breast tissue. Breast Cancer Res Treat 1992; 22(1):31-38.
35. Ruan W, Newman CB, Kleinberg DL. Intact and amino-terminally shortened forms of insulin-like growth factor I induce mammary gland differentiation and development. Proc Natl Acad Sci USA 1992; 89(22):10872-10876.
36. Ruan W, Catanese V, Wieczorek R et al. Estradiol enhances the stimulatory effect of insulin-like growth factor-I (IGF-I) on mammary development and growth hormone-induced IGF-I messenger ribonucleic acid. Endocrinology 1995; 136(3):1296-1302.
37. Wu Y, Cui K, Miyoshi K et al. Reduced circulating insulin-like growth factor I levels delay the onset of chemically and genetically induced mammary tumors. Cancer Res 2003; 63(15):4384-4388.
38. Hankinson SE, Willett WC, Colditz GA et al. Circulating concentrations of insulin-like growth factor-I and risk of breast cancer. Lancet 1998; 351(9113):1393-1396.
39. Cullen KJ, Yee D, Sly WS et al. Insulin-like growth factor receptor expression and function in human breast cancer. Cancer Res 1990; 50(1):48-53.
40. Pezzino V, Papa V, Milazzo G et al. Insulin-like growth factor-I (IGF-I) receptors in breast cancer. Ann NY Acad Sci 1996; 784:189-201.
41. Turner BC, Haffty BG, Narayanan L et al. Insulin-like growth factor-I receptor overexpression mediates cellular radioresistance and local breast cancer recurrence after lumpectomy and radiation. Cancer Res 1997; 57(15):3079-3083.
42. Rocha RL, Hilsenbeck SG, Jackson JG et al. Insulin-like growth factor binding protein-3 and insulin receptor substrate-1 in breast cancer: Correlation with clinical parameters and disease-free survival. Clin Cancer Res 1997; 3(1):103-109.
43. Lee AV, Jackson JG, Gooch JL et al. Enhancement of insulin-like growth factor signaling in human breast cancer: Estrogen regulation of insulin receptor substrate-1 expression in vitro and in vivo. Mol Endocrinol 1999; 13(5):787-796.
44. Hoang CD, Zhang X, Scott PD et al. Selective activation of insulin receptor substrate-1 and -2 in pleural mesothelioma cells: Association with distinct malignant phenotypes. Cancer Res 2004; 64(20):7479-7485.
45. Zhang X, Kamaraju S, Hakuno F et al. Motility response to insulin-like growth factor-I (IGF-I) in MCF-7 cells is associated with IRS-2 activation and integrin expression. Breast Cancer Res Treat 2004; 83(2):161-170.
46. Nagle JA, Ma Z, Byrne MA et al. Involvement of insulin receptor substrate 2 in mammary tumor metastasis. Mol Cell Biol 2004; 24(22):9726-9735.
47. Perks CM, Bowen S, Gill ZP et al. Differential IGF-independent effects of insulin-like growth factor binding proteins (1-6) on apoptosis of breast epithelial cells. J Cell Biochem 1999; 75(4):652-664.
48. Oh Y, Muller HL, Lamson G et al. Insulin-like growth factor (IGF)-independent action of IGF-binding protein-3 in Hs578T human breast cancer cells. Cell surface binding and growth inhibition. J Biol Chem 1993; 268(20):14964-14971.
49. Al-Hajj M, Wicha MS, Benito-Hernandez A et al. Prospective identification of tumorigenic breast cancer cells. Proc Natl Acad Sci USA 2003; 100(7):3983-3988.
50. Baselga J, Norton L. Focus on breast cancer. Cancer Cell 2002; 1(4):319-322.
51. Tamoxifen for early breast cancer: An overview of the randomised trials. Early Breast Cancer Trialists' Collaborative Group. Lancet 1998; 351(9114):1451-1467.
52. Rubin R, Baserga R. Insulin-like growth factor-I receptor. Its role in cell proliferation, apoptosis, and tumorigenicity. Lab Invest 1995; 73(3):311-331.
53. Gooch JL, Van Den Berg CL, Yee D. Insulin-like growth factor (IGF)-I rescues breast cancer cells from chemotherapy-induced cell death—proliferative and anti-apoptotic effects. Breast Cancer Res Treat 1999; 56(1):1-10.
54. Xie SP, Pirianov G, Colston KW. Vitamin D analogues suppress IGF-I signalling and promote apoptosis in breast cancer cells. Eur J Cancer 1999; 35(12):1717-1723.
55. Kulik G, Klippel A, Weber MJ. Antiapoptotic signalling by the insulin-like growth factor I receptor, phosphatidylinositol 3-kinase, and Akt. Mol Cell Biol 1997; 17(3):1595-1606.
56. Kothari S, Cizeau J, McMillan-Ward E et al. BNIP3 plays a role in hypoxic cell death in human epithelial cells that is inhibited by growth factors EGF and IGF. Oncogene 2003; 22(30):4734-4744.
57. Peruzzi F, Prisco M, Dews M et al. Multiple signaling pathways of the insulin-like growth factor 1 receptor in protection from apoptosis. Mol Cell Biol 1999; 19(10):7203-7215.
58. Zha J, Harada H, Yang E et al. Serine phosphorylation of death agonist BAD in response to survival factor results in binding to 14-3-3 not BCL-X(L). Cell 1996; 87(4):619-628.
59. Singleton JR, Dixit VM, Feldman EL. Type I insulin-like growth factor receptor activation regulates apoptotic proteins. J Biol Chem 1996; 271(50):31791-31794.
60. Dunn SE, Hardman RA, Kari FW et al. Insulin-like growth factor 1 (IGF-1) alters drug sensitivity of HBL100 human breast cancer cells by inhibition of apoptosis induced by diverse anticancer drugs. Cancer Res 1997; 57(13):2687-2693.

61. Lu Y, Zi X, Pollak M. Molecular mechanisms underlying IGF-I-induced attenuation of the growth-inhibitory activity of trastuzumab (Herceptin) on SKBR3 breast cancer cells. Int J Cancer 2004; 108(3):334-341.
62. Mamay CL, Mingo-Sion AM, Wolf DM et al. An inhibitory function for JNK in the regulation of IGF-I signaling in breast cancer. Oncogene 2003; 22(4):602-614.
63. Heron-Milhavet L, Karas M, Goldsmith CM et al. Insulin-like growth factor-I (IGF-I) receptor activation rescues UV-damaged cells through a p38 signaling pathway. Potential role of the IGF-I receptor in DNA repair. J Biol Chem 2001; 276(21):18185-18192.
64. Heron-Milhavet L, LeRoith D. Insulin-like growth factor I induces MDM2-dependent degradation of p53 via the p38 MAPK pathway in response to DNA damage. J Biol Chem 2002; 277(18):15600-15606.
65. Shahrabani-Gargir L, Pandita TK, Werner H. Ataxia-telangiectasia mutated gene controls insulin-like growth factor I receptor gene expression in a deoxyribonucleic acid damage response pathway via mechanisms involving zinc-finger transcription factors Sp1 and WT1. Endocrinology 2004; 145(12):5679-5687.
66. Kuhn C, Hurwitz SA, Kumar MG et al. Activation of the insulin-like growth factor-1 receptor promotes the survival of human keratinocytes following ultraviolet B irradiation. Int J Cancer 1999; 80(3):431-438.
67. Decraene D, Agostinis P, Bouillon R et al. Insulin-like growth factor-1-mediated AKT activation postpones the onset of ultraviolet B-induced apoptosis, providing more time for cyclobutane thymine dimer removal in primary human keratinocytes. J Biol Chem 2002; 277(36):32587-32595.
68. Trojanek J, Ho T, Del Valle L et al. Role of the insulin-like growth factor I/insulin receptor substrate 1 axis in Rad51 trafficking and DNA repair by homologous recombination. Mol Cell Biol 2003; 23(21):7510-7524.
69. Jayanth VR, Belfi CA, Swick AR et al. Insulin and insulin-like growth factor-1 (IGF-1) inhibit repair of potentially lethal radiation damage and chromosome aberrations and alter DNA repair kinetics in plateau-phase A549 cells. Radiat Res 1995; 143(2):165-174.
70. Holdaway IM, Mason BH, Lethaby AE et al. Serum insulin-like growth factor-I and insulin-like growth factor binding protein-3 following chemotherapy for advanced breast cancer. ANZ J Surg 2003; 73(11):905-908.
71. Kajdaniuk D, Marek B. Influence of adjuvant chemotherapy with cyclophosphamide, methotrexate and 5-fluorouracil on plasma insulin-like growth factor-I and chosen hormones in breast cancer premenopausal patients. J Clin Pharm Ther 2000; 25(1):67-72.
72. Formelli F. Quality control for HPLC assay and surrogate end point biomarkers from the fenretinide (4-HPR) breast cancer prevention trial. J Cell Biochem Suppl 2000; 34:73-79.
73. Yang LM, Tin UC, Wu K et al. Role of retinoid receptors in the prevention and treatment of breast cancer. J Mammary Gland Biol Neoplasia 1999; 4(4):377-388.
74. Chrzanowski K, Bielawska A, Palka J. Proline analogue of melphalan as a prodrug susceptible to the action of prolidase in breast cancer MDA-MB 231 cells. Farmaco 2003; 58(11):1113-1119.
75. Gronbaek H, Tanos V, Meirow D et al. Effects of tamoxifen on insulin-like growth factors, IGF binding proteins and IGFBP-3 proteolysis in breast cancer patients. Anticancer Res 2003; 23(3C):2815-2820.
76. Ferrari L, Bajetta E, Martinetti A et al. Could exemestane affect insulin-like growth factors, interleukin 6 and bone metabolism in postmenopausal advanced breast cancer patients after failure on aminoglutethimide, anastrozole or letrozole? Int J Oncol 2003; 22(5):1081-1089.
77. Torrisi R, Baglietto L, Johansson H et al. Effect of raloxifene on IGF-I and IGFBP-3 in postmenopausal women with breast cancer. Br J Cancer 2001; 85(12):1838-1841.
78. Nielsen TO, Andrews HN, Cheang M et al. Expression of the insulin-like growth factor I receptor and urokinase plasminogen activator in breast cancer is associated with poor survival: Potential for intervention with 17-allylamino geldanamycin. Cancer Res 2004; 64(1):286-291.
79. Jerome L, Shiry L, Leyland-Jones B. Anti-insulin-like growth factor strategies in breast cancer. Semin Oncol 2004; 31(1 Suppl 3):54-63.
80. Byron SA, Yee D. Potential therapeutic strategies to interrupt insulin-like growth factor signaling in breast cancer. Semin Oncol 2003; 30(5 Suppl 16):125-132.
81. Szepeshazi K, Schally AV, Armatis P et al. Antagonists of GHRH decrease production of GH and IGF-I in MXT mouse mammary cancers and inhibit tumor growth. Endocrinology 2001; 142(10):4371-4378.
82. Kahan Z, Varga JL, Schally AV et al. Antagonists of growth hormone-releasing hormone arrest the growth of MDA-MB-468 estrogen-independent human breast cancers in nude mice. Breast Cancer Res Treat 2000; 60(1):71-79.
83. Ingle JN, Suman VJ, Kardinal CG et al. A randomized trial of tamoxifen alone or combined with octreotide in the treatment of women with metastatic breast carcinoma. Cancer 1999; 85(6):1284-1292.

84. Friend KE. Cancer and the potential place for growth hormone receptor antagonist therapy. Growth Horm IGF Res 2001; 11(Suppl A):S121-123.
85. Van den Berg CL, Cox GN, Stroh CA et al. Polyethylene glycol conjugated insulin-like growth factor binding protein-1 (IGFBP-1) inhibits growth of breast cancer in athymic mice. Eur J Cancer 1997; 33(7):1108-1113.
86. Yee D, Jackson JG, Kozelsky TW et al. Insulin-like growth factor binding protein-1 (IGFBP-1) expression inhibits IGF-I action in MCF-7 breast cancer cells. Cell Growth and Differentiation 1994; 5:73-77.
87. Zi X, Zhang J, Agarwal R et al. Silibinin up-regulates insulin-like growth factor-binding protein 3 expression and inhibits proliferation of androgen-independent prostate cancer cells. Cancer Res 2000; 60(20):5617-5620.
88. D'Ambrosio C, Ferber A, Resnicoff M et al. A soluble insulin-like growth factor I receptor that induces apoptosis of tumor cells in vivo and inhibits tumorigenesis. Cancer Res 1996; 56(17):4013-4020.
89. Samani AA, Chevet E, Fallavollita L et al. Loss of tumorigenicity and metastatic potential in carcinoma cells expressing the extracellular domain of the type 1 insulin-like growth factor receptor. Cancer Res 2004; 64(10):3380-3385.
90. Goya M, Miyamoto S, Nagai K et al. Growth inhibition of human prostate cancer cells in human adult bone implanted into nonobese diabetic/severe combined immunodeficient mice by a ligand-specific antibody to human insulin-like growth factors. Cancer Res 2004; 64(17):6252-6258.
91. Arteaga CL, Kitten LJ, Coronado EB et al. Blockade of the type I somatomedin receptor inhibits growth of human breast cancer cells in athymic mice. J Clin Invest 1989; 84(5):1418-1423.
92. Li SL, Liang SJ, Guo N et al. Single-chain antibodies against human insulin-like growth factor I receptor: Expression, purification, and effect on tumor growth. Cancer Immunol Immunother 2000; 49(4-5):243-252.
93. Sachdev D, Li SL, Hartell JS et al. A chimeric humanized single-chain antibody against the type I insulin-like growth factor (IGF) receptor renders breast cancer cells refractory to the mitogenic effects of IGF-I. Cancer Res 2003; 63(3):627-635.
94. Ye JJ, Liang SJ, Guo N et al. Combined effects of tamoxifen and a chimeric humanized single chain antibody against the type I IGF receptor on breast tumor growth in vivo. Horm Metab Res 2003; 35(11-12):836-842.
95. Maloney EK, McLaughlin JL, Dagdigian NE et al. An anti-insulin-like growth factor I receptor antibody that is a potent inhibitor of cancer cell proliferation. Cancer Res 2003; 63(16):5073-5083.
96. Burtrum D, Zhu Z, Lu D et al. A fully human monoclonal antibody to the insulin-like growth factor I receptor blocks ligand-dependent signaling and inhibits human tumor growth in vivo. Cancer Res 2003; 63(24):8912-8921.
97. Blum G, Gazit A, Levitzki A. Development of new insulin-like growth factor-1 receptor kinase inhibitors using catechol mimics. J Biol Chem 2003; 278(42):40442-40454.
98. Parrizas M, Gazit A, Levitzki A et al. Specific inhibition of insulin-like growth factor-1 and insulin receptor tyrosine kinase activity and biological function by tyrphostins. Endocrinology 1997; 138(4):1427-1433.
99. Mitsiades CS, Mitsiades NS, McMullan CJ et al. Inhibition of the insulin-like growth factor receptor-1 tyrosine kinase activity as a therapeutic strategy for multiple myeloma, other hematologic malignancies, and solid tumors. Cancer Cell 2004; 5(3):221-230.
100. Garcia-Echeverria C, Pearson MA, Marti A et al. In vivo antitumor activity of NVP-AEW541-A novel, potent, and selective inhibitor of the IGF-IR kinase. Cancer Cell 2004; 5(3):231-239.
101. Salatino M, Schillaci R, Proietti CJ et al. Inhibition of in vivo breast cancer growth by antisense oligodeoxynucleotides to type I insulin-like growth factor receptor mRNA involves inactivation of ErbBs, PI-3K/Akt and p42/p44 MAPK signaling pathways but not modulation of progesterone receptor activity. Oncogene 2004; 23(30):5161-5174.
102. Beech DJ, Parekh N, Pang Y. Insulin-like growth factor-I receptor antagonism results in increased cytotoxicity of breast cancer cells to doxorubicin and taxol. Oncol Rep 2001; 8(2):325-329.
103. Benini S, Manara MC, Baldini N et al. Inhibition of insulin-like growth factor I receptor increases the antitumor activity of doxorubicin and vincristine against Ewing's sarcoma cells. Clin Cancer Res 2001; 7(6):1790-1797.
104. Camirand A, Pollak M. Cotargeting IGF-1R and c-kit: Synergistic inhibition of proliferation and induction of apoptosis in H 209 small cell lung cancer cells. Br J Cancer 2004; 90(9):1825-1829.
105. Baserga R. Targeting the IGF-1 receptor: From rags to riches. Eur J Cancer 2004; 40(14):2013-2015.
106. Albain KS, Green SJ, Ravdin PM et al. Adjuvant chemohormonal therapy for primary breast cancer should be sequential instead of concurrent: Initial results from intergroup trial 0100 (SWOG-8814). Proc ASCO 2002, (Abs#:143).

CHAPTER 8

EGF Receptor in Breast Cancer Chemoresistance

Robert B. Dickson* and T.B. Deb

Introduction

Acquisition of resistance of breast cancer to chemotherapy is commonly associated with progression of the disease to increased metastatic spread. Although early studies on this problem examined the possible roles of chemotherapy drug metabolism and efflux by resistant breast cancer cells, more recent work has implicated aberrant growth factor receptor and signal transduction pathways in the process for such commonly used drugs as the anthracyclines. Cell mutability, DNA repair defects, and loss of DNA damage checkpoint controls certainly must play key roles in the ability of cancer cells to evolve such resistance mechanisms.[1,2]

Identification of specific growth factor-receptor pathways involved in chemo-resistance has allowed the possibility of targeting its specific determinants to enhance effectiveness of tumor eradication. To date, the most clear data on growth factor pathway involvement involve the EGF receptor family, a group of four related receptors: EGF receptor (ErbB), ErbB2, ErbB3, and ErbB4. The family forms heterodimeric and homodimeric associations, with ErbB2 and ErbB homodimers and heterodimers most likely to bear poor prognostic relationships to breast cancer. Specifically, their overexpression is associated with the estrogen receptor negative form of the disease, which is intrinsically the most clinically aggressive form. ErbB2 is now widely accepted as a poor prognostic biomarker of breast cancer, with implications for therapy choice.[4-6] Specifically, ErbB2 expression appears to indicate that a tumor will be refractory to therapy with anthracyclines and will be sensitive to an ErbB2 targeted therapy (herceptin, or trastuzumab).[7-9]

EGFR Signal Transduction

Multiple cellular signal transduction pathways emanate from EGF receptor complexes with its family members. EGFR has also been shown to be transactivated by ligand stimulated, G-protein coupled receptors. However, four dominant pathways govern cell proliferation and survival, key determinants of chemoresistance: the PI-3 kinase/AKT pathway, JAK-STAT pathway, phospholipase Cγ, the Shc-Grb2-SOS-Ras MAPK pathway, and cytoplasmic tyrosine kinases (c-Src, c-Yes). Phosphorylated tyrosine residues on EGF receptor family members serve as docking/activation sites for these various signal transduction cascades. Other signal transduction pathways also exist, but are beyond the scope of the current, brief review.[2,3]

The functions of each of the major EGF receptor family signal transduction pathways noted above are varied. The PI-3K pathway, possibly the most polyfunctional of all, utilizes Akt to regulate p53, cyclin D_1, nitrous oxide synthetase, FKHR, NF-κB, Fas-L, Bad, and

*Contributing Author: Robert B. Dickson—Georgetown University Medical Center, 3970 Reservoir Road NW, Washington, DC 20007-2197, U.S.A. Email: dicksonr@georgetown.edu

Breast Cancer Chemosensitivity, edited by Dihua Yu and Mien-Chie Hung.

mTor. These pathways modulate cell proliferation, cell survival, and angiogenesis. Phospholipase Cγ regulates cellular calcium influx and protein kinase C family enzymes, and activation of transcription factors STAT and STAT5 have strong influences on breast cancer progression. The cytoplasmic tyrosine kinase c-Src regulates cell transforming effects of EGF receptor family, particularly as they relate to ErbB2. Finally, the Ras-MAPK pathway is especially important in regulating cell proliferation.[2,3] It is important to note that cross-talk between these signaling mechanisms is also observed in many breast cancer cells, with potential amplification of signals.

A Central Role for Akt in Chemoresistance

As noted in the prior section, Akt is a key mediator of PI-3K for important determinants of oncogenesis and chemoresistance. Akt, a three gene family, is commonly gene-amplified or overexpressed in multiple cancers, including breast cancer. In ovarian cancer, expression of Akt 2 confers cancer cell-resistance to paclitaxel.[10] In MCF-7 breast cancer cells expressing ErbB3, ErbB2 transfection causes PI-3K-dependent activation of Akt and polychemotherapy resistance (paclitaxel, doxorubicin, 5-fluorouracil, etoposids, and camptothecin).[12-14] Resistance was reversed by cell transfection with dominant negative forms of PI-3K and Akt. While PI-3K is able to dock and become activated through interactions with EGF receptor, ErbB3, and ErbB4, ErbB2 does not possess the cognate protein-protein docking site for PI-3K, and so ErbB2 must interact with a heterotypic partner in order to function through the PI-3K-Akt pathway.[3] This would imply that although ErbB2 may be the most important EGF receptor family member biomarker for chemoresistance in breast cancer, the actual signal transduction pathway of interest for resistance to proapoptotic therapies passes directly through *other* ErbB family members to reach Akt.

Akt is known to be activated by growth factors and other stimuli, through both PI-3K -dependent and independent mechanisms.[15-18] This kinase is a ubiquitous lipid kinase and a upstream effector of Akt.[19] It has also been implicated in a wide variety of cellular functions: cell survival and antiapoptosis,[20,21] growth and proliferation,[22,23] differentiation,[24,25] cytoskeletal rearrangement,[26] translocation of glucose transporter GLUT4[27,28] and membrane ruffling.[29] Following cell exposure to growth factors, PI-3K produces phosphatidylinositol 3,4-biphosphates and phosphatidyl inositol 3,4,5-triphosphates, at the plasma membrane. These phospholipids serve as binding anchors for the pleckstrin homology domain of Akt. This results in accumulation of Akt on the inner surface of the plasma membrane.[10,30-32] There Akt is phosphorylated at Ser-473 and Thr-308 to become fully activated.[33] There also exists, a PI-3K-independent mechanism(s) of Akt activation, involving calmodulin kinase kinase, which directly phosphorylates Akt, independent of calcium.[15] Calmodulin, an allosteric regulator of calmodulin kinases, also regulates Akt activation. This regulation is independent of calmodulin kinase kinase, and has been reported for neuronal cell survival[34,35] and GLUT4 translocation in 3T3-L1 adipocytes.[36] A consensus sequence in the p110 catalytic subunit of PI-3K is a likely binding site of calmodulin, but no biochemical data currently exist to support this idea.[37]

Since mammary epithelial cells can release calcium from intracellular stores in response to growth factors and survival ligands,[38,39] our group has examined the role calcium plays in survival of these cells. We reported that EGF-induced survival of c-Myc-overexpressing mammary carcinoma cells is mediated by activation of PI-3K/Akt kinase.[40] We next searched for a specific survival mechanism(s), downstream of EGFR, which may be a therapeutic target(s) in breast carcinoma. Our studies[41] identified a calcium/calmodulin-dependent Akt activation and survival mechanism in these cells. We found that EGF-induced Akt activation is mediated by calmodulin, resulting in cell survival. Calmodulin did not exert its effects directly at the PI-3K level. Rather, we found that an EGF-dependent complex forms between calmodulin and Akt. It appears that this mechanism serves transport Akt to the plasma membrane for its subsequent activation by a PI-3K-dependent mechanism. Cellular treatment with calmodulin antagonists leads to apoptotic cell death in tumorigenic mammary carcinoma cells. Recent other studies further confirm the important link provided by Ca^{++}-CaM in growth factor survival signaling.

For example, a recent study[42] using PTEN-deficient tumor cells as a screen for novel inhibitors of PI-3K/Akt survival signaling identified multiple antagonists of calmodulin signaling. A second, completely independent line of investigation used a proteomic scan of human proteins to identify Akt-1 as a direct binding partner of calmodulin.[43] Thus interference with CaM-Akt interaction could represent a novel therapeutic target in the EGF pathway.

EGFR as a Target for Therapy in Chemoresistant Tumors

As described above, multiple therapeutic approaches may be possible, targeting EGF receptor family members and their survival signaling pathways. While ErbB2-directed therapy is now well established, the areas relating to targeting of Akt and EGF receptor are still rapidly emerging.[44,45] EGF receptor-targeted therapies began with the isolation by Dr. John Mendelsohn's group and therapeutic application of antibodies that blocked human EGF ligand-receptor interactions. An early application of these antibodies in the MDA MD-468 human breast cancer cell line demonstrated their anti-cancer activity for breast carcinoma.[46,47] Since these early studies, the approach has been further validated in animal models, the antibodies humanized, and clinical trials are now underway.[47] In addition, small molecule drugs directed toward the tyrosine kinase domain of the EGF receptor are now moving forward in clinical trials.[47] For the latter class, both EGF receptor-selective drugs, as well as pan-EGF receptor family drugs are now in use. Dosing and scheduling regimens are now being tested to optimize the various therapeutic approaches.[47] A particularly exciting recent development was the identification of a subset of EGF therapy-responsive lung cancer tumors bearing particular EGF receptor kinase-domain mutations. However, these mutations have not yet been detected in human breast cancer.[47]

In terms of the monoclonal antibody approach, there are now in clinical trial three anti-EGF-directed drugs (ABX-EGF, EMD-7200, and h-R3), and one ErbB2-directed drug (Pertuzamab), in addition to the already marketed Cetuximab (against EGF receptor) and Trastuzumab (against ErbB2). The EGF receptor monoclonal antibody derived antagonists have not been particularly effective against breast cancer, so their clinical trials focus on cancers of the colon, head and neck, kidney, cervix, lung, and pancreas. In contrast, the small molecule inhibitors, particularly the pan EGF family tyrosine kinase inhibitors may have more promise in breast cancer. The currently available drugs in this class are the following: Gefitinib (EGF receptor selective) Erlotinib (EGF receptor), Lapatinib (EGF receptor and ErbB2), C1-1033 (all EGF receptor family members), and EKB-569 (EGF receptor, ErbB2, VEGF receptor). Among these, Lapatinib is currently in Phase III trial from breast cancer, while the others are either openly marketed (Gefitinib, Erlotinib) or in clinical trials in other disease sites (lung cancer, pancreatic, skin, colon).[47]

Current clinical trials are now using pre and post therapy biopsies in order to more rapidly begin to evaluate therapeutic efficacy and mechanism of action of EGF receptor antagonists in breast cancer therapy. Lapatinib, a reversible inhibitor of the EGF receptor and ErbB2 tyrosine kinases, has been recently examined for its effects on patient tumors by sequential biopsy. Responsive tumors showed inhibition of phosphorylation on EGF receptors, ErbB2, Erk1/2, Akt, cyclin D_1, and transforming growth factor alpha. Clinical responses were also associated with increased apoptotis as determined by TUNEL staining.[48]

As the diversity of monoclonal antibodies against the EGF receptor continue to be studied for their efficacy and mechanism of action, it is clear that multiple classes are emerging. For example, while most therapeutic antibodies applied so far to therapy block receptor ligand interactions, this need not be the only approach to a therapeutically effective mechanism of action. For example, Pertuzumab is a recombinant, humanized monoclonal antibody that binds ErbB2 and blocks its ability to dimerize with other family members, such as the EGF receptor. Thus, this antibody is fundamentally different from other approaches with anti EGF receptor-directed antibodies or small molecule drugs, and potentially targets downstream survival signaling pathways. A Phase I trial, recently demonstrated good toleration and pharmacokinetics and some partial responses.[49]

Conclusions and Future Directions

In summary, although the EGF receptor has been under study for many years, its potential as a target of therapy for breast cancer is only just beginning to be appreciated. Although ErbB2 may stand at the top of a key growth factor-dependent pathway driving breast cancer progression, other pathway elements, such as EGF receptor, PI-3K and Akt may eventually yield key targets for therapy. However, in order for the field to most rapidly move forward, researchers must take advantage of new in vitro and in vivo assay methodologies. In particular, relevant animal models for drug testing have been particularly problematic, slowing progress. Some of these hurdles may now be resolving with new advances in molecular imaging of tumors for rapid assays that compliment classical pathology approaches.[50,51]

References

1. Dickson RB, Lippman ME. Drug and Hormonal Resistance in Breast Cancer: Cellular and Molecular Mechanisms. New York: Ellis Horwood, 1995:1-440.
2. Dickson RB, Pestell RG, Lippman ME. Molecular biology of breast cancer. In: DeVita V, Hellman S, Rosenberg S, eds. Cancer: Principles and Practice of Oncology. 7th ed. Philadelphia: JB Lippincott, 2005:1399-1414.
3. Siegel PM, Muller WJ. Tyrosine kinase and signal transduction in mouse mammary tumorigenesis. In: Dickson RB, Solomon DS, eds. Hormones and Growth Factors in Development and Neoplasia. New York: Wily Liss, 1999:397-419.
4. Nicholson S, Halcrow P, Sainsbury JR et al. Epidermal growth factor receptor (EGFr) status associated with failure of primary endocrine therapy in elderly postmenopausal patients with breast cancer. Br J Cancer 1988; 58:810-814.
5. Nicholson S, Sainsbury JR, Halcrow P et al. Expression of epidermal growth factor receptors associated with lack of response to endocrine therapy in recurrent breast cancer. Lancet 1989; 28:182-185.
6. Nicholson S, Richard J, Sainsbury C et al. Epidermal growth factor receptor (EGFr); results of a 6 year follow-up study in operable breast cancer with emphasis on the node negative subgroup. Br J Cancer 1991; 63:146-150.
7. Pegram MD, Slamon DJ. Combination therapy with trastuzumab (Herceptin) and cisplatin for chemoresistant metastatic breast cancer: Evidence for receptor-enhanced chemosensitivity. Semin Oncol 1999; 26:89-95.
8. Hayes DF. Prognostic and predictive factors for breast cancer: Translating technology to oncology. J Clin Oncol 2005; 23:1596-1599.
9. Pegram MD, Pietras R, Bajamonde A et al. Targeted therapy: Wave of the future. J Clin Oncol 2005; 23:1776-1781.
10. Mitsuuchi Y, Johnson SW, Selvakumaran M et al. The phosphatidylinositol 3-kinase/AKT signal transduction pathway plays a critical role in the expression of p21WAF1/CIP1/SDI1 induced by cisplatin and paclitaxel. Cancer Res 2002; 60:5390-5394.
11. Kim D, Dan HC, Park S et al. AKT/PKB signaling mechanisms in cancer and chemoresistance. Front Biosci 2005; 10:975-987.
12. Yu D. Mechanisms of ErbB2-mediated pactitaxel resistance and trastuzumab-mediated paclitaxel sensitization in ErbB2-overexpressing breast cancers. Semin Oncol 2001; 28:12-17.
13. Jin W, Wu L, Liang K et al. Roles of the PI-3K and MEK pathways in Ras-mediated chemoresistance in breast cancer cells. Br J Cancer 2003; 89:185-191.
14. Knuefermann C, Lu Y, Liu B et al. HER2/PI-3K/Akt activation leads to a multidrug resistance in human breast adenocarcinoma cells. Oncogene 2003; 22:3205-3212.
15. Yano S, Tokumitsu H, Soderling TR. Calcium promotes cell survival through CaM-K kinase activation of the protein-kinase-B pathway. Nature 1998; 396:584-587.
16. Datta SR, Brunet A, Greenberg ME. Cellular survival: A play in three Akts. Genes Dev 1999; 13:2905-2927.
17. Datta K, Bellacosa A, Chan TO et al. Akt is a direct target of the phosphatidylinositol 3-kinase. Activation by growth factors, v-src and v-Ha-ras, in Sf9 and mammalian cells. J Biol Chem 1996; 271:30835-30839.
18. Filippa N, Sable CL, Filloux C et al. Mechanism of protein kinase B activation by cyclic AMP-dependent protein kinase. Mol Cell Biol 1999; 19:4989-5000.
19. Cantley LC. The phosphoinositide 3-kinase pathway. Science 2002; 296:1655-1657.
20. Ahmed NN, Grimes HL, Bellacosa A et al. Transduction of interleukin-2 antiapoptotic and proliferative signals via Akt protein kinase. Proc Natl Acad Sci USA 1997; 94:3627-3632.

21. Yao R, Cooper GM. Requirement for phosphatidylinositol-3 kinase in the prevention of apoptosis by nerve growth factor. Science 1995; 267:2003-2006.
22. Valius M, Kazlauskas A. Phospholipase C-gamma 1 and phosphatidylinositol 3 kinase are the downstream mediators of the PDGF receptor's mitogenic signal. Cell 1993; 73:321-334.
23. Auger KR, Serunian LA, Soltoff SP et al. PDGF-dependent tyrosine phosphorylation stimulates production of novel polyphosphoinositides in intact cells. Cell 1989; 57:167-175.
24. Kimura K, Hattori S, Kabuyama Y et al. Neurite outgrowth of PC12 cells is suppressed by wortmannin, a specific inhibitor of phosphatidylinositol 3-kinase. J Biol Chem 1994; 269:18961-18967.
25. Kaliman P, Vinals F, Testar X et al. Phosphatidylinositol 3-kinase inhibitors block differentiation of skeletal muscle cells. J Biol Chem 1996; 271:19146-19151.
26. Rodriguez-Viciana P, Warne PH, Khwaja A et al. Role of phosphoinositide 3-OH kinase in cell transformation and control of the actin cytoskeleton by Ras. Cell 1997; 89:457-467.
27. Kamohara S, Hayashi H, Todaka M et al. Platelet-derived growth factor triggers translocation of the insulin-regulatable glucose transporter (type 4) predominantly through phosphatidylinositol 3-kinase binding sites on the receptor. Proc Natl Acad Sci USA 1995; 92:1077-1081.
28. Clarke JF, Young PW, Yonezawa K et al. Inhibition of the translocation of GLUT1 and GLUT4 in 3T3-L1 cells by the phosphatidylinositol 3-kinase inhibitor, wortmannin. Biochem J 1994; 300:631-635.
29. Wennstrom S, Siegbahn A, Yokote K et al. Membrane ruffling and chemotaxis transduced by the PDGF beta-receptor require the binding site for phosphatidylinositol 3' kinase. Oncogene 1994; 9:651-660.
30. James SR, Downes CP, Gigg R et al. Specific binding of the Akt-1 protein kinase to phosphatidylinositol 3,4,5-trisphosphate without subsequent activation. Biochem J 1996; 315:709-713.
31. Frech M, Andjelkovic M, Ingley E et al. High affinity binding of inositol phosphates and phosphoinositides to the pleckstrin homology domain of RAC/protein kinase B and their influence on kinase activity. J Biol Chem 1997; 272:8474-8481.
32. Franke TF, Yang SI, Chan TO et al. The protein kinase encoded by the Akt proto-oncogene is a target of the PDGF-activated phosphatidylinositol 3-kinase. Cell 1995; 81:727-736.
33. Alessi DR, Andjelkovic M, Caudwell B et al. Mechanism of activation of protein kinase B by insulin and IGF-1. EMBO J 1996; 15:6541-6551.
34. Egea J, Espinet C, Soler RM et al. Neuronal survival induced by neurotrophins requires calmodulin. J Cell Biol 2001; 154:585-597.
35. Cheng A, Wang S, Yang D et al. Calmodulin mediates brain-derived neurotrophic factor cell survival signaling upstream of Akt kinase in embryonic neocortical neurons. J Biol Chem 2003; 278:7591-7599.
36. Yang C, Watson RT, Elmendorf JS et al. Calmodulin antagonists inhibit insulin-stimulated GLUT4 (glucose transporter 4) translocation by preventing the formation of phosphatidylinositol 3,4,5-trisphosphate in 3T3L1 adipocytes. Mol Endocrinol 2000; 14:317-326.
37. Fischer R, Julsgart J, Berchtold MW. High affinity calmodulin target sequence in the signalling molecule PI 3-kinase. FEBS Lett 1998; 425:175-177.
38. Ichikawa J, Furuya K, Miyata S et al. EGF enhances Ca(2+) mobilization and capacitative Ca(2+) entry in mouse mammary epithelial cells. Cell Biochem Funct 2000; 18:215-225.
39. Ichikawa J, Kiyohara T. Suppression of EGF-induced cell proliferation by the blockade of Ca2+ mobilization and capacitative Ca2+ entry in mouse mammary epithelial cells. Cell Biochem Funct 2001; 19:213-219.
40. Ramljak D, Coticchia CM, Nishanian TG et al. Epidermal growth factor inhibition of c-Myc-mediated apoptosis through Akt and Erk involves Bcl-xL upregulation in mammary epithelial cells. Exp Cell Res 2003; 287:397-410.
41. Deb TB, Coticchia CM, Dickson RB. Calmodulin-mediated activation of Akt regulates survival of c-Myc-overexpressing mouse mammary carcinoma cells. J Biol Chem 2004; 279:38903-38911.
42. Kau TR, Schroeder F, Ramaswamy S et al. A chemical genetic screen identifies inhibitors of regulated nuclear export of a Forkhead transcription factor in PTEN-deficient tumor cells. Cancer Cell 2003; 463-476.
43. Shen X, Valencia CA, Szostak J et al. Scanning the human proteome for calmodulin-binding proteins. Proc Natl Acad Sci USA 2005; 102:5969-5974.
44. Schmidt M, Lichtner RB. EGF receptor targeting in therapy-resistant human tumors. Drug Resist Updat 2002; 5:11-18.
45. Harris TK. PDK1 and PKB/Akt: Ideal targets for development of new strategies to structurebased drug design. IUBMB Life 2003; 55:117-1126.

46. Ennis BW, Valverius E, Bates SE et al. Anti-EGF receptor antibodies inhibit the autocrine stimulated growth of MDA-MB-468 human breast cancer cells. Molec Endocrinol 1989; 3:1830-1838.
47. Baselga J, Arteaga CL. Critical update and emerging trends in epidermal growth factor receptor targeting in cancer. J Clin Oncol 2005; 23:2445-2459.
48. Spector NL, Xia W, Burris IIIrd H et al. Study of the biologic effects of lapatinib, a reversible inhibitor of ErbB1 and ErbB2 tyrosine kinases, on tumor growth and survival pathways in patients with advanced malignancies. J Clin Oncol 2005; 23:2502-2512.
49. Agus DB, Gordon MS, Taylor C et al. Phase I clinical study of pertuzumab, a novel HER dimerization inhibitor, in patients with advanced cancer. J Clin Oncol 2005; 23:2534-2543.
50. Blumenthal RD. Chemosensitivity Volume 1: In Vitro Assays. Totowa: Human Press, 2005:1-231.
51. Blumenthal RD. Chemosensitivity Volume 2: In Vitro Models, Imaging, and Molecular Regulators. Totowa: Human Press, 2005:1-442.

Chapter 9

Molecular Mechanisms of ErbB2-Mediated Breast Cancer Chemoresistance

Ming Tan and Dihua Yu*

Introduction

The *erbB2* (also known as HER2 or *neu*) gene encodes a 185-kDa transmembrane glycoprotein, which belongs to the epidermal growth factor receptor (EGFR) family. ErbB2 is a receptor tyrosine kinase with intrinsic tyrosine kinase activity. The mammalian EGFR family comprises four receptors (EGFR, ErbB2, ErbB3, and ErbB4), which are derived from a series of gene duplications early in vertebrate evolution and are 40%–45% identical.[1] ErbB2 is the only EGFR family member for which no ligand has been found. This may be explained by the unique structure of the ErbB2 extracellular domain, which is not favorable for ligand binding.[2,3] Since ErbB2 extracellular domain is always in the open conformation, ErbB2 is the preferred binding partner of all ErbB receptors even as a monomer.[2-4] The binding of ErbB2 to other ErbB receptors results in increased signaling potency of the dimerized receptors through several means, including increased ligand affinity, increased coupling efficiency to signaling molecules, and decreased rate of receptor internalization.[5-8]

ErbB2 plays an important role in human malignancies. The *erbB2* gene is amplified or overexpressed in approximately 30% of human breast cancers[9] and in many other cancer types, including ovarian,[9] stomach,[10] bladder,[11] salivary,[12] and lung carcinomas.[13] Overwhelming evidence from numerous studies indicates that amplification or overexpression of ErbB2 disrupts normal cell-control mechanisms and gives rise of aggressive tumor cells.[14] Patients with ErbB2-overexpressing breast cancer have substantially lower overall survival rates and shorter disease-free intervals than patients whose cancer does not overexpress ErbB2. Moreover, overexpression of ErbB2 leads to increased breast cancer metastasis.[15-17] The important roles of ErbB2 in cancer progression render it a highly attractive target for therapeutic interventions of breast cancer.[18,19] The humanized ErbB2-targeting antibody trastuzumab (Herceptin, from Genentech)[20] was approved for the treatment of ErbB2-overexpressing breast cancers in 1998. In both phase II and phase III clinical trials, the antibody has shown remarkable therapeutic efficacy when given in combination with chemotherapeutic agents.[21,22] Trastuzumab represents an excellent example of the ErbB2-targeting therapies. Other efforts to develop ErbB2-targeting cancer therapies also have yielded promising results, such as E1A gene therapy,[23,24] single-chain antibodies,[25,26] and tyrosine kinase inhibitors,[27] just to name a few.

These novel targeted cancer therapies provide exciting new hope and opportunity for fighting breast cancers. However, the majority of patients with breast carcinoma still receive chemotherapy as a critical component of multimodality treatment.[28] Thus, understanding the effect and mechanisms of ErbB2 on chemosensitivity is important for anticancer agent selection and individualization of patient treatment, which are critical to the success of treatment of breast cancers.

*Corresponding Author: Dihua Yu—The University of Texas M.D. Anderson Cancer Center, Houston, Texas 77030, U.S.A. Email: dyu@mdanderson.org

Breast Cancer Chemosensitivity, edited by Dihua Yu and Mien-Chie Hung.

ErbB2 and Chemoresistance

Although the role and the mechanisms of ErbB2 overexpression on chemosensitivity still require intensive investigation, findings of many clinical and laboratory studies suggest that ErbB2 overexpression leads to increased chemoresistance to certain chemotherapeutic agents.

Clinical Studies

In a clinical study of breast cancer patients receiving adjuvant chemotherapy (cyclophosphamide, methotrexate, 5-fluorouracil, and prednisone [CMFP]), those with ErbB2-negative tumors showed a significantly greater rate of disease free survival in response to therapy than patients with ErbB2-positive tumors, indicating that overexpression of ErbB2 may play a role in resistance to chemotherapy.[29] In another study, it was found that tumors that overexpress ErbB2 are less responsive to cyclophosphamide, methotrexate, and 5-fluorouracil (CMF) adjuvant therapy regimen than tumors that express a normal amount of ErbB2 protein.[30] ErbB2 overexpression was also shown to have predictive value in epirubicin therapy in patients with advanced breast cancer.[31] It has been reported that tumors coexpressing ErbB2 and Ras proteins were less responsive to tamoxifen and CMF regimens than those expressing low levels of ErbB2 and Ras.[32] Recently, several groups reported that an elevated serum ErbB2 level predicted an unfavorable response to hormonal therapy or chemotherapy in patients with advanced metastatic breast cancer.[33-36] Supporting data also came from the clinical trials in which trastuzumab was combined with chemotherapeutic agents. In a phase III first-line study, the combination of trastuzumab and chemotherapy produced significantly greater clinical benefit than chemotherapy alone.[37] Similarly, a randomized phase II study comparing trastuzumab plus docetaxel with docetaxel alone in 188 patients with metastatic breast cancer yielded a statistically significant difference in terms of overall response rate, time to treatment progression and overall survival in favor of the combination.[38] These results suggest that ErbB2 overexpression plays a role in inducing chemoresistance.

Laboratory Findings

Data from laboratory studies have provided more direct evidence that ErbB2-overexpressing breast cancer cells are more resistant to certain chemotherapeutic agents than cells that do not overexpress ErbB2. A panel of human breast cancer cell lines expressing ErbB2 at different levels was tested for their sensitivity to paclitaxel and docetaxel.[39] Higher expression of ErbB2 in these cell lines correlated well with resistance to the drugs, and downregulation of cell-surface ErbB2 using an anti-ErbB2 monoclonal antibody significantly sensitized the cell lines to paclitaxel. The results indicate that overexpression of ErbB2 renders human breast cancer cells resistant to paclitaxel.[39] In another study, MDA-MB-435 human breast cancer cells[40] were stably transfected with the *erbB2* gene that led to increased ErbB2 expression and no change in the expression of the multidrug resistance gene *mdr1*, but the ErbB2-overexpressing transfectants were more resistant to paclitaxel and docetaxel than their parental cells.[41] This leads to the conclusion that overexpression of ErbB2 can lead to intrinsic paclitaxel and docetaxel resistance independent of the *mdr1*-mediated multidrug resistance mechanism. These findings are supported by results of other independent studies using multiple breast cancer cell lines that express high levels of ErbB2 protein.[42-46] Moreover, downregulation of ErbB2 with the antisense oligonucleotides in ErbB2-overexpressing BT-474 breast cancer cells suppressed ErbB2 overexpression by 60.5% and subsequently increased the sensitivity of these cells to adriamycin and paclitaxel 20.8- and 10.8-fold, respectively.[47] Furthermore, several other ErbB2-targeting approaches that downregulate ErbB2, such as humanized anti-ErbB2 antibodies,[48,49] tyrosine kinase inhibitors,[50] and adenovirus type 5 E1A,[51,52] all led to sensitization to certain chemotherapeutic agents in cultured cancer cells or in animal models. Taken together, these laboratory findings clearly indicate that ErbB2 overexpression is linked to resistance to particular chemotherapeutic agents.

The Existing Controversy

Despite the supporting evidence just described, current clinical and experimental data on the effect of ErbB2 overexpression on chemosensitivity remain controversial. Results of several clinical studies have suggested that ErbB2 overexpression does not necessarily lead to chemoresistance. One clinical study showed that ErbB2 expression does not predict response to docetaxel or sequential methotrexate and 5-fluorouracil in advanced breast cancers.[53] Another study showed that ErbB2 expression was not significantly associated with tumor response to neoadjuvant treatment with fluorouracil, adriamycin and cyclophoshamide (FAC) in 329 cases of breast cancer.[54] In yet another study, ErbB2-overexpressing breast cancers responded better to doxorubicin than did breast cancers expressing low levels of ErbB2.[55]

These contradictory clinical observations may be partly explained by intrinsic differences in the design of the studies. Since clinical studies are complicated processes, numerous factors may affect the outcome of the investigations. For example, amplification and overexpression of ErbB2 can be detected by fluorescence in situ hybridization, immunohistochemistry, or enzyme-linked immunosorbent assay on tumor tissue samples.[56] However, use of nonstandardized assay, subjectivity of the assay performer, limitations of techniques, differences in antibodies or DNA probes used, and differences in tissue treatment procedure have resulted in discordance in determining ErbB2 expression levels. In other words, the way in which the ErbB2 expression level was determined and defined is a very important factor that may affect the outcome of a study. Other factors that may affect outcome include the timing of treatment (neoadjuvant or adjuvant); the type of regimen (e.g., FAC or CMF); the treatment status of the patient (previously treated or untreated); patient's age (younger or older), menopausal status, and ethnic background; estrogen-receptor status of the tumor (positive or negative); and presence or absence of other genetic alterations that may interact with the ErbB2 receptor. Thus, any of these factors may lead to discordance in the outcomes of investigations on the role of ErbB2 in modulating chemosensitivity.[57]

The controversy over the role of ErbB2 in chemosensitivity also exists in laboratory studies. For example, In a laboratory study in which *erbB2*-transfected breast and ovarian cancer cells were used to determine the effect of ErbB2 overexpression on chemosensitivity, ErbB2 overexpression was found to be insufficient to induce intrinsic resistance to drugs, including paclitaxel.[58] Although this study involved the use of a series of *erbB2*-transfected breast cancer cells, including *erbB2*-transfected MDA-MB-435 breast cancer cells, it yielded results that apparently disagreed with those of an earlier study showing that ErbB2 can confer paclitaxel resistance in *erbB2*-transfected MDA-MB-435 cells.[41] The discrepancy may be explained partly by differences in ErbB2 expression levels between the *erbB2*-transfected MDA-MB-435 cells used in the two studies. In particular, the *erbB2* transfectants that showed paclitaxel resistance expressed very high levels of ErbB2 protein, similar to those expressed in the SKBR3 breast cancer cell line, which was established from an ErbB2-overexpressing primary breast tumor.[41] However, the *erbB2* transfectants that showed no paclitaxel resistance produced less than one-third of the level of ErbB2 protein expressed by SKBR3 cells.[58,57] Based on the data from these two independent studies, we suggest that ErbB2 overexpression must reach a threshold level in breast cancer cells before they become resistant to paclitaxel. This might provide a reasonable explanation for the discrepancy between the findings of the two studies.

Molecular Mechanisms of ErbB2-Mediated Chemoresistance

The role of ErbB2 in chemoresistance is a problem of great clinical importance, and the observational data are abundant, as described above. Because of the controversy and the complexity of this problem, however, our knowledge on the molecular mechanisms of how ErbB2 confers chemoresistance is still limited. It is generally believed that breast cancer cells overexpressing ErbB2 are intrinsically resistant to DNA-damaging agents such as cisplatin as the result of an altered cell-cycle checkpoint, altered DNA repair mechanisms, and altered apoptosis responses.[59,60]

Apoptosis is a predominant mechanism by which cancer chemotherapeutic agents kill cells.[61,62] The failure of cancer cells to detect chemotherapeutic agent-induced damage and to activate apoptosis may lead to multidrug resistance. The results from our studies indicate that overexpression of ErbB2 renders human breast cancer cells resistant to paclitaxel.[39,41] While studying the molecular mechanisms of ErbB2-mediated paclitaxel resistance, we found that treatment of MDA-MB-435 breast cancer cells with paclitaxel caused them to undergo apoptosis, which was inhibited in their paired transfectants overexpressing ErbB2. Further investigation showed that paclitaxel-treatment induced a premature activation of $p34^{Cdc2}$ kinase, the mitotic serine/threonine kinase that binds to cyclin B and also plays an important role in cancer cell apoptosis.[63] The premature activation of $p34^{Cdc2}$ led to mitotic catastrophe, i.e., apoptosis. A chemical inhibitor of $p34^{Cdc2}$ and a dominant-negative mutant of $p34^{Cdc2}$ blocked paclitaxel-induced apoptosis in these cells, indicating that a premature Cdc2 activation is a prerequisite for paclitaxel-induced apoptosis in MDA-MB-435 cells. We demonstrated that overexpression of ErbB2 in MDA-MB-435 cells transcriptionally upregulates $p21^{Cip1}$, which associates with $p34^{Cdc2}$, inhibits paclitaxel-mediated $p34^{Cdc2}$ activation, delays cell entrance to G_2/M phase, and thereby inhibits paclitaxel-induced apoptosis. Therefore, upregulation of $p21^{Cip1}$ by ErbB2 inhibits $p34^{Cdc2}$ and deregulates G_2/M checkpoint that contributes to resistance to paclitaxel-induced apoptosis in ErbB2-overexpressing breast cancer cells.[64] In addition, we found that phosphorylation on tyrosine (Tyr)15 of Cdc2, the crucial inhibitory phosphorylation site of Cdc2, is elevated in ErbB2-overexpressing breast cancer cells and primary tumors independent of Wee1, Cdc25C, and $p21^{Cip1}$. We showed that ErbB2 binds to and colocalizes with cyclin B-Cdc2 complexes and can directly and specifically phosphorylate Cdc2 on Tyr15. Increased Cdc2-Tyr15 phosphorylation in ErbB2-overexpressing cells corresponds with delayed M phase entry and reduced sensitivity to paclitaxel-induced apoptosis. Expression of a kinase-defective ErbB2 in MDA-MB-435 breast cancer cells or expression of a nonphosphorylatable Cdc2Y15F mutant in ErbB2 gene transfectant of the MDA-MB-435 cells render the cells sensitive to paclitaxel-induced apoptosis. These data indicate that the increased Cdc2-Tyr15 phosphorylation by ErbB2 tyrosine kinase may be a pertinent cell-cycle checkpoint defect that is involved in paclitaxel resistance in ErbB2-overexpressing breast cancers. Taken together, these findings indicate that ErbB2 overexpression can confer breast cancer cell resistance to paclitaxel-induced apoptosis by inhibiting Cdc2 activation through at least two mechanisms: (1) ErbB2 kinase-independent p21 upregulation and (2) ErbB2 kinase-dependent direct phosphorylation of Cdc2 on Tyr15 (Fig. 1). This model is further supported by results of a study on patient samples showing that ErbB2 overexpression is correlated with $p21^{Cip1}$ upregulation and with increased $p34^{Cdc2}$-Tyr15 phosphorylation in breast tumors.[65]

Other molecular mechanisms may also underlie ErbB2-mediated paclitaxel resistance. In addition, the molecular mechanisms of ErbB2-mediated resistance to different chemotherapeutic agents could be different. For example, other molecular mechanisms of ErbB2-mediated chemoresistance may involve activation of the PI3K/Akt pathway by ErbB2, which leads to increased cancer cell survival,[46] the estrogen receptor-ErbB2 cross-talk in ErbB2-positive breast cancer cells,[66,67] and coexpression of ErbB2 with other ErbB family receptors.[44] Although hints of how ErbB2-overexpressing breast tumors evade chemotherapeutic agent-induced apoptosis and develop chemoresistance are beginning to surface, it is obvious that many questions remain to be answered.

Targeting ErbB2 to Overcome Chemoresistance

Numerous lines of laboratory and clinical evidence indicate that ErbB2-targeting agents can sensitize the response of tumor cells to chemotherapeutic agents; therefore, developing ErbB2-targeting strategies to improve the therapy of ErbB2-overexpressing breast cancer remains a high priority. During the last decade, several exciting techniques have been developed to target ErbB2. Although some are still under investigation, many studies have shown that

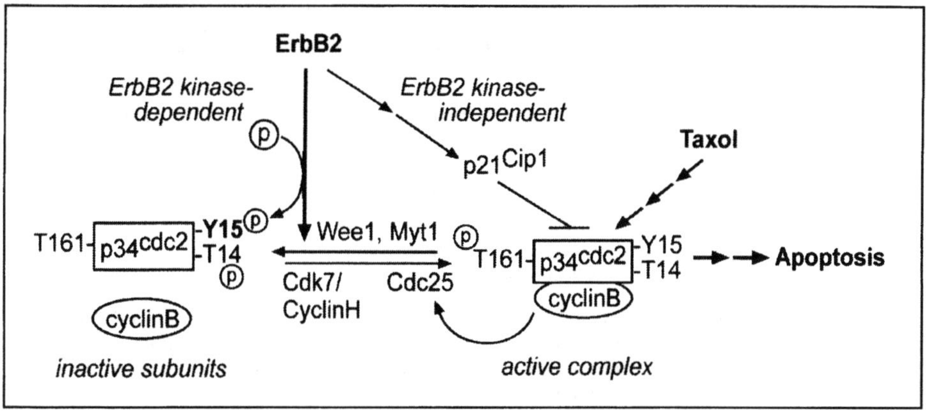

Figure 1. Model of inhibition of $p34^{Cdc2}$ activation and apoptosis by ErbB2. $P34^{Cdc2}$ remains inactive without binding to cyclin B as it is phosphorylated on Thr14 by Myt1 and Tyr15 by Wee1. Activation of $p34^{Cdc2}$ occurs by accumulation and binding of cyclin B, dephosphorylation of Thr14 and Tyr15 by Cdc25C, and phosphorylation of Thr161 by Cdk7/cyclin H. The ErbB2 tyrosine kinase inhibits $p34^{Cdc2}$ activity by directly phosphorylating Cdc2 on Tyr15, which requires ErbB2 kinase activity. On the other hand, ErbB2 can upregulate $p21^{Cip1}$, which inhibits $p34^{Cdc2}$ activities independent of the ErbB2 kinase activity. Since paclitaxel (Taxol)-induced apoptosis requires the activation of $p34^{Cdc2}$, overexpression of ErbB2 can confer resistance to paclitaxel-induced apoptosis by inhibiting $p34^{Cdc2}$ activation through at least two mechanisms, direct phosphorylation of Cdc2-Tyr15 and upregulation of $p21^{Cip1}$. (Figure reprinted from the article by Tan et al, Molecular Cell 9:993-1004, ©2002 Elsevier, with permission).

these ErbB2-targeting techniques not only inhibit tumor growth, but also lead to chemosensitization of ErbB2-overexpressing cancer cells (Table 1).

One of the most successful example is Trastuzumab, which is a humanized antibody that binds to the extracellular domain of ErbB2.[20] Recent studies have shown that, in addition to inhibition of ErbB2 signaling, trastuzumab has other functions, such as activation of the *PTEN* tumor suppressor gene,[68] induction of $p27^{Kip1}$, and induction of G_1 cell cycle arrest.[69,70] Trastuzumab has demonstrated tumor-inhibitory and chemosensitizing effects for paclitaxel and several other chemotherapeutic agents in preclinical studies and in phase II and phase III clinical trials.[21,22,48,71] These results represent an excellent example of anti-ErbB2 antibody-mediated chemosensitization. On the basis of our understanding of the mechanisms of ErbB2-mediated paclitaxel resistance, we investigated the mechanisms by which trastuzumab enhances the antitumor effects of paclitaxel in vitro and in vivo. We found that treatment of ErbB2-overexpressing cells with trastuzumab can inhibit ErbB2-mediated Cdc2-Tyr15 phosphorylation and $p21^{Cip1}$ upregulation, which allows effective $p34^{Cdc2}$ activation and induction of apoptosis upon paclitaxel treatment.[49,65]

In addition, the past 15 years have witnessed the development of several effective ErbB2-tageting strategies. These include, but not limited to, adenovirus type 5 E1A protein (please refer to the chapter by Liao and Hung for more details),[51,72-75,76] ErbB2-specific tyrosine kinase inhibitors such as emodin,[50] HKI-272,[27] and GW572016,[77] anti-ErbB2 intracellular single-chain antibodies (sFv),[25,26,78,79] ErbB2-targeting antisense oligonucleotide,[47,80-83] rationally designed anti-ErbB2 peptide mimetics (AHNP),[84-86] all-trans retinoic acid (ATRA) and fenretinide (4-HPR).[87] These ErbB2-tageting strategies either have been shown to have chemosensitization effect or are currently under testing for chemosensitization in ErbB2-overexpressing breast cancer cell lines, animal models, or clinical trials.[57]

These approaches discussed above have already shown promise in overcoming ErbB2-mediated chemoresistance in either laboratory or clinical studies. Other new technologies are also under development. For example, ErbB2-targeting small interfering RNAs (siRNAs)

Table 1. Possible ways to overcome ErbB2-mediated chemoresistance

ErbB2-Targetng Agents	ErbB2-Targeting Mechanism	Chemosensitization Efficacy
Anti-ErbB2 antibodies		
Trastuzumab (Herceptin)	Binds to the extracellular domain of ErbB2 and downregulates the cell membrane ErbB2, block ErbB2 initiated cell signaling	Demonstrated in animal and in Phase II and Phase III clinical trials
Pertuzumab (Omnitarg)	Prevent HER2 homodimer or heterodimer formation and inhibits ErbB2 downstream signaling pathways	Phase Ib studies are planned with pertuzumab in combination with docetaxel in breast cancer
Adenovirus type 5 E1A	Transcriptionally repress erbB2 gene expression	Tested in Phase I clinical trial
Tyrosine kinase inhibitors	Inhibits the tyrosine kinase activity of ErbB2 receptor tyrosine kinase	Tested in laboratory studies
Retinoic acid	Downregulation of erbB2 mRNA and protein expression	Tested in laboratory studies
ErbB2 antisense oligonucleotides	Block erbB2 transcription	Tested in laboratory studies
Small anti-ErbB2 peptide mimic	Binds to the extracellular domain of ErbB2 and downregulates the cell membrane ErbB2, block ErbB2 initiated cell signaling	Tested in laboratory studies
SiRNA	Block ErbB2 transcription	To be tested

silence *erbB2* mRNA by using double-stranded RNA oligonucleotides.[88-90] One study showed that ovarian cancer cells and breast cancer cells infected with a retrovirus expressing anti-ErbB2 siRNA exhibited effective downregulation of ErbB2 expression and slower proliferation, increased apoptosis, increased G_0/G_1 arrest, and decreased tumor growth in mouse models.[89] On the basis of these results, ErbB2-targeting siRNA may have great potential in overcoming ErbB2-mediated chemoresistance in ErbB2-overexpressing breast cancer cells. Additionally, attempts are being made to modulate existing chemotherapeutic agents so that they can overcome resistance.[91] These combined efforts held great potential to improve the therapies for patients with ErbB2-overexpressing breast tumors.

Future Investigation

Understanding the role of ErbB2 in chemoresistance is important and has significant clinical relevance. Although many studies have been done during the last decade, our current knowledge on the role of ErbB2 in cancer chemosensitivity is still limited. To develop more effective treatment for patients with ErbB2-overexpressing breast cancers, scientists and physicians need to team up to pursue the following and other related issues.

Well Designed and Performed Laboratory and Clinical Studies

Current data from our studies and those of others indicate that ErbB2 renders breast cancers resistant to DNA-damaging agents such as cisplatin[60] and to the microtubule-stabilizing agents taxenes.[39,41] To better understand the role of ErbB2 in chemosensitivity, we must determine whether ErbB2 overexpression also renders breast cancers resistant to other chemotherapeutic agents frequently used for treating such cancers. This should be investigated (1) in carefully designed laboratory studies that use ErbB2-overexpressing breast cancer cell lines that express ErbB2 at levels similar to those detected in primary breast tumors and compare between cell lines having similar genetic backgrounds, and (2) in well-controlled and -defined large-scale clinical trials that can adequately assess the impact of the various factors important for the chemoresponse of tumors.

Other Possible Mechanisms for ErbB2-Mediated Chemoresistance

Many mechanisms of drug resistance in various tumor cells are well established, including enhanced drug metabolism, reduced drug accumulation, drug target amplification, and repair of damaged targets and apoptosis resistance.[61] Our data indicate that in breast cancer cells, ErbB2 can block paclitaxel-induced apoptosis by upregulating $p21^{cip1}$ and by directly phosphorylating the inhibitory tyrosine on $p34^{Cdc2}$. Both mechanisms lead to the inhibition of $p34^{Cdc2}$ kinase, which is required for paclitaxel-induced apoptosis. It has also been reported that chemotherapeutic agents-induced apoptosis depends on a balance between cell-cycle checkpoints and DNA-repair mechanisms. In addition, downstream signaling of the ErbB2 receptor tyrosine kinase, e.g., Akt activation, also plays a critical role in ErbB2-mediated chemoresistance.[46] However, the downstream signal-transduction cascades by which ErbB2 affects chemosensitivity still require more investigation. Most ErbB2-overexpressing breast cancers also express other ErbB receptors, which can form heterodimers with ErbB2. An interconnected network of ErbB signaling pathways may be activated by the overexpression of ErbB2, and that determines tumor cell response to chemotherapeutic stress. It is, therefore, important to perform in-depth investigations into the role of ErbB2 receptor signaling in the regulation of stress-responsive genes. A better understanding of other possible mechanisms underlying ErbB2-mediated chemoresistance is also critical to the development of better alternative therapeutic strategies to overcome intrinsic chemoresistance.

Tailoring Therapeutic Strategies to Individual Patients

One of the important aspects of future research efforts on ErbB2-mediated chemoresistance is to identify other unknown genetic alterations that interact with ErbB2 and contribute to ErbB2-mediated chemoresistance. ErbB2-mediated chemoresistance is a very complicated problem, as it may be specific to tumor, cell, chemotherapeutic agent, regimen, timing, population, or patient age.[57] With today's powerful genomic, proteomic, kinomic, and tissue microarray technologies, it has become feasible to identify factors that interact with ErbB2 and contribute to ErbB2-mediated chemoresistance from tumors of individual patient. This information can then be used to develop therapeutic strategies that are tailored for individual patient and are based on the ErbB2 biology and any identified relevant factors. These individually tailored therapeutic strategies will greatly maximize the therapeutic benefit for individual breast cancer patients.

New Approaches for Overcoming ErbB2-Mediated Chemoresistance

We have discussed several promising strategies for overcoming ErbB2-mediate chemoresistance (Table 1). New and more effective strategies are still in demand and can be developed. Recently, nanotechnology has emerged as a promising strategy for drug delivery. Nanotechnology-based delivery systems can be used to achieve cellular or tissue targeted drug delivery, to improve drug bioavailability, to sustain drug effect in target tissue, to solubilize drugs for intravascular delivery, and to improve the stability of therapeutic agents.[92] With nanotechnology-based delivery systems, chemotherapeutic agents such as paclitaxel, doxorubicin, and 5-fluorouracil, have been shown to achieve improved efficacy to inhibit tumor growth in vitro and in vivo.[93,94] In addition, nanotechnology-based system has been shown to potentially improve the antitumor efficacy of ErbB2-targeting agents, such as trastuzumab.[95] It is foreseeable that by using these new techniques, which will produce highly efficient chemotherapeutic agents and ErbB2-targeting strategies, we may conquer ErbB2-mediated chemoresistance more effectively.

Acknowledgements

We apologize to those whose work could not be cited owing to space limitations. We thank Katherine Hale for editorial assistance. D. Yu's laboratory is supported by NIH grants RO1-CA60488, RO1-CA109570, 1PO1-CA099031-1 project 4, and the Department of Defense grant DAMD17-02-1-0462.

References

1. Stein RA, Staros JV. Evolutionary analysis of the ErbB receptor and ligand families. J Mol Evol 2000; 50(5):397-412.
2. Cho HS, Mason K, Ramyar KX et al. Structure of the extracellular region of HER2 alone and in complex with the Herceptin Fab. Nature 2003; 421(6924):756-760.
3. Garrett TP, McKern NM, Lou M et al. The crystal structure of a truncated ErbB2 ectodomain reveals an active conformation, poised to interact with other ErbB receptors. Mol Cell 2003; 11(2):495-505.
4. Graus-Porta D, Beerli RR, Daly JM et al. ErbB-2, the preferred heterodimerization partner of all ErbB receptors, is a mediator of lateral signaling. EMBO J 1997; 16:1647-1655.
5. Pinkas-Kramarski R, Soussan L, Waterman H et al. Diversification of Neu differentiation factor and epidermal growth factor signaling by combinatorial receptor interactions. EMBO J 1996; 15:2452-2467.
6. Riese IInd DJ, van Raaij TM, Plowman GD et al. The cellular response to neuregulins is governed by complex interactions of the erbB receptor family. Mol Cell Biol 1995; 15(10):5770-5776.
7. Karunagaran D, Tzahar E, Beerli RR et al. ErbB-2 is a common auxiliary subunit of NDF and EGF receptors: Implications for breast cancer. EMBO J 1996; 15:254-264.
8. Jones JT, Akita RW, Sliwkowski MX. Binding specificities and affinities of egf domains for ErbB receptors. FEBS Lett 1999; 447(2-3):227-231.
9. Slamon DJ, Godolphin W, Jones LA et al. Studies of the HER-2/neu proto-oncogene in human breast and ovarian cancer. Science 1989; 244(4905):707-712.
10. Lemoine NR, Jain S, Silvestre F et al. Amplification and overexpression of the EGF receptor and c-erbB-2 proto-oncogenes in human stomach cancer. Br J Cancer 1991; 64(1):79-83.
11. Sauter G, Moch H, Moore D et al. Heterogeneity of erbB-2 gene amplification in bladder cancer. Cancer Res 1993; 53(10 Suppl):2199-2203.
12. Stenman G, Sandros J, Nordkvist A et al. Expression of the ERBB2 protein in benign and malignant salivary gland tumors. Genes Chromosomes Cancer 1991; 3(2):128-135.
13. Tateishi M, Ishida T, Mitsudomi T et al. Prognostic value of c-erbB-2 protein expression in human lung adenocarcinoma and squamous cell carcinoma. Eur J Cancer 1991; 27(11):1372-1375.
14. Hung M-C, Schechter AL, Chevray P-YM et al. Molecular cloning of the neu gene: absence of gross structural alteration in oncogenic alleles. Proc Natl Acad Sci USA 1986; 83:261-264.
15. Tan M, Yao J, Yu D. Overexpression of the c-erbB-2 gene enhanced intrinsic metastatic potential in human breast cancer cells without increasing their transformation abilities. Cancer Res 1997; 57:1199-1205.
16. Moody SE, Sarkisian CJ, Hahn KT et al. Conditional activation of Neu in the mammary epithelium of transgenic mice results in reversible pulmonary metastasis. Cancer Cell 2002; 2(6):451-461.
17. Holbro T, Civenni G, Hynes NE. The ErbB receptors and their role in cancer progression. Exp Cell Res 2003; 284(1):99-110.
18. Yarden Y, Sliwkowski MX. Untangling the ErbB signalling network. Nat Rev Mol Cell Biol 2001; 2(2):127-137.
19. Shawver LK, Slamon D, Ullrich A. Smart drugs: Tyrosine kinase inhibitors in cancer therapy. Cancer Cell 2002; 1(2):117-123.
20. Carter P, Presta L, Gorman CM et al. Humanization of an anti-p185HER2 antibody for human cancer therapy. Proc Natl Acad Sci USA 1992; 89(10):4285-4289.
21. Pegram MD, Lipton A, Hayes DF et al. Phase II study of receptor-enhanced chemosensitivity using recombinant humanized anti-p185HER2/neu monoclonal antibody plus cisplatin in patients with HER2/neu-overexpressing metastatic breast cancer refractory to chemotherapy treatment. J Clin Oncol 1998; 16(8):2659-2671.

22. Dickman S. Antibodies stage a comeback in cancer treatment. Science 1998; 280:1196-1197.
23. Hortobagyi GN, Ueno NT, Xia W et al. Cationic liposome-mediated E1A gene transfer to human breast and ovarian cancer cells and its biologic effects: A phase I clinical trial. J Clin Oncol 2001; 19(14):3422-3433.
24. Madhusudan S, Tamir A, Bates N et al. A multicenter Phase I gene therapy clinical trial involving intraperitoneal administration of E1A-lipid complex in patients with recurrent epithelial ovarian cancer overexpressing HER-2/neu oncogene. Clin Cancer Res 2004; 10(9):2986-2996.
25. Alvarez RD, Barnes MN, Gomez-Navarro J et al. A cancer gene therapy approach utilizing an anti-erbB-2 single-chain antibody-encoding adenovirus (AD21): A phase I trial. Clin Cancer Res 2000; 6(8):3081-3087.
26. Azemar M, Djahansouzi S, Jager E et al. Regression of cutaneous tumor lesions in patients intratumorally injected with a recombinant single-chain antibody-toxin targeted to ErbB2/HER2. Breast Cancer Res Treat 2003; 82(3):155-164.
27. Rabindran SK, Discafani CM, Rosfjord EC et al. Antitumor activity of HKI-272, an orally active, irreversible inhibitor of the HER-2 tyrosine kinase. Cancer Res 2004; 64(11):3958-3965.
28. Buchholz TA, Hunt KK, Whitman GJ et al. Neoadjuvant chemotherapy for breast carcinoma: Multidisciplinary considerations of benefits and risks. Cancer 2003; 98(6):1150-1160.
29. Allred DC, Clark GM, Tandon AK et al. Her-2/neu in node-negative breast cancer: prognostic significance of overexpression influenced by the presence of in situ carcinoma. J Clin Oncol 1992; 10:599-605.
30. Gusterson BA, Gelber RD, Goldhirsch A et al. Prognostic importance of c-erbB-2 expression in breast cancer. J Clin Oncol 1992; 10(7):1049-1056.
31. Jarvinen TA, Holli K, Kuukasjarvi T et al. Predictive value of topoisomerase II alpha and other prognostic factors for epirubicin chemotherapy in advanced breast cancer. Br J Cancer 1998; 77:2267-2273.
32. Giani C, Finocchiaro G. Mutation rate of the CDKN2 gene in malignant gliomas. Cancer Res 1994; 54(24):6338-6339.
33. Fehm T, Maimonis P, Katalinic A et al. The prognostic significance of c-erbB-2 serum protein in metastatic breast cancer. Oncology 1998; 55:33-38.
34. Colomer R, Montero S, Lluch A et al. Circulating HER2 extracellular domain and resistance to chemotherapy in advanced breast cancer. Clin Cancer Res 2000; 6:2356-2362.
35. Classen S, Kopp R, Possinger K et al. Clinical relevance of soluble c-erbB-2 for patients with metastatic breast cancer predicting the response to second-line hormone or chemotherapy. Tumour Biol 2002; 23(2):70-75.
36. Colomer R, Llombart-Cussac A, Lluch A et al. Biweekly paclitaxel plus gemcitabine in advanced breast cancer: Phase II trial and predictive value of HER2 extracellular domain. Ann Oncol 2004; 15(2):201-206.
37. Slamon DJ, Leyland-Jones B, Shak S et al. Use of chemotherapy plus a monoclonal antibody against HER2 for metastatic breast cancer that overexpresses HER2. N Engl J Med 2001; 344(11):783-792.
38. Extra JM, Cognetti F, Chan S et al. First-line trastuzumab (Herceptin') plus docetaxel versus docetaxel alone in women with HER2-positive metastatic breast cancer (MBC): Results from a randomised phase II trial (M77001). Br Cancer Res Treat 2003; 82(suppl 1):S47.
39. Yu D, Liu B, Jing T et al. Overexpression of both p185c-erbB2 and p170mdr-1 renders breast cancer cells highly resistant to taxol. Oncogene 1998; 16(16):2087-2094.
40. Sellappan S, Grijalva R, Zhou X et al. Lineage infidelity of MDA-MB-435 cells: Expression of melanocyte proteins in a breast cancer cell line. Cancer Res 2004; 64(10):3479-3485.
41. Yu D, Liu B, Tan M et al. Overexpression of c-erbB-2/neu in breast cancer cells confers increased resistance to Taxol via mdr-1-independent mechanisms. Oncogene 1996; 13:1359-1365.
42. Sabbatini AR, Basolo F, Valentini P et al. Induction of multidrug resistance (MDR) by transfection of MCF-10A cell line with c-Ha-ras and c-erbB-2 oncogenes. Int J Cancer 1994; 59(2):208-211.
43. Ciardiello F, Caputo R, Pomatico G et al. Resistance to taxanes is induced by c-erbB-2 overexpression in human MCF-10A mammary epithelial cells and is blocked by combined treatment with an antisense oligonucleotide targeting type I protein kinase A. Int J Cancer 2000; 85(5):710-715.
44. Chen X, Yeung TK, Wang Z. Enhanced drug resistance in cells coexpressing ErbB2 with EGF receptor or ErbB3. Biochem Biophys Res Commun 2000; 277(3):757-763.
45. Witters LM, Santala SM, Engle L et al. Decreased response to paclitaxel versus docetaxel in HER-2/neu transfected human breast cancer cells. Am J Clin Oncol 2003; 26(1):50-54.
46. Knuefermann C, Lu Y, Liu B et al. HER2/PI-3K/Akt activation leads to a multidrug resistance in human breast adenocarcinoma cells. Oncogene 2003; 22(21):3205-3212.

47. Tanabe K, Kim R, Inoue H et al. Antisense Bcl-2 and HER-2 oligonucleotide treatment of breast cancer cells enhances their sensitivity to anticancer drugs. Int J Oncol 2003; 22(4):875-881.
48. Baselga J, Norton L, Albanell J et al. Recombinant humanized anti-HER2 antibody (Herceptin®) enhances the antitumor activity of paclitaxel and doxorubicin against HER2/neu overexpressing human breast cancer xenografts. Cancer Res 1998; 58:2825-2831.
49. Lee S, Yang W, Lan KH et al. Enhanced sensitization to taxol-induced apoptosis by herceptin pretreatment in ErbB2-overexpressing breast cancer cells. Cancer Res 2002; 62(20):5703-5710.
50. Zhang L, Lau YK, Xia W et al. Tyrosine kinase inhibitor emodin suppresses growth of HER-2/neu-overexpressing breast cancer cells in athymic mice and sensitizes these cells to the inhibitory effect of paclitaxel. Clin Cancer Res 1999; 5:343-353.
51. Ueno NT, Yu D, Hung MC. Chemosensitization of HER-2/neu-overexpressing human breast cancer cells to paclitaxol (Taxol) by adenovirus type 5 E1A. Oncogene 1997; 15:953-960.
52. Ueno NT, Bartholomeusz C, Xia W et al. Systemic gene therapy in human xenograft tumor models by liposomal delivery of the E1A gene. Cancer Res 2002; 62(22):6712-6716.
53. Sjostrom J, Collan J, von Boguslawski K et al. C-erbB-2 expression does not predict response to docetaxel or sequential methotrexate and 5-fluorouracil in advanced breast cancer. Eur J Cancer 2002; 38(4):535-542.
54. Rozan S, Vincent-Salomon A, Zafrani B et al. No significant predictive value of c-erbB-2 or p53 expression regarding sensitivity to primary chemotherapy or radiotherapy in breast cancer. Int J Cancer 1998; 79:27-33.
55. Porter-Jordan K, Lippman ME. Overview of the biologic markers of breast cancer. Breast Cancer 1994; 8:73-100.
56. Ross JS, Fletcher JA, Bloom KJ et al. HER-2/neu testing in breast cancer. Am J Clin Pathol 2003; 120(Suppl):S53-71.
57. Yu D, Hung M-C. Role of erbB2 in breast cancer chemosensitivity. BioEssays 2000; 22:673-680.
58. Pegram MD, Finn RS, Arzoo K et al. The effect of HER2/neu overexpression on chemotherapeutic drug sensitivity in human breast and ovarian cancer cells. Oncogene 1997; 15:537-547.
59. Pietras RJ, Pegram MD, Finn RS et al. Remission of human breast cancer xenografts on therapy with humanized monoclonal antibody to HER-2 receptor and DNA-reactive drugs. Oncogene 1998; 17:2235-2249.
60. Alaoui-Jamali MA, Paterson J, Al Moustafa AE et al. The role of erbB-2 tyrosine kinase receptor in cellular intrinsic chemoresistance: Mechanisms and implications. Biochem Cell Biol 1997; 75:315-325.
61. Fisher DE. Apoptosis in cancer therapy: Crossing the threshold. Cell 1994; 78:539-542.
62. Wang CY, Cusack Jr JC, Liu R et al. Control of inducible chemoresistance: Enhanced anti-tumor therapy through increased apoptosis by inhibition of NF-kappaB. Nat Med 1999; 5(4):412-417.
63. Gao CY, Zelenka PS. Induction of cyclin B and H1 kinase activity in apoptotic PC12 cells. Exp Cell Res 1995; 219:612-618.
64. Yu D, Jing T, Liu B et al. Overexpression of ErbB2 blocks Taxol-induced apoptosis by upregulation of p21Cip1, which inhibits p34Cdc2 kinase. Mol Cell 1998; 2(5):581-591.
65. Yang W, Klos KS, Zhou X et al. ErbB2 overexpression in human breast carcinoma is correlated with p21Cip1 up-regulation and tyrosine-15 hyperphosphorylation of p34Cdc2: Poor responsiveness to chemotherapy with cyclophoshamide methotrexate, and 5-fluorouracil is associated with Erb2 overexpression and with p21Cip1 overexpression. Cancer 2003; 98(6):1123-1130.
66. Osborne CK, Bardou V, Hopp TA et al. Role of the estrogen receptor coactivator AIB1 (SRC-3) and HER-2/neu in tamoxifen resistance in breast cancer. J Natl Cancer Inst 2003; 95(5):353-361.
67. Shou J, Massarweh S, Osborne CK et al. Mechanisms of tamoxifen resistance: Increased estrogen receptor-HER2/neu cross-talk in ER/HER2-positive breast cancer. J Natl Cancer Inst 2004; 96(12):926-935.
68. Nagata Y, Lan KH, Zhou X et al. PTEN activation contributes to tumor inhibition by trastuzumab, and loss of PTEN predicts trastuzumab resistance in patients. Cancer Cell 2004; 6(2):117-127.
69. Le XF, Claret FX, Lammayot A et al. The role of cyclin-dependent kinase inhibitor p27Kip1 in anti-HER2 antibody-induced G1 cell cycle arrest and tumor growth inhibition. J Biol Chem 2003; 278(26):23441-23450.
70. Nahta R, Takahashi T, Ueno NT et al. P27(kip1) down-regulation is associated with trastuzumab resistance in breast cancer cells. Cancer Res 2004; 64(11):3981-3986.
71. Baselga J, Tripathy D, Mendelsohn J et al. Phase II study of weekly intravenous recombinant humanized anti-p185HER2 monoclonal antibody in patients with HER2/neu-overexpressing metastatic breast cancer. J Clin Oncol 1996; 14(3):737-744.
72. Yu D, Scorsone K, Hung M-C. Adenovirus type 5 E1A gene products act as transformation suppressors of the neu oncogene. Mol Cell Biol 1991; 11(3):1745-1750.

73. Yu DH, Hamada JI, Zhang H et al. Mechanisms of c-erbB2/neu oncogene-induced metastasis and repression of metastatic properties by adenovirus 5 E1A gene products. Oncogene 1992; 7:2263-2270.
74. Yu D, Suen T-C, Yan D-H et al. Transcriptional repression of the neu protooncogene by the adenovirus 5 E1A gene products. Proc Natl Acad Sci USA 1990; 87(12):4499-4503.
75. Yu D, Matin A, Xia W et al. Liposome-mediated in vivo E1A gene transfer suppressed dissemination of ovarian cancer cells that overexpress HER-2/neu. Oncogene 1995; 11:1383-1388.
76. Liao Y, Zou YY, Xia WY et al. Enhanced paclitaxel cytotoxicity and prolonged animal survival rate by a nonviral-mediated systemic delivery of E1A gene in orthotopic xenograft human breast cancer. Cancer Gene Ther 2004; 11(9):594-602.
77. Spector NL, Xia W, Burris IIIrd H et al. Study of the biological effects of lapatinib, a reversible inhibitor of ErbB1 and ErbB2 tyrosine kinases, on tumor growth and survival pathways in patients with advanced malignancies. J Clin Oncol 2005; (Epub ahead of print).
78. Deshane J, Loechel F, Conry RM et al. Intracellular single-chain antibody directed against erbB2 down-regulates cell surface erbB2 and exhibits a selective anti-proliferative effect in erbB2 overexpressing cancer cell lines. Gene Ther 1994; 1(5):332-337.
79. Barnes MN, Deshane JS, Siegal GP et al. Novel gene therapy strategy to accomplish growth factor modulation induces enhanced tumor cell chemosensitivity. Clin Cancer Res 1996; 2:1089-1095.
80. Rait AS, Pirollo KF, Xiang L et al. Tumor-targeting, systemically delivered antisense HER-2 chemosensitizes human breast cancer xenografts irrespective of HER-2 levels. Mol Med 2002; 8(8):475-486.
81. Rait AS, Pirollo KF, Ulick D et al. HER-2-targeted antisense oligonucleotide results in sensitization of head and neck cancer cells to chemotherapeutic agents. Ann NY Acad Sci 2003; 1002:78-89.
82. Waterhouse DN, Dragowska WH, Gelmon KA et al. Pharmacodynamic behavior of liposomal antisense oligonucleotides targeting Her-2/neu and vascular endothelial growth factor in an ascitic MDA435/LCC6 human breast cancer model. Cancer Biol Ther 2004; 3(2):197-204.
83. Funato T, Kozawa K, Fujimaki S et al. Increased sensitivity to cisplatin in gastric cancer by antisense inhibition of the her-2/neu (c-erbB-2) gene. Chemotherapy 2001; 47(4):297-303.
84. Park BW, Berezov A, Wu CW et al. Rationally designed anti-HER2/neu peptide mimetic disables $p185^{HER2/neu}$ tyrosine kinases in vitro and in vivo. Nat Biotech 2000; 18:194-198.
85. Berezov A, Chen J, Liu Q et al. Disabling receptor ensembles with rationally designed interface peptidomimetics. J Biol Chem 2002; 277(31):28330-28339.
86. Murali R, Liu Q, Cheng X et al. Antibody like peptidomimetics as large scale immunodetection probes. Cell Mol Biol 2003; 49(2):209-216.
87. Grunt T, Dittrich E, Ofterdinger M et al. Effects of retinoic acid and fenretinide on the c-erbB-2 expression, growth and cisplatin sensitivity of breast cancer cells. Br J Cancer 1998; 78:79-87.
88. Mellinghoff IK, Vivanco I, Kwon A et al. HER2/neu kinase-dependent modulation of androgen receptor function through effects on DNA binding and stability. Cancer Cell 2004; 6(5):517-527.
89. Yang G, Cai KQ, Thompson-Lanza JA et al. Inhibition of breast and ovarian tumor growth through multiple signaling pathways by using retrovirus-mediated small interfering RNA against Her-2/neu gene expression. J Biol Chem 2004; 279(6):4339-4345.
90. Choudhury A, Charo J, Parapuram SK et al. Small interfering RNA (siRNA) inhibits the expression of the Her2/neu gene, upregulates HLA class I and induces apoptosis of Her2/neu positive tumor cell lines. Int J Cancer 2004; 108(1):71-77.
91. Hayes DF, Henderson IC, Shapiro CL. Treatment of metastatic breast cancer: Present and future prospects. Seminars in Oncol 1995; 22(2):5-21.
92. Panyam J, Labhasetwar V. Biodegradable nanoparticles for drug and gene delivery to cells and tissue. Adv Drug Deliv Rev 2003; 55(3):329-347.
93. BrannonPeppas L, Blanchette JO. Nanoparticle and targeted systems for cancer therapy. Adv Drug Deliv Rev 2004; 56(11):1649-1659.
94. Garber K. Improved Paclitaxel formulation hints at new chemotherapy approach. J Natl Cancer Inst 2004; 96(2):90-91.
95. Ito A, Kuga Y, Honda H et al. Magnetite nanoparticle-loaded anti-HER2 immunoliposomes for combination of antibody therapy with hyperthermia. Cancer Lett 2004; 212(2):167-175.

Chapter 10

Estrogen Receptors in Resistance to Hormone Therapy

Matthew H. Herynk and Suzanne A.W. Fuqua*

Abstract

Estrogen and its receptors α and β (ERα and ERβ) play a major role in tumor progression and approximately two-thirds of breast cancers express these functional receptors. Thus, the ER is a major target for current and developing therapies. Although most ER-positive tumors initially respond to hormonal therapies such as tamoxifen, many tumors will eventually become resistant to tamoxifen induced growth inhibition. This chapter will discuss molecular mechanisms that contribute to hormonal resistance of current therapies including ERα mutations, the roles of proliferation and apoptosis in tumor homeostasis and receptor coregulator proteins. Additionally, the role of nonclassical ERα signaling through growth factor receptors and the subsequent downstream-initiated signaling, and the role of the progesterone receptors will be discussed.

Introduction

Aberrant estrogenic signaling has been associated with a number of human cancers including breast,[1] colon[2] and ovarian[3] cancers. Estrogen and its receptors, ERs α and β, play a particularly important role in the growth and progression of breast cancer.[4,5] As early as the 1800s it was shown that oophorectomy of premenopausal women with metastatic breast cancer caused tumor regression in approximately one-third of patients.[6,7] However it was not until the later half of the 20th century that the molecular mechanisms behind this therapy were discovered. In the 1950s, Jenson and Jacobson demonstrated that estrogen was targeted to specific tissues,[8,9] and ERα was purified in the following decade.[10] However, it was not until the 1980s that ERα was actually cloned[11-14] and the second estrogen receptor, ERβ, was cloned in the mid 1990s.[15-17] Because ERα was discovered and cloned much earlier than ERβ, much more is known about this receptor, thus this chapter will focus mainly on ERα.

It has been demonstrated that the majority of human breast cancers express ERα. Because of the important role of the ERs in breast cancer growth, targeting this receptor has proven to be an effective therapy for ER-positive breast cancer. Selective estrogen receptor modulators (SERMS), such as tamoxifen and raloxifene, inhibit the ability of estrogen to bind to the ERs through competitive inhibition, and have been shown to display agonist or antagonist activity in a tissue-type specific manner.[18] The pure antiestrogen fulvestrant (ICI182,780) binds with the ERs resulting in antagonistic effects and downregulation of ER.[19,20] A newer class of drugs called the aromatase inhibitors (AI), seek to block estrogen signaling by inhibiting the production of estrogen compounds from testosterone precursors.

*Corresponding Author: Suzanne A.W. Fuqua—Breast Center, Baylor College of Medicine, One Baylor Plaza, BCM 600, Houston, Texas, U.S.A. 77030. Email: sfuqua@breastcenter.tmc.edu

Breast Cancer Chemosensitivity, edited by Dihua Yu and Mien-Chie Hung.

Clinically, the AIs are demonstrating increased efficacy and reduced side effects when compared with tamoxifen,[21-24] the most commonly prescribed antiestrogen. Targeting the ER with tamoxifen has proven to be an effective therapy for many patients with ER-positive breast cancer. However, the majority of tumors initially responding to tamoxifen will develop hormone resistance (HR) and relapse within five years. A number of different molecular mechanisms have been implicated in HR and will be discussed in this chapter.

Receptor Structure and Function

ERs α and β belong to the nuclear hormone receptor superfamily.[14,25,26] The ERs contain six structural domains and several defined functional domains (Fig. 1A). ERα contains a ligand-independent, amino-terminal activation function (AF-1) domain, a DNA-binding domain, a hinge region, and the hormone binding domain with its associated ligand-dependent AF-2 region. ERβ shares a similar structure but lacks the AF-1 domain (Fig. 1B), thus ERβ has significantly reduced ligand-independent activity. Binding of hormone to the ER results in reduced binding of the coregulatory repressor protein complexes with histone deacetylase complexes (HDACs), and the subsequent recruitment of coactivator proteins with histone acetyl transferase (HAT) activity, leading to increased transcriptional activity. Thus, altered association with these coregulatory molecules modulates and "fine-tunes" ER transcriptional activity.

When the crystal structures of ERα bound to various ligands were solved, the structural mechanisms behind the agonist action of SERMS was firmly established. Hormone binding induces helix 12 in the receptor to reposition itself over the ligand binding pocket forming a tight cover over the bound ligand.[27-32] This conformational change stabilizes helix 12 allowing

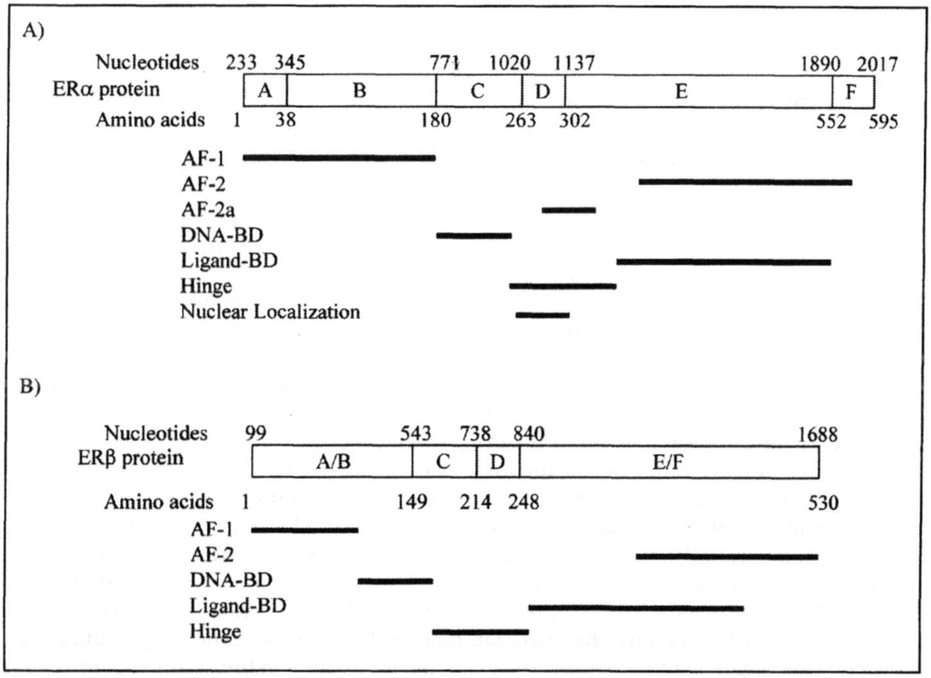

Figure 1. Domains of the estrogen receptors. Exons are numbered in the corresponding blocked region with the nucleotide number above. ATG start codon and the TAG stop codon are shown below. The protein domains are labeled A-F, nucleotide numbers corresponding to the start of each domain are above with amino acid numbers below. Relative positions of some of the known functional domains are represented by solid bars below. A) ERα, B) ERβ.

for the recruitment of receptor coactivator proteins.[33,34] In contrast, when ER is bound to partial agonists or antagonists such as tamoxifen, raloxifene, or faslodex, the "bulky" side chain of the compound prevents helix 12 from forming the agonist bound structure, thus inhibiting coactivator recruitment.[31,32,35-38] Inhibitory compounds without "bulky" side chains, such as genestein, inhibit full ER activation by stabilizing nonproductive conformations of the ligand-binding pocket.[36] Stabilization of helix 12 over the ligand binding pocket allows coactivator recruitment, while the antagonist bound conformation of helix 12 prevents coactivator recruitment, thereby preventing transcriptional activity.

Does Estrogen Receptors α or β Expression Predict Response to Therapy?

Approximately two-thirds of invasive breast cancers express ER α and β isoforms.[39-44] It is clear that ERα expression is an important prognostic factor and a positive predictor of response to endocrine therapy.[45-47] The prognostic value of ERβ is much less clear. While the majority of studies suggest that ERβ protein expression is associated with a more favorable outcome correlating with know prognostic factors,[48-51] a few studies have demonstrated that ERβ expression is associated with a poor clinical outcome.[41,52,53] However, it must be noted that the majority of studies suggesting a less favorable outcome with ERβ expression, have only analyzed RNA expression levels. Additionally, multiple studies have demonstrated that ERβ protein expression is associated with a more favorable response to tamoxifen[54-56] and low ERβ expression predicts tamoxifen resistance.[57] Thus, while ERα is an accepted prognostic and predictive factor, the value of ERβ protein expression has not been fully elucidated. However, direct protein analyses have suggested that like ERα, ERβ is an indicator of a more favorable clinical outcome.

Mechanisms of Resistance to Hormonal Therapies

Estrogen Receptor α Mutations

While mutations in ERs α and β directly causing HR is an attractive hypothesis, to date, clinical data supporting ER mutations as a major means of HR do not exist. One study from the mid 1990s estimated that only 1% of primary invasive breast tumors exhibit missense mutations.[58] Currently, approximately 19 separate mutations have been identified in ERα while even fewer have been found in ERβ (for a complete see review ref. 59). Karnik et al identified an ERα 437stop mutation in 1 of 5 metastatic breast tumors.[60] A similar mutation ERα 417stop was identified in the tamoxifen resistant T47DCO cell line.[61] While we found mutations in 3 of 30 metastatic lesions, only Y537N demonstrated tamoxifen resistance.[62] Although a number of experimental mutations have been made in ERα and a few result in tamoxifen resistance including, G400V[63] and L540Q,[49,64] only a handful of mutations have been identified in breast cancer patient samples, thus clinical evidence that ERα mutations in tumors play a fundamental role in HR is lacking.

While the laboratory value of ER mutations is immense (see ref. 59), the clinical value of ER mutations is only beginning to be elucidated. Recently we identified an ERα somatic A to G transition in 30% of the premalignant breast lesions studied. This mutation changes the 303 lysine to arginine (K303R) resulting in a receptor that is hypersensitive to estrogen leading to increased transcriptional activity at subphysiological estrogen concentrations.[65] Additionally, this K303R mutant receptor has increased binding to coactivator proteins.[65] In contrast to the previously identified mutations, this mutation that has been identified in a large number of patient samples suggesting a potential role for this mutant receptor in breast tumorigenesis and progression. What role, if any, the K303R ERα mutant has in HR remains to be elucidated.

Tumor Homeostasis (Proliferation and Apoptosis Signals)

Tumor homeostasis requires a balance between proliferation and apoptosis. ERα signaling has been implicated in a number of different processes involved in tumor homeostasis. For instance, upregulation of growth signals, or downregulation of apoptotic signals, could lead to an inability of a cell to respond to antiestrogens in an antagonistic manner, thus contributing to HR. Many genes involved in cell cycle regulation such as cyclin A1, cyclin D1, and the E2F1 transcription factor have been identified as estrogen-upregulated genes.[66-68] Furthermore, tamoxifen can function as an agonist to upregulate genes promoting cell cycle progression, including c-fos, c-myc, cyclin A2, and E2F1.[69] Estrogen has also been shown to upregulate the antiapoptotic genes bcl-2 and bcl-xl,[70] thereby protecting a cell from death-inducing signals. Overexpression of HER-2 in MCF-7 cells leads to tamoxifen resistance and an upregulation of bcl-2 and bcl-xl,[71] suggesting a role for these proteins in blocking cell death following tamoxifen treatment. Additionally, Teeck et al has demonstrated that long-term tamoxifen treatment in vitro leads to a reduction in proapoptotic genes and increased antiapoptotic genes.[72] This antiapoptotic effect extended to other drugs such as the topoisomerase inhibitor, etoposide.[72] Both estrogen and tamoxifen have been shown to upregulate genes involved in tumor homeostasis, thereby altering normal growth controls.

One of the more studied estrogen mediated growth control proteins is cyclin D1, a key regulatory subunit for the cyclin dependent kinases, cdk4 and cdk6. Recently, retrospective analysis of patients with long-term clinical follow-up demonstrated that high cyclin D1 levels were associated with a worse overall survival in tamoxifen-treated patients,[73] suggesting a role for cyclin D1 in HR. Increased levels of cyclin D1 may titrate out the cell cycle inhibitors Cip/kip, thus "forcing" the cell through G1 progression and bypassing normal control mechanisms.[74] Additionally, cyclin D1 can form a complex with ERα and coactivators thereby affecting ERα signaling.[75,76] Cyclin D1 has also been shown to stimulate estrogen-independent ERα activity, thus providing a mechanism for estrogen-independent proliferation in the presence of high cyclin D1 levels.[77] It is clear that ERα signaling plays a key role in the balance between life and death signals which can significantly affect the cellular response to tamoxifen. We are just beginning to dissect the molecular mechanisms behind the roles of proliferation and apoptosis in HR.

Coactivators and Corepressors

Coregulator proteins directly affect the activity of ERα, thus it is not surprising that these proteins have been implicated in breast cancer progression and HR. In fact, the expression of many of these coregulator molecules have been shown to be altered during breast tumorigenesis.[78-80] However, one study was not able to find altered expression of coregulators in de novo tamoxifen-resistant tumors,[81] suggesting a role in tumorigenesis, but not HR. One such altered protein, the coactivator SRC-1, has been shown to increase the tamoxifen agonist activity in certain cells.[82] Furthermore, in endometrial cells, where tamoxifen acts as an agonist, tamoxifen treatment recruits SRC-1 to ERα.[83] These data showing tissue-type specificity of tamoxifen agonist effects clearly demonstrate a role for SRC-1 in tamoxifen-stimulated ERα transactivation. Additionally, clinical studies have shown that high SRC-1 levels may correlate with a favorable response to tamoxifen treatment in women with recurrent breast cancer,[84] an apparent contradiction to the previously mentioned in vitro studies. Although the exact role of SRC-1 in tamoxifen induced ERα activity is undecided, it is clear that SRC-1 plays a key role in tamoxifen signaling through ERα.

Gene amplification of AIB1/SRC3 has been demonstrated in a number of breast and ovarian cancer cell lines, as well as breast cancer biopsies.[85,86] Increased AIB1 resulted in estrogen-independent growth and tamoxifen resistance.[87] These effects may be mediated through AIB1's interactions with the E2F1 transcription factor, thereby leading to increased cell proliferation.[87] Analysis of clinical samples has revealed that patients with elevated levels of AIB1 and HER-2 did not respond well to tamoxifen therapy.[88] However, elevated AIB1 levels are associated with a better prognosis in patients not receiving adjuvant tamoxifen.[88]

In addition to a role for coactivators in HR, the corepressor NCoR1 is associated with tamoxifen resistance.[89,90] Tamoxifen-bound ERα can recruit corepressors such as NCoR1 and

SMRT.[82,84,91-94] Hence, if corepressor expression is reduced, then it cannot be recruited to repress the activity of ERα. The effects of NCoR1 recruitment occur in a tissue-type specific manner, hence leading to agonist activity of tamoxifen in some cells,[91] but not others.[95] Numerous in vitro studies have visibly displayed a role for coactivator upregulation or corepressor downregulation leading to HR. However, clinical evidence demonstrating a role for coregulator proteins in HR is not as clear, and their role is just beginning to materialize.

Estrogen Receptor Phosphorylation

Phosphorylation of ERα has been shown to be an important mechanism of ligand-independent activation. Mutational analysis has shown that serines 104 and 106 are phosphorylated by cyclin A-CDK2.[96-98] However, the role of phosphorylation at these serines is not clearly defined. Serine 118 is phosphorylated by the MAPKs Erk 1/2 and Cdk7, as well as unknown kinases.[99-102] Additionally, S118 phosphorylation can be induced by a number of different ligands including, estrogen, tamoxifen, ICI 164,384, EGF, IGF, and TPA.[99-105] While some studies have shown that S118 phosphorylation can lead to ligand-independent activity,[101] other studies have shown that S118 phosphorylation is not sufficient for ERα-induced transactivation.[102] However, many studies have shown that phosphorylation of S118 is required for the full transcriptional activation of ERα.[97,101,102,104,106] Casein kinase II, AKT2, and pp90rsk1 have all been shown to phosphorylate S167,[97,104,107,108] in response to various ligands including estrogen, EGF and PMA.[104,109] Mutational analysis has demonstrated reduced transactivation of S167A mutants,[97,104,107,108] demonstrating an important role for S167 phosphorylation in ERα transcactiavtion. Protein Kinase A (PKA) phoshporylates S236 which lies within the DNA-binding domain.[110] Futhermore, S236 phosphorylation is important for ERα dimerization.[110] Recently, PKA has also been shown to phosphorylate S305 inducing a switch between the antagonistic to the agonistic effects of tamoxifen, thus leading to tamoxifen induced ERα transactivation.[111] The Src family kinases c-Src and Lck have been shown to phosphorylate Y537 of ERα.[112] Y537 lies at the amino-terminal cap of helix 12 and forms a hydrogen bond with N348 upon ligand binding,[113,114] thus stabilizing the ligand bound conformation. Additionally, experimental mutations (Y537S and Y537E) in ERα that mimic this hydrogen bond lead to a constitutively-active receptor.[114-117] Interestingly, Y537 is the only phosphorylation site of ERα that has been found to be mutated in patient samples.[62] It is clear that numerous nonestrogen mediated events can lead to phosphorylation of ERα, hence affecting ERα transactivation and possibly HR.

Signal Transduction

As mentioned in the previous section, a number of different kinases have been shown to significantly affect the activity of ERα, and additional clinical and laboratory data suggest a significant role for several of these kinases in the development of HR. One of the most studied kinases in HR is the EGFR family of receptor tyrosine kinase. Analysis of clinical samples has shown that elevated HER-2 or EGFR expression in a pretreatment biopsy correlated with a reduced response to tamoxifen.[118] Additionally, a number of groups have found increased levels of HER-2 and EGFR in tamoxifen-resistant MCF-7 breast cancer cells.[53,119,120] HER-2 overexpressing MCF-7 cell line form tamoxifen-resistant tumors when grown as xenografts in nude mice.[121] Ropero et al found that cytostatic levels of tamoxifen were able to upregulate HER-2 mRNA and protein levels by 66% and 49%, respectively.[122] This recent data asks the question, are HER-2 cells selected by tamoxifen treatment, or is HER-2 expression stimulated by tamoxifen and simply a byproduct of treatment? The fact that tamoxifen-resistance in breast cancer cells can be reversed with EGFR/HER-2 tyrosine kinase inhibitors[123-125] demonstrates an active role for HER-2 in the development of HR. Recently, HER-2 overexpression has been correlated with overexpression of the ERα coactivator AIB1, and poor disease-free survival for patients receiving adjuvant tamoxifen.[88] Additionally, HER-2 status was correlated with expression of the ERα coactivators PEA3, AIB1, and SRC1.[126] However, in this study, only

SRC1 and PEA3 were significantly associated with disease recurrence on endocrine treatment.[126] Collectively, these data demonstrate an important role for EGFR and HER-2 signaling in the development of HR in ERα-positive breast cancer.

MAPK activation can contribute to estrogen-induced proliferation,[127] estrogen sensitivity,[128] and cell survival,[129] therefore it is not surprising that MAPK activation has been implicated in HR. Gee et al analyzed breast cancer patient samples and found that activated MAPKs, Erk-1/2, are biomarkers for a shorter response to tamoxifen.[130] Furthermore, blocking Erk-1/2 activity restored the inhibitory effect of tamoxifen on ERα mediated transcription and cell proliferation in tamoxifen-resistant, HER-2 overexpressing MCF-7 cells.[125] However, activation of the upstream activator of Erk-1/2 MEK1, was insufficient to induce tamoxifen resistance in MCF-7 cells.[131] Additionally, MAPK can affect ERα through the downstream effector of MAPK, pp90rsk1,[105] thus effecting ERα phosphorylation indirectly. In addition to MAPK-induced phosphorylation of ERα, estrogen signaling can also induce the activation of MAPK,[132] thereby suggesting the existence of a two-way crosstalk and feedback loops between these pathways. Although it is clear that MAPK activation can lead to phosphorylation and possible activation of ERα, the molecular mechanisms of MAPK- induced HR remain to be defined.

Activation of the PI3K/AKT pathway has also been implicated in HR.[133] Tamoxifen-resistant ER-negative HER-2 overexpressing MCF-7 cells demonstrate increased ligand-induced (estrogen, EGF, heregulin, and tamoxifen) AKT phosphorylation.[124] Additionally, overexpression of AKT3 in MCF-7 cells leads to estrogen-independent tumors in nude mice.[134] Furthermore, these AKT3-overexpressing tumors were growth stimulated with tamoxifen and growth inhibited with estrogen,[134] thus demonstrating an in vivo role for AKT3 in changing tamoxifen from an antagonist to an agonist. Shoman et al examined protein expression of the AKT regulator, PTEN (phosphatase and tensin homolog deleted on chromosome ten), in tamoxifen-treated ERα-positive breast cancer patients and found that reduced PTEN protein expression was associated with shorter disease-free survival.[135] These in vitro and in vivo experiments, as well as analysis of patient samples, demonstrate a potential role for the PI3K/AKT pathway in HR.

Much less defined are the roles of PKA and PKC-induced phosphorylation of ERα. Activation of both proteins was shown to modestly increase ERα induced transactivation, while estrogen stimulation combined with PKA or PKC activation resulted in a synergistic activation of ERα transactivation.[136] Activation of PKA, or downregulation of the PKA negative regulator, PKA-R1α, and subsequent phosphorylation of ERα at S305 resulted in tamoxifen resistance.[111] Additionally, in clinical samples, it was found that downregulation of PKA-R1α prior to treatment, was associated with tamoxifen resistance.[111] While disrupting normal signaling in these pathways may contribute to tamoxifen resistance, their role has yet to be defined.

Collectively these data help to elucidate the crosstalk between kinase signal transduction pathways and ERα. This coordinated crosstalk may significantly contribute to the cell-type and tissue specific effects of SERMS. A full understanding of the complex interactions between signaling pathways will undoubtedly lead to newer, more effective single and/or combination therapies for the treatment of HR breast cancer.

Interactions with Other Transcription Factors

Functional studies of the AF-1 and AF-2 domains have revealed differential transcriptional activity, many of which are dependent upon the cellular and promoter context.[137] Because the AF-2 domain is the ligand-dependent domain, it is not surprising that when AF-2 is the dominant activity within a cell, tamoxifen can act as a pure antagonist.[138-140] In contrast, when the ligand-independent AF-1 domain is the dominant activity, then tamoxifen behaves as a partial agonist and can stimulate transcriptional activity. While ERα and ERβ demonstrate similar AF-2 transactivation, ERβ has reduced AF-1 activity.[141-145]

Additional studies have demonstrated that ERs α and β can differentially interact with other cellular promoters and with differential ligand-induced activity. For instance, it has been demonstrated that Seratonin-1A can be upregulated through nuclear factor-κB (NF-κB) induced by ERα signaling, but not ERβ signaling.[146] Paech et al have demonstrated that estrogen-induced ERα can upregulate AP-1 activity.[147] In contrast, ERβ AP-1 activity was reduced following estrogen stimulation.[147] Although the receptors displayed differential responses to estrogen, the antiestrogens tamoxifen, raloxifene, and ICI 164,384 all stimulated AP-1 activity[147] to varying amounts, dependent upon the cell-type.[148] Additionally, ERα has been shown to bind with the Sp1 transcriptional activator protein[149] leading to an upregulation of a number of genes including c-myc, heat shock protein 27, and transforming growth factor α, to name a few.[150] In MCF-7 and MDA-MB-231 cells, estrogen, tamoxifen, and fulvestrant, all increased ERα-induced SP1 activity.[149] In contrast, tamoxifen stimulated AP-1 activity through ERβ only in the MCF-7 cells.[149] These cell-type specific effects involve a number of different factors, of which, the molecular identification will undoubtedly lead to new targeted therapies to counteract the partial agonist action of antiestrogens.

Progesterone Receptors A and B

The progesterone receptors A and B (PRA and PRB) isoforms have a strong prognostic and predictive value for response to tamoxifen and longer time to HR.[151-153] Retrospective data supports the concept that ER-positive, but PR-negative tumors are relatively HR.[154] Current data from the ATAC trial has shown a major benefit for the aromatase inhibitor anastrazole in ER+/PR- subgroup while only a modest advantage was seen in the tamoxifen-sensitive ER+/PR+ subgroup when compared with tamoxifen.[155] It has been hypothesized that PR status might be used to select first-line therapy.[156] Tamoxifen followed by an aromatase inhibitor might be the best therapy in ER-/PR+ tumors, while ER+/PR- tumors might receive and initial therapy of aromatase inhibitors because of their relative resistance to tamoxifen.[156] Additionally, we have recently demonstrated that tamoxifen treated, PR-positive tumors with a high PR-A/PR-B ratio were 2.76 times more likely to relapse on tamoxifen than patients with lower ratios.[157] The PR-B isoform is generally a transcriptional activator while the PR-A isoform can act as a dominant negative repressor for ERα and PR-B.[158-161] Thus, an increase in the relative amounts of PR-A can act to repress ERα signaling, hence affecting the response to tamoxifen.

Future Directions

The development of HR in breast cancer is a major clinical problem. This chapter has outlined many of the mechanisms currently thought to be involved in the development of HR following endocrine therapy. It is clear that HR can develop via multiple mechanisms either directly affecting ER signaling such as coregulator proteins, or independent of ER such as hormone-independent cell cycle progression. While we learn more about the mechanisms involved in the development of HR, we also learn that there will not be one simple answer to this devastating problem. The recent introduction of new estrogen/ER targeting therapies is showing some promise for the treatment of recurrent breast cancer. As these newer therapies inhibit ER via different mechanisms of action, it will be necessary to know the reasons behind HR for the accurate treatment of recurrent breast cancer. For instance, the aromatase inhibitor letrozole, was more effective than tamoxifen when tumors coexpressed ERα, and EGFR or HER-2.[162] Additionally, therapies developed to inhibit the molecular crosstalk between ER and other signal transduction pathways may be necessary to freely overcome HR. Important strides are already being demonstrated in the treatment of HR caused by HER-2 overexpression by using specific inhibitors to HER-2 and EGFR to reverse HR.[123-125] The key to targeting HR will be identifying the molecular mechanisms involved in HR in individual patients, and developing therapeutic agents to specifically target those mechanisms. This will require an emergence of accurate diagnostic biomarkers.

One exciting approach to identify individual proteins and signal transduction pathways involved in HR is the technique of microarray expression profiling. Recently, these tools have been used to identify a set of genes associated with clinical outcome in breast cancer patients treated with tamoxifen.[163] Interestingly, the set of genes identified that were predictive of outcome in these patients was different than the genes that were prognostic in untreated patients.[164] These large-scale gene identification approaches allow for the simultaneous analysis of complex signaling pathways, enabling scientists to identify hidden patterns in gene expression.

The problem of HR is complex involving a number of different mechanisms. Understanding the molecular aspects of HR will allow for the development and use of newer, targeted therapies to combat this problem. Additionally, the use of large-scale proteomic and genomic approaches will undoubtedly aid in the identification of new targets for combating HR.

Acknowledgements

This work was supported by grants from the NIH/NCI to SAWF (R01CA72038), and the Department of Defense Fellowship (DAMD17-02-1-0417) to MHH.

References

1. Russo IH, Russo J. Role of hormones in mammary cancer initiation and progression. J Mammary Gland Biol Neoplasia 1998; 3(1):49-61.
2. Di Leo A, Messa C, Cavallini A et al. Estrogens and colorectal cancer. Curr Drug Targets Immune Endocr Metabol Disord 2001; 1(1):1-12.
3. Pujol P, Rey JM, Nirde P et al. Differential expression of estrogen receptor-alpha and -beta messenger RNAs as a potential marker of ovarian carcinogenesis. Cancer Res 1998; 58(23):5367-5373.
4. Feigelson HS, Ross RK, Yu MC et al. Genetic susceptibility to cancer from exogenous and endogenous exposures. J Cell Biochem Suppl 1996; 25:15-22.
5. Colditz GA. Relationship between estrogen levels, use of hormone replacement therapy, and breast cancer. J Natl Cancer Inst 1998; 90(11):814-823.
6. Beatson C. On the treatment of inoperable cases of carcinoma of the mamma: Suggestions for a new method of treatment, with illustrative cases. Lancet 1896; 2:104-107.
7. Boyd S. On oophorectomy in cancer of the breast. Br Med J 1900; 2:1161-1167.
8. Jensen E, Jacobson HI. Fate of steroidal estrogens in target tissues. In: Pincus G, EP V, eds. Biological activities of steroids in relation to cancer. New York: Academic Press, 1960:161-174.
9. Jensen E, Jacobson HI. Basic guides to the mechanism of estrogen action. Recent Prog Horm Res 1962; 18:387-414.
10. Toft D, Gorski J. A receptor molecule for estrogens: Isolation from the rat uterus and preliminary characterization. Proc Natl Acad Sci USA 1966; 55(6):1574-1581.
11. Green S, Kumar V, Krust A et al. Structural and functional domains of the estrogen receptor. Cold Spring Harb Symp Quant Biol 1986; 51(Pt 2):751-758.
12. Green S, Walter P, Kumar V et al. Human oestrogen receptor cDNA: Sequence, expression and homology to v-erb-A. Nature 1986; 320(6058):134-139.
13. Greene GL, Gilna P, Waterfield M et al. Sequence and expression of human estrogen receptor complementary DNA. Science 1986; 231(4742):1150-1154.
14. Walter P, Green S, Greene G et al. Cloning of the human estrogen receptor cDNA. Proc Natl Acad Sci USA 1985; 82(23):7889-7893.
15. Enmark E, Pelto-Huikko M, Grandien K et al. Human estrogen receptor beta-gene structure, chromosomal localization, and expression pattern. J Clin Endocrinol Metab 1997; 82(12):4258-4265.
16. Mosselman S, Polman J, Dijkema R. ER beta: Identification and characterization of a novel human estrogen receptor. FEBS Lett 1996; 392(1):49-53.
17. Ogawa S, Inoue S, Watanabe T et al. The complete primary structure of human estrogen receptor beta (hER beta) and its heterodimerization with ER alpha in vivo and in vitro. Biochem Biophys Res Commun 1998; 243(1):122-126.
18. Jordan VC, Gapstur S, Morrow M. Selective estrogen receptor modulation and reduction in risk of breast cancer, osteoporosis, and coronary heart disease. J Natl Cancer Inst 2001; 93(19):1449-1457.
19. Osborne CK, Pippen J, Jones SE et al. Double-blind, randomized trial comparing the efficacy and tolerability of fulvestrant versus anastrozole in postmenopausal women with advanced breast cancer progressing on prior endocrine therapy: Results of a North American trial. J Clin Oncol 2002; 20(16):3386-3395.
20. Howell A, Robertson JF, Quaresma Albano J et al. Fulvestrant, formerly ICI 182,780, is as effective as anastrozole in postmenopausal women with advanced breast cancer progressing after prior endocrine treatment. J Clin Oncol 2002; 20(16):3396-3403.

21. Mouridsen H, Gershanovich M, Sun Y et al. Superior efficacy of letrozole versus tamoxifen as first-line therapy for postmenopausal women with advanced breast cancer: Results of a phase III study of the International Letrozole Breast Cancer Group. J Clin Oncol 2001; 19(10):2596-2606.
22. Nabholtz JM, Buzdar A, Pollak M et al. Anastrozole is superior to tamoxifen as first-line therapy for advanced breast cancer in postmenopausal women: Results of a North American multicenter randomized trial. Arimidex Study Group. J Clin Oncol 2000; 18(22):3758-3767.
23. Bonneterre J, Thurlimann B, Robertson JF et al. Anastrozole versus tamoxifen as first-line therapy for advanced breast cancer in 668 postmenopausal women: Results of the Tamoxifen or Arimidex Randomized Group Efficacy and Tolerability study. J Clin Oncol 2000; 18(22):3748-3757.
24. Baum M, Budzar AU, Cuzick J et al. Anastrozole alone or in combination with tamoxifen versus tamoxifen alone for adjuvant treatment of postmenopausal women with early breast cancer: First results of the ATAC randomised trial. Lancet 2002; 359(9324):2131-2139.
25. Kuiper GG, Enmark E, Pelto-Huikko M et al. Cloning of a novel receptor expressed in rat prostate and ovary. Proc Natl Acad Sci USA 1996; 93(12):5925-5930.
26. Misrahi M, Atger M, d'Auriol L et al. Complete amino acid sequence of the human progesterone receptor deduced from cloned cDNA. Biochem Biophys Res Commun 1987; 143(2):740-748.
27. Steinmetz AC, Renaud JP, Moras D. Binding of ligands and activation of transcription by nuclear receptors. Annu Rev Biophys Biomol Struct 2001; 30:329-359.
28. Weatherman RV, Fletterick RJ, Scanlan TS. Nuclear-receptor ligands and ligand-binding domains. Annu Rev Biochem 1999; 68:559-581.
29. Tanenbaum DM, Wang Y, Williams SP et al. Crystallographic comparison of the estrogen and progesterone receptor's ligand binding domains. Proc Natl Acad Sci USA 1998; 95(11):5998-6003.
30. Glass CK, Rosenfeld MG. The coregulator exchange in transcriptional functions of nuclear receptors. Genes Dev 2000; 14(2):121-141.
31. Shiau AK, Barstad D, Loria PM et al. The structural basis of estrogen receptor/coactivator recognition and the antagonism of this interaction by tamoxifen. Cell 1998; 95(7):927-937.
32. Brzozowski AM, Pike AC, Dauter Z et al. Molecular basis of agonism and antagonism in the oestrogen receptor. Nature 1997; 389(6652):753-758.
33. Moras D, Gronemeyer H. The nuclear receptor ligand-binding domain: Structure and function. Curr Opin Cell Biol 1998; 10(3):384-391.
34. Feng W, Ribeiro RC, Wagner RL et al. Hormone-dependent coactivator binding to a hydrophobic cleft on nuclear receptors. Science 1998; 280(5370):1747-1749.
35. Shiau AK, Barstad D, Radek JT et al. Structural characterization of a subtype-selective ligand reveals a novel mode of estrogen receptor antagonism. Nat Struct Biol 2002; 9(5):359-364.
36. Pike AC, Brzozowski AM, Hubbard RE et al. Structure of the ligand-binding domain of oestrogen receptor beta in the presence of a partial agonist and a full antagonist. Embo J 1999; 18(17):4608-4618.
37. Pike AC, Brzozowski AM, Walton J et al. Structural aspects of agonism and antagonism in the oestrogen receptor. Biochem Soc Trans 2000; 28(4):396-400.
38. Pike AC, Brzozowski AM, Walton J et al. Structural insights into the mode of action of a pure antiestrogen. Structure (Camb) 2001; 9(2):145-153.
39. Harvey JM, Clark GM, Osborne CK et al. Estrogen receptor status by immunohistochemistry is superior to the ligand-binding assay for predicting response to adjuvant endocrine therapy in breast cancer. J Clin Oncol 1999; 17(5):1474-1481.
40. Fuqua SA, Schiff R, Parra I et al. Estrogen receptor beta protein in human breast cancer: Correlation with clinical tumor parameters. Cancer Res 2003; 63(10):2434-2439.
41. Speirs V, Parkes AT, Kerin MJ et al. Coexpression of estrogen receptor alpha and beta: Poor prognostic factors in human breast cancer? Cancer Res 1999; 59(3):525-528.
42. Jarvinen TA, Pelto-Huikko M, Holli K et al. Estrogen receptor beta is coexpressed with ERalpha and PR and associated with nodal status, grade, and proliferation rate in breast cancer. Am J Pathol 2000; 156(1):29-35.
43. Speirs V, Kerin MJ. Prognostic significance of oestrogen receptor beta in breast cancer. Br J Surg 2000; 87(4):405-409.
44. Skliris GP, Carder PJ, Lansdown MR et al. Immunohistochemical detection of ERbeta in breast cancer: Towards more detailed receptor profiling? Br J Cancer 2001; 84(8):1095-1098.
45. Tamoxifen for early breast cancer: An overview of the randomised trials. Early Breast Cancer Trialists' Collaborative Group. Lancet 1998; 351(9114):1451-1467.
46. Clark GM, McGuire WL. Steroid receptors and other prognostic factors in primary breast cancer. Semin Oncol 1988; 15(2 Suppl 1):20-25.
47. Knight WA, Livingston RB, Gregory EJ et al. Estrogen receptor as an independent prognostic factor for early recurrence in breast cancer. Cancer Res 1977; 37(12):4669-4671.

48. Katzenellenbogen BS, Montano MM, Ediger TR et al. Estrogen receptors: Selective ligands, partners, and distinctive pharmacology. Recent Prog Horm Res 2000; 55:163-193,(discussion 194-165).
49. Montano MM, Ekena K, Krueger KD et al. Human estrogen receptor ligand activity inversion mutants: Receptors that interpret antiestrogens as estrogens and estrogens as antiestrogens and discriminate among different antiestrogens. Mol Endocrinol 1996; 10(3):230-242.
50. Endoh H, Maruyama K, Masuhiro Y et al. Purification and identification of p68 RNA helicase acting as a transcriptional coactivator specific for the activation function 1 of human estrogen receptor alpha. Mol Cell Biol 1999; 19(8):5363-5372.
51. McKenna NJ, Lanz RB, O'Malley BW. Nuclear receptor coregulators: Cellular and molecular biology. Endocr Rev 1999; 20(3):321-344.
52. Dotzlaw H, Leygue E, Watson PH et al. Estrogen receptor-beta messenger RNA expression in human breast tumor biopsies: Relationship to steroid receptor status and regulation by progestins. Cancer Res 1999; 59(3):529-532.
53. Knowlden JM, Hutcheson IR, Jones HE et al. Elevated levels of epidermal growth factor receptor/ c-erbB2 heterodimers mediate an autocrine growth regulatory pathway in tamoxifen-resistant MCF-7 cells. Endocrinology 2003; 144(3):1032-1044.
54. Hopp T, Weiss H, Parra I et al. Low levels of estrogen receptor beta protein predicts resistance to tamoxifen therapy in breast cancer. Clin Cancer Res 2004; 10(22).
55. Mann S, Laucirica R, Carlson N et al. Estrogen receptor beta expression in invasive breast cancer. Hum Pathol 2001; 32(1):113-118.
56. Iwase H, Zhang Z, Omoto Y et al. Clinical significance of the expression of estrogen receptors alpha and beta for endocrine therapy of breast cancer. Cancer Chemother Pharmacol 2003; 52(Suppl 1):S34-38.
57. Esslimani-Sahla M, Simony-Lafontaine J, Kramar A et al. Estrogen receptor beta (ER beta) level but not its ER beta cx variant helps to predict tamoxifen resistance in breast cancer. Clin Cancer Res 2004; 10(17):5769-5776.
58. Roodi N, Bailey LR, Kao WY et al. Estrogen receptor gene analysis in estrogen receptor-positive and receptor-negative primary breast cancer. J Natl Cancer Inst 1995; 87(6):446-451.
59. Herynk MH, Fuqua SA. Estrogen receptor mutations in human disease. Endocr Rev 2004.
60. Karnik PS, Kulkarni S, Liu XP et al. Estrogen receptor mutations in tamoxifen-resistant breast cancer. Cancer Res 1994; 54(2):349-353.
61. Graham IInd ML, Krett NL, Miller LA et al. T47DCO cells, genetically unstable and containing estrogen receptor mutations, are a model for the progression of breast cancers to hormone resistance. Cancer Res 1990; 50(19):6208-6217.
62. Zhang QX, Borg A, Wolf DM et al. An estrogen receptor mutant with strong hormone-independent activity from a metastatic breast cancer. Cancer Res 1997; 57(7):1244-1249.
63. Jiang SY, Langan-Fahey SM, Stella AL et al. Point mutation of estrogen receptor (ER) in the ligand-binding domain changes the pharmacology of antiestrogens in ER-negative breast cancer cells stably expressing complementary DNAs for ER. Mol Endocrinol 1992; 6(12):2167-2174.
64. Ince BA, Schodin DJ, Shapiro DJ et al. Repression of endogenous estrogen receptor activity in MCF-7 human breast cancer cells by dominant negative estrogen receptors. Endocrinology 1995; 136(8):3194-3199.
65. Fuqua SA, Wiltschke C, Zhang QX et al. A hypersensitive estrogen receptor-alpha mutation in premalignant breast lesions. Cancer Res 2000; 60(15):4026-4029.
66. Soulez M, Parker MG. Identification of novel oestrogen receptor target genes in human ZR75-1 breast cancer cells by expression profiling. J Mol Endocrinol 2001; 27(3):259-274.
67. Hayashi SI, Eguchi H, Tanimoto K et al. The expression and function of estrogen receptor alpha and beta in human breast cancer and its clinical application. Endocr Relat Cancer 2003; 10(2):193-202.
68. Coser KR, Chesnes J, Hur J et al. Global analysis of ligand sensitivity of estrogen inducible and suppressible genes in MCF7/BUS breast cancer cells by DNA microarray. Proc Natl Acad Sci USA 2003; 100(24):13994-13999.
69. Hodges LC, Cook JD, Lobenhofer EK et al. Tamoxifen functions as a molecular agonist inducing cell cycle-associated genes in breast cancer cells. Mol Cancer Res 2003; 1(4):300-311.
70. Gompel A, Somai S, Chaouat M et al. Hormonal regulation of apoptosis in breast cells and tissues. Steroids 2000; 65(10-11):593-598.
71. Kumar R, Mandal M, Lipton A et al. Overexpression of HER2 modulates bcl-2, bcl-XL, and tamoxifen-induced apoptosis in human MCF-7 breast cancer cells. Clin Cancer Res 1996; 2(7):1215-1219.
72. Treeck O, Zhou R, Diedrich K et al. Tamoxifen long-term treatment in vitro alters the apoptotic response of MCF-7 breast cancer cells. Anticancer Drugs 2004; 15(8):787-793.

73. Stendahl M, Kronblad A, Ryden L et al. Cyclin D1 overexpression is a negative predictive factor for tamoxifen response in postmenopausal breast cancer patients. Br J Cancer 2004; 90(10):1942-1948.
74. Zhou Q, Hopp T, Fuqua SA et al. Cyclin D1 in breast premalignancy and early breast cancer: Implications for prevention and treatment. Cancer Lett 2001; 162(1):3-17.
75. McMahon C, Suthiphongchai T, DiRenzo J et al. P/CAF associates with cyclin D1 and potentiates its activation of the estrogen receptor. Proc Natl Acad Sci USA 1999; 96(10):5382-5387.
76. Lamb J, Ladha MH, McMahon C et al. Regulation of the functional interaction between cyclin D1 and the estrogen receptor. Mol Cell Biol 2000; 20(23):8667-8675.
77. Zwijsen RM, Wientjens E, Klompmaker R et al. CDK-independent activation of estrogen receptor by cyclin D1. Cell 1997; 88(3):405-415.
78. Bautista S, Valles H, Walker RL et al. In breast cancer, amplification of the steroid receptor coactivator gene AIB1 is correlated with estrogen and progesterone receptor positivity. Clin Cancer Res 1998; 4(12):2925-2929.
79. Kurebayashi J, Otsuki T, Kunisue H et al. Expression levels of estrogen receptor-alpha, estrogen receptor-beta, coactivators, and corepressors in breast cancer. Clin Cancer Res 2000; 6(2):512-518.
80. Murphy LC, Simon SL, Parkes A et al. Altered expression of estrogen receptor coregulators during human breast tumorigenesis. Cancer Res 2000; 60(22):6266-6271.
81. Murphy LC, Leygue E, Niu Y et al. Relationship of coregulator and oestrogen receptor isoform expression to de novo tamoxifen resistance in human breast cancer. Br J Cancer 2002; 87(12):1411-1416.
82. Smith CL, Nawaz Z, O'Malley BW. Coactivator and corepressor regulation of the agonist/antagonist activity of the mixed antiestrogen, 4-hydroxytamoxifen. Mol Endocrinol 1997; 11(6):657-666.
83. Shang Y, Brown M. Molecular determinants for the tissue specificity of SERMs. Science 2002; 295(5564):2465-2468.
84. Jepsen K, Hermanson O, Onami TM et al. Combinatorial roles of the nuclear receptor corepressor in transcription and development. Cell 2000; 102(6):753-763.
85. Anzick SL, Kononen J, Walker RL et al. AIB1, a steroid receptor coactivator amplified in breast and ovarian cancer. Science 1997; 277(5328):965-968.
86. List HJ, Reiter R, Singh B et al. Expression of the nuclear coactivator AIB1 in normal and malignant breast tissue. Breast Cancer Res Treat 2001; 68(1):21-28.
87. Louie MC, Zou JX, Rabinovich A et al. ACTR/AIB1 functions as an E2F1 coactivator to promote breast cancer cell proliferation and antiestrogen resistance. Mol Cell Biol 2004; 24(12):5157-5171.
88. Osborne CK, Bardou V, Hopp TA et al. Role of the estrogen receptor coactivator AIB1 (SRC-3) and HER-2/neu in tamoxifen resistance in breast cancer. J Natl Cancer Inst 2003; 95(5):353-361.
89. Girault I, Lerebours F, Amarir S et al. Expression analysis of estrogen receptor alpha coregulators in breast carcinoma: Evidence that NCOR1 expression is predictive of the response to tamoxifen. Clin Cancer Res 2003; 9(4):1259-1266.
90. Graham JD, Bain DL, Richer JK et al. Thoughts on tamoxifen resistant breast cancer. Are coregulators the answer or just a red herring? J Steroid Biochem Mol Biol 2000; 74(5):255-259.
91. Lavinsky RM, Jepsen K, Heinzel T et al. Diverse signaling pathways modulate nuclear receptor recruitment of N-CoR and SMRT complexes. Proc Natl Acad Sci USA 1998; 95(6):2920-2925.
92. Shang Y, Hu X, DiRenzo J et al. Cofactor dynamics and sufficiency in estrogen receptor-regulated transcription. Cell 2000; 103(6):843-852.
93. Zhang X, Jeyakumar M, Petukhov S et al. A nuclear receptor corepressor modulates transcriptional activity of antagonist-occupied steroid hormone receptor. Mol Endocrinol 1998; 12(4):513-524.
94. Jackson TA, Richer JK, Bain DL et al. The partial agonist activity of antagonist-occupied steroid receptors is controlled by a novel hinge domain-binding coactivator L7/SPA and the corepressors N-CoR or SMRT. Mol Endocrinol 1997; 11(6):693-705.
95. Morrison AJ, Herrera RE, Heinsohn EC et al. Dominant-negative nuclear receptor corepressor relieves transcriptional inhibition of retinoic acid receptor but does not alter the agonist/antagonist activities of the tamoxifen-bound estrogen receptor. Mol Endocrinol 2003; 17(8):1543-1554.
96. Kraus WL, Weis KE, Katzenellenbogen BS. Determinants for the repression of estrogen receptor transcriptional activity by ligand-occupied progestin receptors. J Steroid Biochem Mol Biol 1997; 63(4-6):175-188.
97. Le Goff P, Montano MM, Schodin DJ et al. Phosphorylation of the human estrogen receptor. Identification of hormone-regulated sites and examination of their influence on transcriptional activity. J Biol Chem 1994; 269(6):4458-4466.
98. Rogatsky I, Trowbridge JM, Garabedian MJ. Potentiation of human estrogen receptor alpha transcriptional activation through phosphorylation of serines 104 and 106 by the cyclin A-CDK2 complex. J Biol Chem 1999; 274(32):22296-22302.

99. Liu H, Lee ES, Deb Los Reyes A et al. Silencing and reactivation of the selective estrogen receptor modulator-estrogen receptor alpha complex. Cancer Res 2001; 61(9):3632-3639.
100. Karas RH, Gauer EA, Bieber HE et al. Growth factor activation of the estrogen receptor in vascular cells occurs via a mitogen-activated protein kinase-independent pathway. J Clin Invest 1998; 101(12):2851-2861.
101. Kato S, Endoh H, Masuhiro Y et al. Activation of the estrogen receptor through phosphorylation by mitogen-activated protein kinase. Science 1995; 270(5241):1491-1494.
102. Bunone G, Briand PA, Miksicek RJ et al. Activation of the unliganded estrogen receptor by EGF involves the MAP kinase pathway and direct phosphorylation. Embo J 1996; 15(9):2174-2183.
103. Ali S, Metzger D, Bornert JM et al. Modulation of transcriptional activation by ligand-dependent phosphorylation of the human oestrogen receptor A/B region. Embo J 1993; 12(3):1153-1160.
104. Joel PB, Smith J, Sturgill TW et al. pp90rsk1 regulates estrogen receptor-mediated transcription through phosphorylation of Ser-167. Mol Cell Biol 1998; 18(4):1978-1984.
105. Joel PB, Traish AM, Lannigan DA. Estradiol-induced phosphorylation of serine 118 in the estrogen receptor is independent of p42/p44 mitogen-activated protein kinase. J Biol Chem 1998; 273(21):13317-13323.
106. Castano E, Chen CW, Vorojeikina DP et al. The role of phosphorylation in human estrogen receptor function. J Steroid Biochem Mol Biol 1998; 65(1-6):101-110.
107. Clark DE, Poteet-Smith CE, Smith JA et al. Rsk2 allosterically activates estrogen receptor alpha by docking to the hormone-binding domain. Embo J 2001; 20(13):3484-3494.
108. Sun M, Paciga JE, Feldman RI et al. Phosphatidylinositol-3-OH Kinase (PI3K)/AKT2, activated in breast cancer, regulates and is induced by estrogen receptor alpha (ERalpha) via interaction between ERalpha and PI3K. Cancer Res 2001; 61(16):5985-5991.
109. Arnold SF, Obourn JD, Jaffe H et al. Serine 167 is the major estradiol-induced phosphorylation site on the human estrogen receptor. Mol Endocrinol 1994; 8(9):1208-1214.
110. Chen D, Pace PE, Coombes RC et al. Phosphorylation of human estrogen receptor alpha by protein kinase A regulates dimerization. Mol Cell Biol 1999; 19(2):1002-1015.
111. Michalides R, Griekspoor A, Balkenende A et al. Tamoxifen resistance by a conformational arrest of the estrogen receptor alpha after PKA activation in breast cancer. Cancer Cell 2004; 5(6):597-605.
112. Arnold SF, Obourn JD, Jaffe H et al. Phosphorylation of the human estrogen receptor on tyrosine 537 in vivo and by src family tyrosine kinases in vitro. Mol Endocrinol 1995; 9(1):24-33.
113. Carlson KE, Choi I, Gee A et al. Altered ligand binding properties and enhanced stability of a constitutively active estrogen receptor: Evidence that an open pocket conformation is required for ligand interaction. Biochemistry 1997; 36(48):14897-14905.
114. Skafar DF. Formation of a powerful capping motif corresponding to start of "helix 12" in agonist-bound estrogen receptor-alpha contributes to increased constitutive activity of the protein. Cell Biochem Biophys 2000; 33(1):53-62.
115. Weis KE, Ekena K, Thomas JA et al. Constitutively active human estrogen receptors containing amino acid substitutions for tyrosine 537 in the receptor protein. Mol Endocrinol 1996; 10(11):1388-1398.
116. Zhong L, Skafar DF. Mutations of tyrosine 537 in the human estrogen receptor-alpha selectively alter the receptor's affinity for estradiol and the kinetics of the interaction. Biochemistry 2002; 41(13):4209-4217.
117. Schlegel A, Wang C, Katzenellenbogen BS et al. Caveolin-1 potentiates estrogen receptor alpha (ERalpha) signaling. caveolin-1 drives ligand-independent nuclear translocation and activation of ERalpha. J Biol Chem 1999; 274(47):33551-33556.
118. Newby JC, Johnston SR, Smith IE et al. Expression of epidermal growth factor receptor and c-erbB2 during the development of tamoxifen resistance in human breast cancer. Clin Cancer Res 1997; 3(9):1643-1651.
119. Hutcheson IR, Knowlden JM, Madden TA et al. Oestrogen receptor-mediated modulation of the EGFR/MAPK pathway in tamoxifen-resistant MCF-7 cells. Breast Cancer Res Treat 2003; 81(1):81-93.
120. Nicholson RI, Gee JM, Knowlden J et al. The biology of antihormone failure in breast cancer. Breast Cancer Res Treat 2003; 80(Suppl 1):S29-34, (discussion S35).
121. Benz CC, Scott GK, Sarup JC et al. Estrogen-dependent, tamoxifen-resistant tumorigenic growth of MCF-7 cells transfected with HER2/neu. Breast Cancer Res Treat 1993; 24(2):85-95.
122. Ropero S, Menendez JA, Vazquez-Martin A et al. Trastuzumab plus tamoxifen: Anti-proliferative and molecular interactions in breast carcinoma. Breast Cancer Res Treat 2004; 86(2):125-137.
123. Moulder SL, Yakes FM, Muthuswamy SK et al. Epidermal growth factor receptor (HER1) tyrosine kinase inhibitor ZD1839 (Iressa) inhibits HER2/neu (erbB2)-overexpressing breast cancer cells in vitro and in vivo. Cancer Res 2001; 61(24):8887-8895.

124. Shou J, Massarweh S, Osborne CK et al. Mechanisms of tamoxifen resistance: Increased estrogen receptor-HER2/neu cross-talk in ER/HER2-positive breast cancer. J Natl Cancer Inst 2004; 96(12):926-935.
125. Kurokawa H, Lenferink AE, Simpson JF et al. Inhibition of HER2/neu (erbB-2) and mitogen-activated protein kinases enhances tamoxifen action against HER2-overexpressing, tamoxifen-resistant breast cancer cells. Cancer Res 2000; 60(20):5887-5894.
126. Fleming FJ, Myers E, Kelly G et al. Expression of SRC-1, AIB1, and PEA3 in HER2 mediated endocrine resistant breast cancer; a predictive role for SRC-1. J Clin Pathol 2004; 57(10):1069-1074.
127. Castoria G, Barone MV, Di Domenico M et al. Nontranscriptional action of oestradiol and progestin triggers DNA synthesis. Embo J 1999; 18(9):2500-2510.
128. Santen RJ, Song RX, McPherson R et al. The role of mitogen-activated protein (MAP) kinase in breast cancer. J Steroid Biochem Mol Biol 2002; 80(2):239-256.
129. Oh AS, Lorant LA, Holloway JN et al. Hyperactivation of MAPK induces loss of ERalpha expression in breast cancer cells. Mol Endocrinol 2001; 15(8):1344-1359.
130. Gee JM, Robertson JF, Ellis IO et al. Phosphorylation of ERK1/2 mitogen-activated protein kinase is associated with poor response to anti-hormonal therapy and decreased patient survival in clinical breast cancer. Int J Cancer 2001; 95(4):247-254.
131. Atanaskova N, Keshamouni VG, Krueger JS et al. MAP kinase/estrogen receptor cross-talk enhances estrogen-mediated signaling and tumor growth but does not confer tamoxifen resistance. Oncogene 2002; 21(25):4000-4008.
132. Driggers PH, Segars JH. Estrogen action and cytoplasmic signaling pathways. Part II: The role of growth factors and phosphorylation in estrogen signaling. Trends Endocrinol Metab 2002; 13(10):422-427.
133. Campbell RA, Bhat-Nakshatri P, Patel NM et al. Phosphatidylinositol 3-kinase/AKT-mediated activation of estrogen receptor alpha: A new model for anti-estrogen resistance. J Biol Chem 2001; 276(13):9817-9824.
134. Faridi J, Wang L, Endemann G et al. Expression of constitutively active Akt-3 in MCF-7 breast cancer cells reverses the estrogen and tamoxifen responsivity of these cells in vivo. Clin Cancer Res 2003; 9(8):2933-2939.
135. Shoman N, Klassen S, McFadden A et al. Reduced PTEN expression predicts relapse in patients with breast carcinoma treated by tamoxifen. Mod Pathol 2004.
136. Cho H, Katzenellenbogen BS. Synergistic activation of estrogen receptor-mediated transcription by estradiol and protein kinase activators. Mol Endocrinol 1993; 7(3):441-452.
137. Bocquel MT, Kumar V, Stricker C et al. The contribution of the N- and C-terminal regions of steroid receptors to activation of transcription is both receptor and cell-specific. Nucleic Acids Res 1989; 17(7):2581-2595.
138. Tzukerman MT, Esty A, Santiso-Mere D et al. Human estrogen receptor transactivational capacity is determined by both cellular and promoter context and mediated by two functionally distinct intramolecular regions. Mol Endocrinol 1994; 8(1):21-30.
139. McDonnell DP, Clemm DL, Hermann T et al. Analysis of estrogen receptor function in vitro reveals three distinct classes of antiestrogens. Mol Endocrinol 1995; 9(6):659-669.
140. Fujita T, Kobayashi Y, Wada O et al. Full activation of estrogen receptor alpha activation function-1 induces proliferation of breast cancer cells. J Biol Chem 2003; 278(29):26704-26714.
141. Hall JM, McDonnell DP. The estrogen receptor beta-isoform (ERbeta) of the human estrogen receptor modulates ERalpha transcriptional activity and is a key regulator of the cellular response to estrogens and antiestrogens. Endocrinology 1999; 140(12):5566-5578.
142. McInerney EM, Weis KE, Sun J et al. Transcription activation by the human estrogen receptor subtype beta (ER beta) studied with ER beta and ER alpha receptor chimeras. Endocrinology 1998; 139(11):4513-4522.
143. Hyder SM, Chiappetta C, Stancel GM. Interaction of human estrogen receptors alpha and beta with the same naturally occurring estrogen response elements. Biochem Pharmacol 1999; 57(6):597-601.
144. Delaunay F, Pettersson K, Tujague M et al. Functional differences between the amino-terminal domains of estrogen receptors alpha and beta. Mol Pharmacol 2000; 58(3):584-590.
145. Cowley SM, Parker MG. A comparison of transcriptional activation by ER alpha and ER beta. J Steroid Biochem Mol Biol 1999; 69(1-6):165-175.
146. Wissink S, van der Burg B, Katzenellenbogen BS et al. Synergistic activation of the serotonin-1A receptor by nuclear factor-kappa B and estrogen. Mol Endocrinol 2001; 15(4):543-552.
147. Paech K, Webb P, Kuiper GG et al. Differential ligand activation of estrogen receptors ERalpha and ERbeta at AP1 sites. Science 1997; 277(5331):1508-1510.

148. Webb P, Lopez GN, Uht RM et al. Tamoxifen activation of the estrogen receptor/AP-1 pathway: Potential origin for the cell-specific estrogen-like effects of antiestrogens. Mol Endocrinol 1995; 9(4):443-456.
149. Saville B, Wormke M, Wang F et al. Ligand-, cell-, and estrogen receptor subtype (alpha/beta)-dependent activation at GC-rich (Sp1) promoter elements. J Biol Chem 2000; 275(8):5379-5387.
150. Safe S. Transcriptional activation of genes by 17 beta-estradiol through estrogen receptor-Sp1 interactions. Vitam Horm 2001; 62:231-252.
151. Gelbfish GA, Davidson AL, Kopel S et al. Relationship of estrogen and progesterone receptors to prognosis in breast cancer. Ann Surg 1988; 207(1):75-79.
152. Stonelake PS, Baker PG, Gillespie WM et al. Steroid receptors, pS2 and cathepsin D in early clinically node-negative breast cancer. Eur J Cancer 1994; 30A(1):5-11.
153. Clark GM, McGuire WL, Hubay CA et al. The importance of estrogen and progesterone receptor in primary breast cancer. Prog Clin Biol Res 1983; 132E:183-190.
154. Bardou VJ, Arpino G, Elledge RM et al. Progesterone receptor status significantly improves outcome prediction over estrogen receptor status alone for adjuvant endocrine therapy in two large breast cancer databases. J Clin Oncol 2003; 21(10):1973-1979.
155. Dowsett M. Analysis of time to recurrence in the ATAC (armidex, tamoxifen, alone or in combination) trial according to estrogen receptor and progesterone receptor status. Paper presented at: San Antonio Breast Cancer Symposium, 2003: San Antonio.
156. Fuqua SA, Cui Y, Lee AV et al. Insights into the role of progesterone receptors in breast cancer. Journal of Clinical Oncology 2005; 23(4).
157. Hopp TA, Weiss HL, Hilsenbeck SG et al. Breast cancer patients with progesterone receptor PR-A-Rich tumors have poorer disease-free survival rates. Clin Cancer Res 2004; 10(8):2751-2760.
158. Vegeto E, Shahbaz MM, Wen DX et al. Human progesterone receptor A form is a cell- and promoter-specific repressor of human progesterone receptor B function. Mol Endocrinol 1993; 7(10):1244-1255.
159. Wen DX, Xu YF, Mais DE et al. The A and B isoforms of the human progesterone receptor operate through distinct signaling pathways within target cells. Mol Cell Biol 1994; 14(12):8356-8364.
160. McDonnell DP, Shahbaz MM, Vegeto E et al. The human progesterone receptor A-form functions as a transcriptional modulator of mineralocorticoid receptor transcriptional activity. J Steroid Biochem Mol Biol 1994; 48(5-6):425-432.
161. Kraus WL, Weis KE, Katzenellenbogen BS. Inhibitory cross-talk between steroid hormone receptors: Differential targeting of estrogen receptor in the repression of its transcriptional activity by agonist- and antagonist-occupied progestin receptors. Mol Cell Biol 1995; 15(4):1847-1857.
162. Ellis MJ, Coop A, Singh B et al. Letrozole is more effective neoadjuvant endocrine therapy than tamoxifen for ErbB-1- and/or ErbB-2-positive, estrogen receptor-positive primary breast cancer: Evidence from a phase III randomized trial. J Clin Oncol 2001; 19(18):3808-3816.
163. Ma XJ, Wang Z, Ryan PD et al. A two-gene expression ratio predicts clinical outcome in breast cancer patients treated with tamoxifen. Cancer Cell 2004; 5(6):607-616.
164. van 't Veer LJ, Dai H, van de Vijver MJ et al. Gene expression profiling predicts clinical outcome of breast cancer. Nature 2002; 415(6871):530-536.

Chapter 11

Novel Approaches for Chemosensitization of Breast Cancer Cells:

The E1A Story

Yong Liao, Dihua Yu and Mien-Chie Hung*

Abstract

The adenoviral E1A-mediated sensitization to a variety of anti-cancer drug-induced apoptosis is a well-established phenomenon on different types of cell systems. However, the mechanisms underlying E1A-mediated chemosensitization are still not fully understood. Recent studies demonstrate that E1A-mediated sensitization to drug-induced apoptosis can occur via multiple pathways; some of which depend on the expression of functional p53 and/or p19ARF proteins, while some are not. In human breast cancer cells with Her-2/neu overexpression, which usually are more resistance to anti-cancer drugs than cells without Her-2/neu overexpression, may be sensitized through E1A-mediated downregulation of Her-2/neu. Alternatively, E1A can induce sensitization to anticancer drugs in cancer cells or normal diploid fibroblast cells through upregulating the expression of caspase proenzymes, or downregulating the activity of a critical survival factor Akt and/or upregulating the activities of a pro-apoptotic kinase p38 and a protein phosphatase PP2A, etc. This review summarizes these progresses and proposes a plausible feed-forward model for E1A-mediated chemosensitization in human breast cancer cells.

Introduction

Breast cancer is the most common malignancy in American and northwestern European women. In 2005, 211,240 new cases of invasive breast cancer will be diagnosed among American women, as well as an estimated 58,490 additional cases of in situ breast cancer, and 39,800 women are expected to die from this disease. Only lung cancer accounts for more cancer deaths in women. Men are also susceptible to the disease, with an estimated 1,690 cases and 460 deaths in 2005 in the United States.[1] Despite recent advances in the treatment of breast cancer, survival rates for patients with metastatic breast cancer remain poor. Chemotherapeutics are the most effective treatment for metastatic tumors. However, the ability of cancer cells to become simultaneously resistant to different drugs—a trait known as multidrug resistance—remains a significant impediment to successful chemotherapy. Three decades of multidrug-resistance research have identified a myriad of ways in which cancer cells can elude chemotherapy, and it has become apparent that resistance exists against every effective drug, even the newest agents.[2] Therefore, the ability to circumvent drug resistance is likely to improve the efficacy of chemotherapy.

*Corresponding Author: Mien-Chie Hung—Department of Molecular and Cellular Oncology, The University of Texas M.D. Anderson Cancer Center, Houston, Texas 77030, U.S.A. Email: mhung@mdanderson.org

Breast Cancer Chemosensitivity, edited by Dihua Yu and Mien-Chie Hung.

The adenoviral type-5 and type-2 early region 1A (E1A) proteins were reported originally as an oncogene that could cooperate with other viral and cellular oncogenes to transform primary culture cells but not established cell lines, as distinct from the type-12 E1A, a potent oncogene that can transform established cell lines.[3] However, E1A has not been associated with human malignancies despite extensive studies. Instead, E1A was shown to suppress experimental metastasis of rodent cells transformed by the ras oncogene[4-6] and the Her-2/neu oncogene[7-9] and metastasis of certain human cancer cell lines.[10,11] In addition, increasing experimental results indicate that E1A can inhibit the tumorigenicity of both the transformed rodent cells and the human cancer cell lines.[7,8,12] Therefore, based on its ability to suppress both tumorigenicity and metastasis, E1A has been considered as a tumor suppressor gene[3,11,13-20] and translated into multiple clinical trials.[3,21-27] In addition to tumor suppressor activities, expression of the E1A gene in stably transfected normal fibroblast and human cancer cells has also been shown to increase sensitivity to the in vitro cytotoxicity of several anticancer drugs (such as etoposide and cisplatin) in normal fibroblasts and sarcoma cells, doxorubincin in colon and hepatocellular carcinoma cells, gemcitabine in hepatocellular and breast cancer cells, and paclitaxel in breast and ovarian cancers.[28-39] In primary mouse embryo fibroblasts, E1A-mediated sensitization to apoptosis induced by ionizing radiation and chemotherapeutic drugs has been reported through a p53-dependent and $p19^{ARF}$-dependent pathway.[28,40-44] Moreover, a p53-independent mechanism for E1A-mediated chemosensitization to the anticancer drugs etoposide and cisplastin has been demonstrated in human cancer cell lines, including Saos-2 osteosarcoma cells.[30,31,42,45] So far, E1A has been shown to mediate sensitization to a wide variety of apoptotic stimuli, including serum starvation, ultraviolet (UV)- and γ-radiation, tumor necrosis factor (TNF)-α, and different categories of anticancer drugs. This review describes a framework for drug-induced apoptosis and summarizes some recent work that improves our understanding of the molecular mechanisms underlying the adenovirus E1A-mediated chemosensitization.

Mechanisms of Apoptosis: Intrinsic versus Extrinsic Apoptotic Pathways

Most chemotherapeutic drugs kill cancer cells by inducing apoptosis, and many similarities exist in cellular response to drug-induced apoptosis, regardless of their primary target.[46-52] Defects in apoptosis signaling contribute to resistance of tumors.[47,48,53,54] Apoptosis, from the Greek word for "falling off" or "dropping off" (as leaves from a tree), is defined by distinct morphological and biochemical changes mediated by a family of cysteine aspartic acid-specific proteases (caspases), which are expressed as inactive precursors or zymogens (pro-caspases) and are proteolytically processed to an active state following an apoptotic stimulus. To date, approximately 14 mammalian caspases have been identified and can be roughly divided into three functional groups: apoptosis initiator (including caspase-2, -9, -8, -10), apoptosis effector (including caspase-3, -6, -7), and cytokine maturation (including caspase-1, -4, -5, -11, -12, -13, -14).[55-57] Two separable pathways lead to caspase activation: the extrinsic pathway and the intrinsic pathway. The extrinsic pathway is initiated by ligation of transmembrane death receptors (Fas, TNF receptor, and TRAIL receptor) with their respective ligands (FasL, TNF, and TRAIL) to activate membrane-proximal caspases (caspase-8 and –10), which in turn cleave and activate effector caspases such as caspase-3 and –7. This pathway can be regulated by c-FLIP, which inhibits upstream initiator caspases, and inhibitor of apoptosis proteins (IAPs), which affect both initiator and effector caspases.[58-60] The intrinsic pathway requires disruption of the mitochondrial membrane and the release of mitochondrial proteins, such as cytochrome *c*. Cytochrome *c*, released from the mitochondrial intermembrane space to cytoplasm, works together with the other two cytosolic protein factors, Apaf-1 (apoptoic protease activating factor-1) and procaspase-9, to promote the assembly of a caspase-activating complex termed the apoptosome, which in return induces activation of caspase-9 and thereby initiates the apoptotic caspase cascade.[56,57,61-63]

The primary regulatory step for mitochondrial-mediated caspase activation (the intrinsic pathway) might be at the level of cytochrome *c* release.[62,63] The known regulators of cytochrome *c* release are Bcl-2 family proteins.[62-65] According to their function in apoptosis, the mammalian Bcl-2 family can be divided into pro-apoptotic and anti-apoptotic members. The pro-apoptotic members include Bax, Bcl-Xs, Bak, Bok/Mtd, which contain 2 or 3 Bcl-2 homology (BH) regions, and molecules such as Bad, Bik/Nbk, Bid, Hrk/DP5, Bim/Bod, and Blk, which contains only the BH3 region. The anti-apoptotic Bcl-2 family members include Bcl-2, Bcl-XL, Bcl-w, A1/Bfl-1, Mcl-1, and Boo/Diva, which contain three or four regions with extensive amino acid sequence similarity to Bcl-2 (BH1-BH4).[65,66] Overexpression of the anti-apoptotic molecules such as Bcl-2 or Bcl-XL blocks cytochrome *c* release in response to a variety of apoptotic stimuli. On the contrary, the pro-apoptotic members of the Bcl-2 family proteins (such as Bax and Bid) promote cytochrome *c* release from the mitochondria. Pro-apoptotic and anti-apoptotic members of the Bcl-2 protein family can physically interact. For example, binding of BH-3 only proteins (e.g., Noxa, Puma, Bad, and Bim) to anti-apoptotic Bcl-2 proteins (e.g., Bcl-2 and Bcl-XL) results in activation of Bax and Bak.[64,65] In addition, there is considerable cross-talk between the extrinsic and intrinsic pathways. For example, caspase-8 can proteolytically activate Bid, which can then facilitate the release of cytochrome *c* and amplifies the apoptotic signal following death receptor activation.[52,56,57,61-63] Most of the anticancer agents either directly induce DNA damage or indirectly induce secondary stress-responsive signaling pathways to trigger apoptosis by activation of the intrinsic apoptotic pathway, and some can simultaneously activate the extrinsic receptor pathway. Therefore, molecules or signaling events that regulate the processes of apoptosis can also affect cellular response to drugs.

Factors and Key Molecules Involved in the Regulation of Apoptosis and Drug Response in Breast Cancer

Her-2/neu

The erbB2 gene (also known as HER-2, neu, and NGL) encodes a 185-kDa transmembrane glycoprotein (ErbB2) that belongs to the epithelial growth factor receptor (EGFR) family and heterodimerizes with other EGFR family members upon ligand stimulation. Similar to EGFR, the Her-2/neu growth factor receptor has intrinsic tyrosine-kinase activities that autophosphorylates and phosphorylates its dimerization partners, resulting in simultaneous activation of several signal transduction pathways and leading to the activation of the PI3K-Akt cell survival pathway and mitogen-activated protein kinase pathways that induce mitogenesis, cell proliferation, cell survival, and genomic instability. Amplification or overexpression of Her-2/neu or both have been detected in about 30% of human breast cancers and in many other types of human malignancies. In addition, overexpression of HER-2/neu in breast cancer has been associated with poor overall survival and has been shown preclinically to enhance malignancy and the metastatic phenotype.[67-69] Although discrepancies exist among different reports, Her-2/neu overexpression seems to have induced chemoresistance in several preclinical studies and a few clinical observations.[69] The predictive value of Her-2/neu overexpression and chemoresistance has been demonstrated in breast and ovarian cancer patients. In general, tumor cells overexpressing Her-2/neu are intrinsically resistant to DNA-damaging agents such as cisplatin. Although the molecular mechanisms by which Her-2/neu induces drug resistance are not yet established, there is evidence that this may be a consequence of altered cell cycle checkpoint and DNA repair mechanisms and dysregulation of apoptotic pathway(s)[67,69] (please refer to section *PI3K/Akt Pathway* and chapters by Yu et al in this book). Drug-induced apoptosis depends on the balance between cell cycle checkpoints and DNA repair mechanisms.[51] Blockade of Her-2/neu signaling using Her-2/neu antagonists, dominant negative mutants, or chemical inhibitors of Her-2/neu tyrosine kinase activity induces cell cycle arrest, inhibits DNA repair, and (or) promotes apoptosis.[22,68] Less understood are downstream signal transduction cascades by which Her-2/neu affects these regulatory mechanisms.

P53

The p53 tumor suppressor gene is the most frequently mutated gene in human tumors, and loss of p53 function can both disable apoptosis and accelerate tumor development in transgenic mice. In addition, functional mutations or altered expression of p53 upstream regulators (such as ATM, Chk2, MDM2, and $p14^{ARF}$) and downstream effectors (PTEN, Bax, Bak, and Apaf-1) are also common in human tumors, including human breast cancer. Mutations in p53 or in the p53 pathway can produce the multidrug resistance phenotype in vitro and in vivo, and reintroduction of wild-type p53 into p53-null tumor cells can reestablish chemosensitivity.[48,70-73] Direct DNA damage that caused by most DNA-damaging drugs and UV or ionizing radiation and overexpression of oncoproteins (e.g., E1A, ras, myc) can activate the intrinsic apoptotic pathway.[52] As a sensor of cellular stress, p53 is a critical initiator of this pathway. P53 can transcriptionally activate pro-apoptotic Bcl-2 family members (such as Bax, Bak, PUMA, and Noxa) and repress anti-apoptotic Bcl-2 proteins (e.g., Bcl-2, Bcl-XL) and IAPs (e.g., Survivin). Additionally, p53 can also transactivate other genes that may contribute to apoptosis, including tumor-suppressor PTEN, apoptosis regulators or inducers such as Apaf-1, PERP, p53AIP1, CD95 and TRAIL receptors (TRAIL-R2/DR5), and genes that leading to an increase in reactive oxygen species (ROS).[71,72,74]

Bcl-2 Family Proteins

As discussed above, Bcl-2 family proteins play a pivotal role in the regulation of the intrinsic apoptotic pathway as guardians of the mitochondria, since these proteins localize particularly to the mitochondrial membrane.[75] Mutations or altered expression of pro-apoptotic or anti-apoptotic Bcl-2 family proteins can drastically alter drug response in experimental systems.[76,77] In addition, several clinical reports have provided support that a high expression level of anti-apoptotic Bcl-2 proteins confers a clinically important chemoresistant phenotype on cancer cells. Likewise, reduced expression of pro-apoptotic Bax levels has been associated with poor response to chemotherapy and shorter overall survival for patients with breast cancers, whereas enhanced expression of Bax protein correlated with a good response to chemotherapy in vivo.[76,78] In addition, chemotherapeutic drugs exert their effect in part by modulating the expression of several members of the Bcl-2 family proteins.[79] For example, paclitaxel, doxorubicin, and thiotepa upregulate several pro-apoptotic Bcl-2 proteins and downregulate anti-apoptotic Bcl-2 proteins. Blockade of Bcl-2 activity or expression by either an anti-Bcl-2 single chain antibody (anti-Bcl-2 sFv) or a novel Bcl-2/Bcl-XL bi-specific anti-sense oligonucleotide has demonstrated remarkable effect in enhancing drug-induced cytotoxicity in breast cancer cells.[79,80]

P-Glycoprotein and MDR

Expression of the multidrug resistance gene (mdr) is one of the most-studied potential mechanisms underlying multidrug resistance. The human mdr gene family encompasses two homologous members, the first of which, called the mdr1 gene, is the best characterized so far. The human mdr1 gene encode a membrane P-170 glycoprotein that, on the basis of its structure, is considered to act as a drug-efflux pump excreting various drugs from cells. Resistance results because increased drug efflux lowers intracellular drug concentrations.[81,82] Analysis of 31 reports published between 1989 and 1996 found that 41% of breast tumors expressed P-170 glycoprotein. P-170 expression increased after therapy and was associated with a greater likelihood of treatment failure, although there was considerable inter-study variability.[2,82] In addition to actively effluxing chemotherapeutic drugs, the P-170 glycoprotein can also protect cells against apoptosis mediated by the death receptor pathway, UV irradiation, and serum starvation.[2,52]

Ceramide

Ceramide, or N-acyl-sphingosine, has been implicated in the acquired drug resistance in breast cancer. Ceramide is a metabolite of sphingomyelin hydrolysis by neutral or acidic sphingomyelinases. Sphingomyelin is the most abundant lipid in the plasma membrane of mammalian cells. Ceramide functions as a second messenger to signaling cascades that promote cell differentiation, proliferation, senescence, and apoptosis.[83] Ceramide is produced in response to a wide range of stimuli, including a long list of chemotherapeutic agents. Several lines of evidence have linked the failure of ceramide production to chemotherapy resistance, including the lack of a ceramide response in drug-resistant cell lines and the abrogation of paclitaxel-induced apoptosis by blocking ceramide synthesis.[79,83] In addition, protein phosphatases activated by ceramide, such as protein phosphatase 2A (PP2A), have been shown to promote inactivation of a number of anti-apoptotic molecules or pro-growth regulators, including Bcl-2, PKC, and Akt.[51,83]

PI3K/Akt Pathway

The PI3K-Akt signaling pathway, a major signaling component downstream of growth factor receptor tyrosine kinases, regulates fundamental cellular processes such as cell survival, growth, and motility—that are critical for tumorigenesis.[84] Indeed, aberrant activation of the PI3K-Akt pathway has been widely implicated in many cancers. For example, the serine/threonine kinase Akt and its family members Akt 2 and 3 are amplified or their activity is constitutively elevated in human carcinomas such as breast, pancreatic, ovarian, brain, prostate, lung, and gastric cancers.[85,86] As a direct downstream target of PI3K, Akt is a key oncogenic survival factor that can phosphorylate and inactivate a panel of critical pro-apoptotic molecules, including Bad, caspase-9, the Forkhead transcription factor FKHRL1 (known to induce expression of pro-apoptotic factors such as Fas ligand), GSK3-β, ASK1 (apoptosis signal-regulating kinase-1), cell cycle inhibitors p21 and p27, and tumor-suppressor TSC2.[71,84,87-93] In addition, we and others found that Akt can also inactivate p53 through phosphorylation and nuclear localization of MDM2.[88,89] Blockage of the Akt pathway by a genetic approach using a dominant-negative Akt (DN-Akt) construct resulted in p53 stabilization and accumulation and had a synergistic effect with the genotoxic anti-cancer drug etoposide in inducing apoptosis in NIH3T3 cells both in vitro and in vivo.[88] Numerous groups have reported the effects of chemotherapy on Akt activity in tumor cell lines. These studies showed that the administration of chemotherapeutic agents commonly results in reduced activity of the PI3K/Akt pathway. Although in certain cases chemotherapy agents increased Akt activity, in most cases such increases in Akt activity were transient and were followed by subsequent decreases in activity.[94,95] In addition, studies performed in vitro and in vivo combining small molecule inhibitors of the PI3K/Akt pathway with standard chemotherapy have successfully attenuated chemotherapeutic resistance.[85,96-98]

NF-κB

NF-κB is a multi-subunit nuclear transcription factor that regulates several cellular functions, including cell growth, differentiation, development, adaptive response to redox balance, and apoptosis.[99] In its inactive form, NF-κB resides in the cytoplasm bound to inhibitory IκB proteins that shield the DNA binding site. External stimuli, such as pro-inflammatory agents, infectious agents, stress, and chemotherapeutic drugs, activate NF-κB by phosphorylation and subsequent degradation of IκB proteins by the IκB kinase (IKK), whereupon NF-κB is transported into the nucleus and transcription of the target genes occur. As a nuclear transcription factor, NF-κB target genes including several anti-apoptotic proteins, e.g., the IAP family of caspase inhibitory proteins, TRAF1 and TRAF2 (which are thought to repress caspase-8 activation), Bfl1/A1, Bcl-XL, FLIP, and inducible nitric oxide synthetase. Interestingly, NF-κB also controls promoter activation of certain pro-apoptotic factors, such as CD95 (Fas) and CD95 ligand (FasL) and TRAIL receptors (TRAIL-R1 and -R2).[99,100] Whether NF-κB targets pro- or anti-apoptotic genes depends on the stimulus-specific signaling pathway activated. Administration of certain types of anticancer drugs induced NF-κB transcriptional activation, whereas inhibition of NF-κB in parallel with chemotherapy strongly enhanced the cytotoxic effect of chemotherapy.[79,80,100] Thus, NF-κB may play an important role in inducible chemoresistance, and inhibition of NF-κB may confer sensitivity to cytotoxic anticancer drugs.

Structures, Biochemical Features, and Associated Cellular Proteins of E1A

The adenovirus type 5 (Ad5) E1A encodes two differentially spliced 12S and 13S mRNAs that give rise to 243 and 289 residue (243R, 289R) proteins, respectively. Each is comprised of two exons and, as splice joining exons 1 and 2 of the 12S mRNA is in frame with the spliced 13S transcript, the encoded proteins are identical apart from a central 46 residue region present in the larger product.[101,102] There are three minor E1A mRNA transcripts from Ad5 (11S, 10S, and 9S) that are expressed primarily at later stages of infection, and their functions are largely unknown.[101,103] The roles of the major 12S and 13S E1A products have been studied extensively and their importance in sensitization to apoptosis induced by anticancer drugs is well-documented and will be discussed here.

Both major E1A products are nuclear, highly acidic, and extensively phosphorylated transcription factors that exert their effects through their functional domains, which are highly conserved in virtually all human adenoviruses and in many adenoviruses of other animal species. These regions in Ad5 include conserved regions 1 and 2 (CR1 and CR2) located in Ad5 between residues 40 to 80 and between residues 120 to 139, respectively, and CR3 is located between residues 140 to 185. All three are encoded within exon 1 of the 289R product. CR1 and CR2 are present in both 289R and 243R; however, CR3 is unique to 289R, as it represents the 46 residues eliminated by splicing of the 12S mRNA[101] (Fig. 1). In addition, a region at the amino terminus, which is not highly conserved, is also of great importance. The first E1A-binding proteins to be detected were pRb tumor suppressor and the related proteins p107 and p130, and a 300-kDa transcriptional coactivator/signal integrating protein p300 and CBP, which is a relative of the cyclic AMP responsive element binding (CREB) protein.[101,104] In addition to pRb and p300/CBP, E1A also interacts with a 400-kDa protein doublet through its amino terminus 25-36 residues, which was recently characterized as a strong SWI2/SNF2 homology, p400, and TRRAP/PAF400, a myc-associated, transcriptional coactivator (TRRAP, transactivation/transformation-domain-associated protein) as well as a component of the p/CAF histone acetyltransferase complex (PAF400).[102] The E1A-p400 complex interaction is mainly linked to the E1A transformation mechanism and is not the focus of the current review.

Binding of the pRb family of proteins occurs primarily through a conserved binding site Leu-X-Cys-X-Glu found in the CR2 domain; however, a minor but critical contact is also made with a portion of CR1. Binding of p300/CREB/p400 requires the amino terminus and a region of CR1. P300 and CBP possess endogenous histone acetyl transferase (HAT) activity, which can be directly inhibited by E1A.[104,105] E1A may serve as a substrate for p300 and CBP to repress the HAT activity of p300/CBP, as it can be acetylated at Lys-239 by p300 and P/CAF.[105,106] The CR3 domain interacts with a number of transcription factors, including the TATA-binding protein (TBP), a critical component of the basal transcription complex, and upstream factors such as ATF members, Sp1, and c-Jun. A relatively uncharacterized 48-kDa proteins, termed CtBP, bind to a region encoded by exon 2, just adjacent to the carboxy terminal nuclear localization signal. CtBP is a transcriptional corepressor that was identified originally by its interaction with a conserved PXDLS motif near the carboxyl terminus of E1A.[107] Various repressor proteins that also contain the PXDLS motif recruit CtBP to promoters, where it represses transcription by subsequent recruitment of histone deacetylases (HDACs).[107] It has been shown that mutants of E1A that mimic the effect of acetylation of the Lys239 at the PXDLS motif of E1A by p300/CBP and P/CAF make E1A defective in CtBP binding. These E1A mutants that mimic the effect of acetylation were also defective in repressing

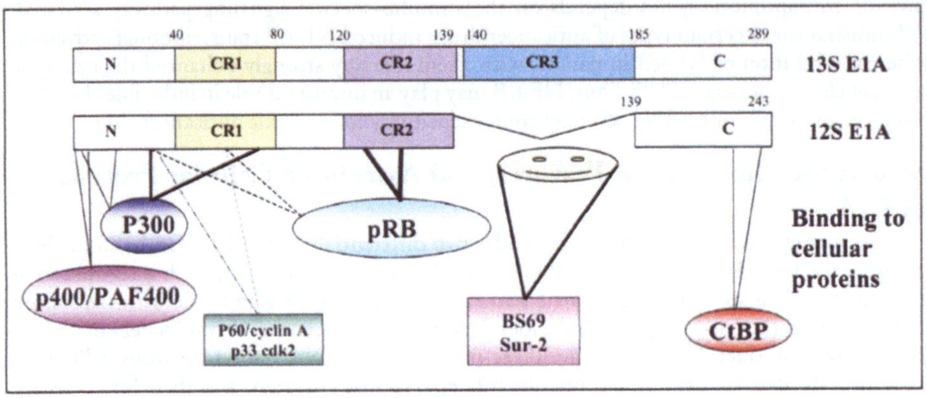

Figure 1. Map of E1A functional domains in 13S and 12S E1A proteins: conserved regions (CR) 1, 2, 3 and consensus sites for binding to cellular proteins.

CREB-stimulated (CBP-dependent) transcriptional activation under certain conditions, suggesting that E1A-mediated transcriptional repression may require interaction with CtBP. In addition, the tumor-suppressor proteins pRb and BRCA1 also recruit CtBP through an adaptor protein CtIP, indicating that CtBP might be involved in the transcriptional repression and/or apoptosis that is regulated by these proteins.[107]

E1A products interact with pRb, p107, and p130 via the major binding site in the CR2 domain and the N-terminal portion of the CR1 domain plays an auxiliary role; however, it is the CR1 domain that competes directly with E2F for access to the pRB pocket, i.e., the E1A CR1 domain can inhibit E2F-pRb complex formation and E1A, by binding with pRB through CR1, releases E2F.[101,108] E2F was originally described as a factor regulating adenovirus E2 transcription; however, it soon became apparent that it is of general importance in the expression of genes encoding DNA synthetic enzymes and regulators of the cell cycle and apoptosis.[109,110] E2F is a sequence-specific, DNA-binding transcription factor that exists as a family of heterodimers containing one of six E2F proteins bound to either DP1 or DP2.[110] A major function of the pRb tumor-suppressor family is to bind to and inactivate E2F. E2F-1, -2, and –3 heterodimers bind to pRb; E2F-4 to pRb, p107 and p130; and E2F-5 to p130 only. E2F6 does not interact with the pRb family of pocket proteins and functions as a negative regulator of E2F-dependent transcription via complexing with chromatin modifiers.[109-111] Binding of E2F to pRb family members involves the "large binding pocket" that overlaps the region required for E1A protein binding. When E1A proteins bind to pRb and its family members, the viral protein competes with E2F for occupancy of the pocket domain, dissociating E2F from the tumor suppressor proteins. When E2F is released, it can activate transcription from promoters that contains its binding site.[108] Exogenous expression of E2F1, E2F2 or E2F3 in quiescent cells results in S-phase entry. In addition, ectopic expression of E2F1 leads to apoptosis in tissue culture cells and transgenic mice.[109]

Chemosensitization by E1a

Ever since the landmark discovery by Lowe and coworkers in 1993[28,40] that the adenovirus E1A can sensitize fibroblasts to apoptosis induced by ionizing radiation and anticancer drugs in a p53-dependent manner, a variety of mechanisms for E1A-mediated chemosensitization have been proposed. It is now recognized that both p53-dependent and –independent pathways, p14^{ARF}-dependent and –independent pathways, and additional pathways may all contribute to E1A-mediated chemosensitization in different cellular contexts. Summarizing these different molecular mechanisms underlying E1A-mediated chemosensitization is the major focus of this review.

Trancriptional Activation of Pro-Apoptotic Molecules: The E2F-1 Pathway

Earlier studies in mouse embryo fibroblasts (MEF) with wild-type p53 or p53 deletion demonstrated that the E1A-mediated sensitization to apoptosis induced by ionizing radiation and several chemotherapeutic agents (5-fluorouracil, etoposide, and doxorubicin) depends on the expression of functional p53 tumor suppressor, as treatment with E1A-expressing cells lacking p53 had little or no effect on viability.[28,40-42,112-118] The stability of p53 protein is increased in cells expressing E1A. Further study showed that stabilization and accumulation of p53 by E1A is through the p19^{ARF} tumor suppressor, since the ability of E1A to induce p53 and its transcriptional targets is severely compromised in ARF-null cells.[44,117,119] ARF (p19^{ARF} in the mouse, p14^{ARF} in human cells) is encoded by an Alternative Reading Frame of the Ink4A tumor suppressor locus. It binds to MDM2, an E3 ubiquitin ligase of p53, and inhibits its E3 ligase activity, thereby allowing p53 to escape MDM2-mediated ubiquitination and degradation and a consequent accumulation of active p53.[120] Reports from White's group showed that E1A inhibited MDM2 transactivation without affecting the expression of p21 or Bax, which resulted in a high level of p53 accumulation and apoptosis by a p300-dependent mechanism.[121] Since p300 is required for MDM2 induction by p53 and subsequent inhibition of p53 stabilization, inhibition of p300 by E1A results in p53 stabilization and causes apoptosis.[121]

A recent report from our group demonstrated that an alternative mechanism may also contribute to the stabilization of p53 by E1A.[122] By screening a yeast two-hybrid library, an MDM2-related p53 binding protein, MDM4, was identified as a direct binding partner of E1A. The NH2-terminal region of MDM4 and the CR1 domain of E1A were required for the interaction between MDM4 and E1A.[122] Since the CR1 domain of E1A is involved in binding with p300, it is not clear whether p300 is also involved in the interaction between MDM4 and E1A. However, we were able to show a tri-complex formation among E1A, MDM4, and p53 that resulted in the stabilization of p53 independent of the p14^{ARF} protein expression.[122] E1A did not affect the p53-MDM2 interaction, but it inhibited MDM2 binding to MDM4, resulting in decreased nuclear exportation of p53.[122] As discussed above, p53 can initiate apoptosis by transcriptionally activating pro-apoptotic Bcl-2 family members and repressing anti-apoptotic bcl-2 proteins and IAPs. In addition, p53 can also transactivate other genes that contribute to apoptosis, including PTEN, Apaf-1, PERP, p53AIP1. Among these p53 target genes, elevated expressions of Apaf-1 and Bax have been reported in response to E1A and partly contributed to E1A-mediated chemosensitzation.[123,124] In baby rat kidney cells, the activity of p53 as a transcription factor is directly correlated with the ability of E1A to induce apoptosis. Whether other p53 target genes are also altered in E1A-expressing cells needs further investigation.

Disrupting RB-E2F heterodimers and subsequent releasing of free E2F-1 by E1A has been suggested to contribute to ARF induction, which is also consistent with the possibility that ARF is an E2F-response gene. Enforced expression of E2F induces p19^{ARF} and, conversely, ARF null cells are resistant to E2F-1-induced apoptosis.[109,110,118,125-130] In addition to the ARF protein, E2F1 can directly activate transcription of p73, a member of the p53 family, in turn leading to activation of p53-responsive target genes and apoptosis. Disruption of the p73 function, by dominant negative p73 mutants or by gene targeting, inhibits E2F-1-induced apoptosis.[109,131] E2F-1 also directly activates the expression of the Apaf-1 gene.[132] Induction of E2F-1 activity results in an increase of mRNA and protein levels of Apaf-1 and a concomitant activation of caspase-9, -3, and -6 and possibly caspase-7. E2F-1-induced apoptosis is significantly reduced by gene disruption of Apaf-1. In addition, RB null embryos exhibit increased levels of Apaf-1, suggesting that Apaf-1 is required for apoptosis induced by E1A-mediated pRb deficiency and subsequent E2F-1 activation.[109,132] Apaf-1 is a also direct transcriptional target of p53, raising the possibility that its transactivation by E2F-1 is indirect. However, E2F-1 binds the Apaf-1 promoter and transactivates the Apaf-1 gene in cells lacking p53, and E2F-1 also activates the deleted version of the Apaf-1 promoter that is not activated by p53, suggesting Apaf-1 is a direct target of E2F-1.[109] In addition to ARF, p53, p73, and Apaf-1, a recent study from Lowe's laboratory also demonstrated that pro-caspase-3, -7, -8, and –9 can also be direct transcriptional targets of E2F-1 and disrupting pRb-E2F-1 complex by E1A, loss of pRb expression by genetic approach, or enforced E2F-1 expression results in the accumulation of caspase proenzymes and sensitization to apoptosis after the release of cytochrome *c* in fibroblasts.[130] Although a number of recent studies have demonstrated that RB inactivation or E2F-1 overexpression leads to apoptosis that is inhibited by loss of p53 but not by loss of ARF, E1A induces expression of these pro-caspase enzymes in cells deficient in either p53 or ARF, suggesting that E1A regulates pro-caspase expression through a p53-independent mechanism.[130]

Interestingly, studies using E2F-1 mutants have demonstrated that although its DNA-binding activity is required, transcriptional transactivation is not necessary for the induction of apoptosis by E2F-1.[129,133,134] Regardless of the fact that expression of E1A induced accumulation of caspase pro-enzymes in human normal diploid fibroblasts, screening the expression levels of caspase pro-enzymes in E1A–expressing human carcinoma cells with epithelial origin, including breast cancer MDA-MB-231, MDA-MB-453, MCF-7, and ovarian cancer 2774, did not observe a unanimous increase of these caspase pro-enzymes.[37] This again suggests that transcriptional upregulation of these caspase pro-enzymes in human cancer cells may not be as critical as that in the normal fibroblast cells for E1A–mediated sensitization to apoptosis. The discrepancy between normal diploid fibroblasts and epithelial carcinoma cells in E1A–mediated sensitization to apoptosis may reflect the nature of the intrinsic difference between normal fibroblasts and carcinoma cells. In addition, the pRb pathway is functionally inactivated in most human cancers; however, loss of pRb expression in human cancer cells may not necessarily result in an increased expression of caspase pro-enzymes, according to our screening of human breast cancer cells and normal skin cells and fibrosarcoma cell lines with different genetic backgrounds in expression of p53, ARF, and pRb proteins (Fig. 2). Intriguingly, noticeable induction of apoptosis in anticancer drug-treated cells expressing E1A mutant incapable of binding to pRB was also observed recently by Chattopadhyay et al,[135] and they also failed to detect the release of E2F-1 in their system, suggesting that additional mechanisms may be involved in E1A-mediated sensitization to apoptosis in human cancer cells.

Transcriptional Inactivation of Receptor Tyrosine Kinases: The Her-2/neu and Axl Pathway

Earlier studies from our group demonstrated that E1A may function as a tumor suppressor for Her-2/neu-overexpressing cancer cells by repressing the expression of Her-2/neu at the transcriptional level.[7-9,17,22,32,136-139] The Her-2/neu promoter contains several positive elements that require the p300/CBP coactivator proteins, which are inhibited by direct E1A interaction. The NH2-terminus and CR1 domain of E1A that is required for E1A to bind with p300 is also required for transcriptional repression of Her-2/neu.[140,141] When introduced into rat B104-1-1 cells transformed by the mutated rat Her-2/neu gene, E1A gene products suppressed the transformed phenotypes and inhibited the metastatic potential of the B104-1-1 cells. Similar results were also observed in SKOV3.ip1 human ovarian cancer cells that overexpress Her-2/neu. The E1A-expressing ovarian cancer cell lines showed decreased p185Her-2/neu expression and reduced malignancy, including a decreased ability to induce tumors in *nu/nu* mice. In addition, preclinical studies using either liposome- or adenovirus-mediated E1A gene transfer found inhibited tumor growth of Her-2/neu-overexpressing breast cancer cells injected into the mammary fat pads of mice and prolonged animal survival compared with controls.[32,137-139] As discussed above, breast tumors that overexpress Her-2/neu are less responsive to treatment with various anticancer agents, such as cyclophosphamide, methotrexate, 5-fluorouracil, epirubicin, paclitaxel and docetaxel, and patients with cancers that overexpress Her2/neu are associated with unfavorable prognosis, shorter relapse time, and lower survival rate.[69,142-144] Therefore, we were prompted to test whether downregulation of Her-2/neu expression by E1A conferred an enhanced sensitivity to anticancer drugs in Her-2/neu overexpressing breast cancer cells. When the E1A gene was transferred into two human breast cancer cell lines that

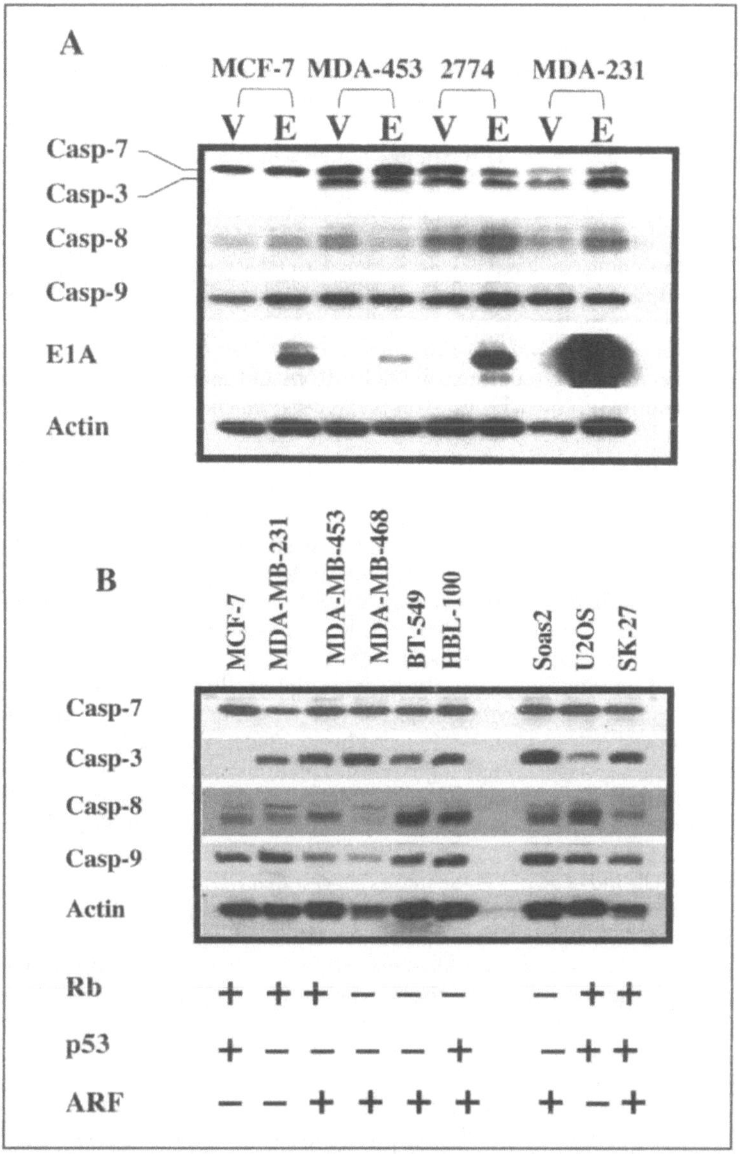

Figure 2. Expression of caspase-3, –7, –8, and –9 proenzymes in human cancer cells. A) Caspase proenzyme expression in E1A stable cells established in human breast cancer cell lines MCF-7, MDA-MB-231, and MDA-MB-453; and ovarian cancer cell line 2774. V: vector control; E: E1A stable cells. B) Procaspase enzyme expression in Rb wild-type breast cancer cells MCF-7, MDA-MB-231, MDA-MB-453; normal foreskin fibroblast cell SK-27; and osteosarcoma U2OS cells versus Rb-deficient cells (MDA-MB-468, BT-549, HBL-100, and Soas2). SK-27, U2OS, and Soas2 cells were obtained from the American Type Culture Collection (ATCC, Manassas, VA). Mouse monoclonal antibodies against human caspase–3 and –7 were from Transduction Laboratories (1:1000; C31720, Lexington, KY) and BD PharMingen (1:1000; 66871A, San Diego, CA), respectively. Rabbit polyclonal antibodies against human caspase-8 and –9 were from Santa Cruz Biotechnology (1:500; SC-7890/H-134, Santa Cruz, CA) and Cell Signaling Technology (1:500; #9502, Beverly, MA), respectively. A) Adapted with permission from reference 37 (Liao Y, Hung MC. Mol Cell Bio 2003; 23:6836).

overexpress Her-2/neu (MDA-MB-453 and MDA-MB-361), levels of Her-2/neu expression in the E1A transfected cells were reduced. The cell proliferation assay and soft agar colony-formation assay indicated a synergistic growth-inhibitory effect following treatment with the combination of E1A gene and paclitaxel in breast cancer cells that overexpress Her-2/neu. A similar synergistic effect was also observed recently following liposome-mediated systemic delivery of the E1A gene via the mouse tail vein in Her-2/neu over-expressing human breast cancer xenograft model.[145] The data indicated that E1A can sensitize Her-2/neu-overexpressing breast cancer cells to paclitaxel through E1A-mediated Her-2/neu repression. This finding has important clinical implications for the development of a novel therapeutic strategy by combining paclitaxel chemotherapy with E1A gene therapy for the treatment of Her-2/neu-overexpressing breast cancers, and a Phase I/II clinical trial is currently underway to determine the therapeutic efficacy of this combination strategy in such cancers. An earlier report by Sabbatini et al demonstrated that expression of Her-2/neu and c-Ha-ras oncogenes induces expression of the P-glycoprotein MDR1 mRNA and confers resistance to doxorubicin in immortalized normal epithelial MCF-10A breast cells, which are negative for the expression of P-glycoprotein.[146] In studies of low Her-2/neu-expressing MDA-MB-435 breast cancer cells, Yu et al observed that expression of Her-2/neu in the cells did not induce P-glycoprotein expression and blocking MDR1 by thioradazine did not sensitize these transfectants to paclitaxel treatment, suggesting that overexpression of Her-2/neu leads to intrinsic paclitaxel resistance independent of the MDR1 mechanism.[142] Further studies by Yu et al suggested that Her-2/neu acts synergistically with MDR1 to confer a higher degree of paclitaxel resistance.[147] It is yet to be determined whether expression of E1A also affects MDR1 expression, in addition to down-regulation of Her-2/neu, for E1A-mediated chemosensitization.

Using a tyrosine kinase differential display (TK-DD) approach, we found that Axl, another oncogenic receptor tyrosine kinase, can also be transcriptionally repressed by the expression of E1A. Similar to the observed downregulation of Her-2/neu, transcriptional repression of Axl by E1A also sensitized cells to apoptosis induced by serum deprivation.[148] Further elucidation of the molecular mechanisms underlying E1A-mediated inhibition of Axl led to the finding that Akt, a critical Axl downstream molecule, was inactivated as a result of E1A expression (see detailed discussions below).[149] However, when we enforced Axl expression in the E1A-expressing cells, E1A-mediated sensitization to apoptosis was abrogated and Akt was reactivated in the presence of the Axl ligand Gas6.[149] These studies suggest that E1A can sensitize cells to apoptosis using alternative pathways in addition to its function in repressing Her-2/neu expression. Indeed, we and others have reported that sensitization to the cytotoxic effects of anticancer drugs (such as cisplatin, adriamycin, gemcitabine, methotrexate, and paclitaxel) by the expression of E1A was also achieved in cells that do not overexpress Her-2/neu.[30,37,39,150-152] In one study, expression of 12S E1A alone or 12S plus 13S E1A did not affect Her-2/neu expression in cancer cell lines expressing a low level of Her-2/neu; however, expression of E1A not only severely reduced the anchorage-independent and tumorigenic growth of these cell lines but also sensitized these cells to the cytotoxic effects of the anticancer drugs.[30] In an E1A-expressing murine melanoma cell system that does not overexpress Her-2/neu, expression of E1A repressed tumor growth in vivo and sensitized the cells to apoptosis induced by either γ-radiation or cisplatin.[150] In addition, recent reports from our group showed that E1A mediated chemosensitization in the low Her-2/neu-expressing breast cancer cell lines MCF-7 and MDA-MB-231 in vitro and in an orthotopic animal model in vivo in a systemic gene therapy setting.[37,39,152] We had previously shown that E1A, through downregulation of Her-2/neu, sensitized cellular response to paclitaxel-induced apoptosis in Her-2/neu overexpressing cells.[32,137,138] In those studies, we could not detect E1A-mediated chemosensitization in the low Her-2/neu-expressing cells. The major reason for this discrepancy was due to the paclitaxel concentrations tested. The Her-2/neu-overexpressing cancer cells are resistant to paclitaxel at a dosage of 10.0 μM; in the presence of E1A, however, they became sensitive at a paclitaxel dose of 1.0 μM. Whereas, the low Her-2/neu-expressing cells, such as MDA-MB-231and MDA-MB-435, are much more sensitive to paclitaxel, even at a concentration of 0.1 μM.

Therefore, in the previous studies, the dose of paclitaxel applied was too high (1 μM) to detect the in vitro E1A-mediated paclitaxel sensitization that was found in the later study (0.01 μM). Therefore, transcriptional repression of Her-2/neu expression is only one of the many mechanisms underlying E1A-mediated chemosensitization and tumor suppression.

Targeting the Intracellular Signaling: The Akt/ASK-MEKK3/p38 Pathway

The breast cancer cell lines MCF-7 and MDA-MB-231 express low levels of Her-2/neu, however, expression of E1A enhanced their sensitivity to apoptosis induced by various anticancer drugs. Because MDA-MB-231 cells express no functional p53 and ARF, whereas MCF-7 cells express an undetectable level of Axl mRNA and protein,[153] we therefore prompted to explore whether additional mechanisms may contribute to E1A-mediated chemosensitization in these cell systems. To determine whether apoptosis-related kinases are involved in E1A-mediated sensitization to apoptosis, we examined the phosphorylation status of the well-known kinases representing different signaling pathways involved in the regulation of apoptosis[154]—p38, Akt, Erk, and JNK—in E1A-expressing MDA-MB-231 and MCF-7 cells (231-E1A and MCF-7-E1A) versus vector-transfected cells (231-Vect, MCF-7-Vect). The phosphorylation levels of JNK1/2 kinases were unchanged in E1A-expressing versus vector control cells in the absence of drug stimuli, whereas upon treatment with paclitaxel, the JNK1/2 kinases were transiently activated and then reduced to the basal level. Kinetically, phosphorylation of JNK1/2 does not correlate with paclitaxel-induced Bcl-2 phosphorylation and PARP cleavage, two hallmarks of paclitaxel-induced apoptosis; therefore, it is unlikely that JNK is the critical player in E1A-mediated chemosensitization in these cell systems. Although the basal phosphorylation level of Erk1/2 was increased in E1A-expressing cells compared with their respective vector controls, blocking of Erk1/2 activity by a specific MEK (a direct upstream kinase of Erk1/2) inhibitor PD98058 alone drastically enhanced spontaneous apoptosis in the absence of paclitaxel in E1A-expressing cells but not in the vector controls, suggesting that Erk is a critical survival factor in stable E1A-expression cells and that elevated Erk1/2 activity do not contribute to E1A-mediated chemosensitization in our system (Liao et al, unpublished data). Elevated expression of phosphorylated p38 was observed in both E1A-expressing MCF-7 and MDA-MB-231 cells, whereas the total level of p38 protein was similar in cells with or without expression of E1A. To test whether alteration of the kinase activity of p38 played a role in E1A-mediated sensititization to drug-induced apoptosis, we compared the kinetics of phosphorylation of p38 with paclitaxel-induced apoptosis in 231-E1A cells, using PARP cleavage and Bcl-2 phosphorylation as apoptotic cell death markers. Kinetically, PARP cleavage and Bcl-2 phosphorylation occurred after increased p38 phosphorylation in 231-E1A cells upon exposure to paclitaxel, suggesting that elevated p38 phosphorylation may be associated with E1A-mediated sensitization to paclitaxel-induced apoptosis. To evaluate whether activation of p38 is required for E1A-mediated sensitization to paclitaxel, we tested whether blocking p38 activity will inhibit E1A-mediated sensitization in 231-E1A cells. We used a specific p38 inhibitor (SB203580) and an IPTG-inducible dominant-negative p38 (DN-p38) mutant to block p38 activation; as expected, blockade of p38 activation in 231-E1A cells by either the p38 inhibitor or the DN-p38 obviously compromised E1A-mediated sensitization to anticancer drug-induced apoptosis, suggesting that p38 activation is required for E1A-mediated sensitization to apoptosis.[37]

Unlike results for p38, the level of phosphorylated Akt was reduced in E1A-expressing 231-E1A and MCF-7-E1A cells. Kinetically, decreased levels of phosphorylated Akt correlated with paclitaxel-induced PARP cleavage and Bcl-2 phosphorylation in 231-E1A cells, suggesting that downregulation of Akt phosphorylation in E1A-expressing cells may also contribute to E1A-mediated chemosensitization. To determine whether downregulation of Akt activity is required for E1A-mediated chemosensitization to paclitaxel, we examined whether activation of Akt by transfection of a constitutively active Akt construct (CA-Akt) would inhibit paclitaxel-induced apoptosis in 231-E1A cells. Reactivation of Akt in E1A-expressing 231-E1A cells repressed p38 phosphorylation in the presence or absence of paclitaxel and reduced

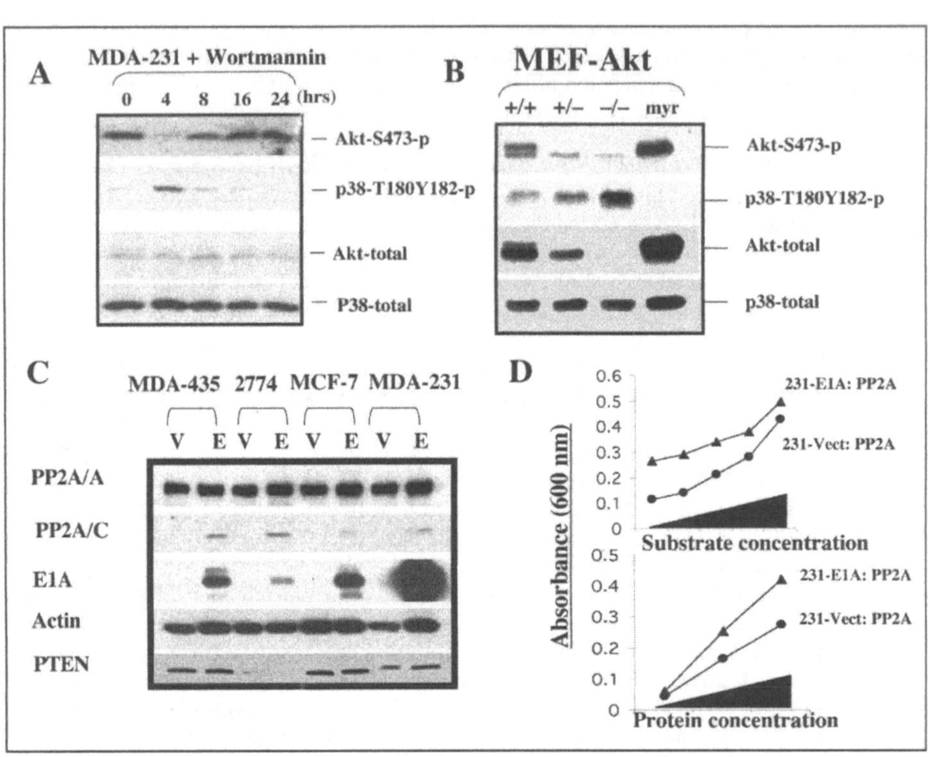

Figure 3. Inhibition of Akt activation by the PI3K inhibitor wortmannin or gene disruption of Akt1 results in p38 activation. A) Stable E1A-expressing or parental MDA-MB-231 cells were serum starved for 24 hours before exposure to 0.1 μM wortmannin. B) Expression of phospho-p38 and phospho-Akt in Akt1 knockout MEF cells and myr-Akt–transfected Rat1 cells. C) E1A upregulates the expression of PP2A/C, the catalytic subunit of PP2A. D) Expression of E1A upregulates protein phosphatase activity of PP2A. A-B and C-D) Adapted with permission from reference 37 (Liao Y, Hung MC. Mol Cell Bio 2003; 23:6836) and reference 152 (Liao Y, Hung MC. Cancer Res 2004; 64:5938), respectively.

paclitaxel-induced apoptosis in the cells. In addition, we also observed that activation of p38 and inactivation of Akt represent a general cellular mechanism in response to different apoptotic stimuli in the presence or absence of E1A.[37]

In an attempt to elucidate whether downregulation of Akt and upregulation of p38 are separate events or are linked events, we sought to determine whether Akt might act upstream of p38. Blockade of Akt activity in MDA-MB-231 cells by wortmannin, a specific PI3K inhibitor, decreased Akt phosphorylation and increased p38 phosphorylation. Similar results were also obtained by using LY2049002, another PI3K inhibitor. These results indicated that Akt phosphorylation is required for repressing p38 activation, that is, that the former is upstream from the latter. This conclusion was further supported by our study on Akt1 knockout mouse embryo fibroblast (MEF) cells and myr-Akt1 transfected stable cells. In that study, we observed that the level of phosphorylated p38 was increased in Akt1 (–/–) MEF cells compared with that in Akt1 (+/+) and Akt1 (+/–) MEF cells. Furthermore, the phospho-p38 protein was undetectable in the myr-Akt1 stable cells, suggesting that activation of p38 is associated with inactivation of Akt or, in other words, that Akt represses p38 activation in a physiological condition (Fig. 3).

Recent reports have demonstrated that ASK1 and MEKK3 are substrates of Akt, and both kinases have been shown to be upstream kinases of p38,[91,155-157] suggesting that Akt may indirectly regulate p38 activity through repression of ASK1 or MEKK3 or both. To test this, we blocked the activity of either ASK1 or MEKK3 by using a kinase-dead, dominant-negative mutant of ASK1 (DN-ASK1) or MEKK3 (DN-MEKK3). As expected, blockade of either ASK1 or MEKK3 activity by DN-ASK1 or DN-MEKK3 repressed p38 phosphorylation and its kinase activity, as measured by phosphorylation of GST-ATF-2 fusion protein in a dose-dependent manner in Akt1 (–/–) MEF cells. This result suggests that Akt can inhibit p38 activation through repression of ASK1 and/or MEKK3 activation. Unlike what we expected, however, the combination of both DN-ASK1 and DN-MEKK3 did not results in an additional reduction of p38 kinase activity or phosphorylation of p38 in these MEF Akt1 (–/–) cells.[37] These data imply that alternative pathway(s) may exist in which Akt acts on p38, though low transfection efficiency may also contribute to the incomplete blockade of ASK1 and MEKK3 activity and subsequent p38 activation in the Akt1 (–/–) cells. The data established that E1A induced p38 activation by repressing Akt. A recent report by Yuan et al[158] demonstrated that, like Akt1, Akt2 can also inhibit p38 activation through phosphorylation of ASK1, which is attributed to Akt2-mediated chemo-resistance; whether E1A also represses Akt2 phosphorylation needs further elucidation. We also observed that p38 activity is repressed in different types of human tumors, which is associated with enhanced Akt activation in human tumor tissue samples.[37]

Akt is a critical downstream target of PI3K and plays a pivotal role in positive regulation of cell survival and cell growth and also in negative regulation of apoptotic cell death, as summarized above.[78,84,86,87,159-163] Inhibition of Akt activation by E1A has important implications in terms of E1A-mediated sensitization to apoptosis. Akt regulates cell survival through phosphorylation of downstream substrates that directly or indirectly control the apoptotic machinery. For example, phosphorylation of IKK by Akt releases NF-κB from inhibition by I-κB (an inhibitor of NF-κB) and increases transcription of pro-survival genes, such as inhibitors of caspases (c-IAP1 and c-IAP2), Bcl-2 homologues (A1/Bfl-1 and Bcl-xL), and adaptor molecules intercepting death signals (e.g., c-FLIP, TRAF-1, and TRAF-2). In return, TRAF-2 can augment TNF-α-induced NF-κB activation.[60,87,164-166] Earlier reports from our laboratory demonstrated that expression of E1A downregulated IKK and NF-κB activities, which was also attributed to E1A-mediated sensitization to apoptosis induced by γ-radiation, TNF-α and anticancer drug gemcitabine.[38,167,168] In addition, E1A has been shown to sensitize cells to TNF-α-induced apoptosis, by downregulating the expression of c-FLIP and preventing its induction by TNF-α.[169] Expression of c-FLIP has been reported to depend on the activity of PI3K/Akt pathway;[164,166] it is not yet clear whether E1A-mediated down-regulation of c-FLIP, IKK and NF-κB activities are resulted from E1A-mediated inhibition of Akt.

Expression of exogenous E2F-1 or induction of endogenous E2F-1 activity by inactivation of pRb was shown to downregulate TRAF-2 protein levels, thus leading to impaired TNF-receptor-mediated NF-κB activation in response to TNF-α, which does not require E2F-1-dependent transactivation.[134] A recent report further showed that E2F-1 can inhibit NF-κB activation by either stabilizing I-κB or competing with p50 for RelA/p65 binding,[170,171] which provide another mechanism for E1A-mediated inhibition of NF-κB activation, though it does not exclude the possibility that Akt may still be involved.

Studies from our group and several others have demonstrated that Akt can also regulate the stability of p53 by phosphorylation and nuclear localization of MDM2.[71,72,88,172] Phosphorylation of MDM2 by Akt promotes nuclear localization of MDM2 and its interaction with p300 and inhibits the interaction of MDM2 with p19^{ARF}, thus increasing p53 degradation through MDM2-mediated ubiquitination.[71,72,88,172] If E1A-mediated repression of Akt activity could also affect p53 stability by this mechanism, it would be another E2F-1-independent pathway for p53 stabilization in E1A-expressing cells, and application of the dominant negative E2F-1 or the E2F-1-null MEF cell system may help to clarify this issue.

Duelli et al[173] have shown that E1A facilitates cytochrome *c* release from the mitochondria, thereby contributing to E1A-mediated sensitization to anticancer drugs; however, the mechanism by which E1A facilitates cytochrome *c* release is unclear. Akt is known to play an important role in maintaining mitochondria integrity and inhibiting the release of cytochrome *c*, and overexpression of Akt confers resistance to paclitaxel by inhibiting paclitaxel-induced cytochrome *c* release.[174] Akt also negatively regulates FasL transcription and translation by phosphorylating members of the Forkhead protein family. In addition, FasL transcriptional activation has been shown to depend on p38;[175] p38 is also involved in the regulation of cytochrome *c* release.[176] Therefore, E1A may alter mitochondrial potential by downregulating Akt and upregulating p38, thereby facilitating the release of cytochrome *c* or the expression of FasL upon treatment with chemotherapeutic drugs such as paclitaxel. Thus, in E1A-expressing cells, a relatively low concentration of drug would be sufficient to trigger apoptosis, regardless of the drug's primary target. Yet, it is not clear whether expression of E1A altered FasL expression.

Moreover, Schmidt et al demonstrated that expression of an active Akt suppresses chemotherapy-induced apoptosis by preventing mammary epithelial cells from undergoing anoikis (from the Greek word for "homelessness"),[177] another form of apoptotic cell death that occurs when epithelial cells lose contact with the extracellular matrix or bind through an inappropriate integrin.[16,18,178-182] Tumor suppressor PTEN has been shown to promote anoikis through its ability to repress Akt activation.[183-185] A recent report by Nagata et al[186] demonstrated that PTEN activation contributes to tumor inhibition by the Her-2/neu targeting antibody trastuzumab (Herceptin). Like constitutive activation of Akt, which confers resistance to chemotherapy and trastuzumab in breast cancer cells,[94] loss of PTEN also predicts trastuzumab resistance in patients due to deregulated PI3K-Akt activation and PI3K inhibitors can rescue loss of PTEN-induced resistance to trastuzumab. E1A is the first protein that has been reported to confer sensitivity to anoikis, and the ability to induce anoikis has also been linked to E1A-mediated tumor suppression and mesenchymal-epithelial transition (MET).[16,18,180,181] Therefore, it would be interesting to test whether down-regulation of Akt by E1A contributes in part to E1A-induced anoikis and MET.

Taken together, these findings indicate that many cellular proteins that interact with and/or are regulated by E1A, such as p53, caspase-8, caspase-9, Bax, NF-κB, p38, are also regulated by Akt. Although not tested, the critical apoptotic regulators Apaf-1 and Bcl-2 may be direct targets of Akt as they both contain the consensus Akt phosphorylation sites. In addition, phenotypes induced by expression of E1A, such as apoptosis, anticancer drug response, anoikis, MET, and tumor formation, are also subject to regulation by Akt. Thus, Akt is not only a pivotal cellular target for the development of anticancer drugs; it is also a critical target of E1A.

Targeting the Protein Phosphatase 2A (PP2A): A Feed-Forward Model

Repression of Akt activation by E1A has also been demonstrated in normal fibroblast IMR90 and Cos-7 cells and linked to the sensitization to apoptosis induced by anticancer drug cisplatin. In addition, expression of E1A can inhibit insulin-mediated Akt activation in the Cos-7 cells.[36] We also observed a similar inhibition effect on Akt activation by insulin in the epithelial breast cancer cell lines MDA-MB-231 and MCF-7 cells, which stably express E1A (unpublished data). However, the mechanisms underlying E1A-mediated repression of Akt activation are not yet known. In Cos-7 cells, E1A was shown to directly repress basal and insulin-stimulated Akt phosphorylation without affecting the Akt protein expression.[36] This raises a question how can E1A, which is not a kinase, represses another kinase without affecting the protein expression or upstream factors or respective protein phosphatases? As discussed above, downregulation of Her-2/neu and/or Axl expression may contribute in part to repression of Akt activation by E1A in cells that overexpress Her-2/neu or Axl. Again, both MDA-MB-231 and MCF-7 cells express low levels of Her-2/neu, and MCF-7 cells have an undetectable level of Axl mRNA and protein.[153] We also failed to detect alterations of PI3K activity or protein expression in E1A-expressing versus control cells. Therefore, the contribution of downregulation of Her-2/neu and Axl by E1A to the repression of Akt activation is minimal in these cells that express low levels of Her-2/neu or Axl.

Phosphorylation of protein kinases is tightly regulated by related protein phosphatases, and two phosphatases, PTEN and PP2A, have been shown to repress Akt activation through dephosphorylation.[71,187-201] To determine whether protein phosphatases were involved in the E1A-mediated downregulation of Akt activation, we measured the alteration of protein phosphatases, such as PTEN and PP2A, in stable E1A-expressing cells versus vector control cells. We did not detect any increased expression of PTEN protein in stable E1A-expressing cells versus the controls. Furthermore, there was no change in the expression level of the PP2A regulatory A subunit PP2A/A; however, we did detect elevated expression of the catalytic subunit of PP2A (PP2A/C) in multiple stable E1A-expressing cells (Fig. 3C).[152] We therefore further measured the PP2A activity and found it was enhanced in E1A-expressing MDA-MB-231 cells (231-E1A) in a dose (protein and substrate concentration)-dependent manner compared with that of the vector control cells (231-Vect), suggesting that E1A enhances the activity of PP2A by upregulating PP2A/C expression (Fig. 3D). In addition, we were able to show that activation of PP2A/C is required for E1A–mediated sensitization to drug–induced apoptosis, since blocking PP2A/C expression using a specific small interfering RNA (siRNA) against PP2A/C reduced drug sensitivity in E1A-expressing cells. Deletion mutation of the conserved domain of E1A, which is required for E1A-mediated upregulation of PP2A/C, also abrogated E1A's ability to sensitize cells to drug-induced apoptosis. We also observed that blockade of PP2A activity by a specific PP2A phosphatase inhibitor (okadaic acid) or repression of PP2A/C expression by a specific siRNA resulted in enhanced Akt phosphorylation and reduced p38 phosphorylation, which further suggests that PP2A regulates Akt and p38 activities.[152] In other words, E1A-induced p38 activation resulted from E1A-mediated upregulation of PP2A activity, which in turn repressed Akt activation.

The core enzyme of PP2A is a dimer consisting of a catalytic subunit (PP2A/C) and a regulatory or structural A subunit (PP2A/A). A third regulatory B subunit (PP2A/B), which determines substrate specificity, can be associated with this core structure.[188,190,192,194,196-198,202] The A and C subunit each exist as two isoforms, whereas the 16 B subunits fall into four families. PP2A A subunits are composed of 15 nonidentical tandem repeats of a 39 amino acid sequence, termed a HEAT motif (named after proteins that contain them: huntingtin, elongation factor, A subunit, TOR kinase). The B subunits bind to repeats 1-10, and the C subunits binds to repeats 11-15 of the A subunit. Recent evidence indicates that PP2A forms stable complexes with protein kinase signaling molecules, indicating that it plays a central regulatory role in signal transduction mediated by reversible protein phosphorylation.[188,190,192,194,196-198] Although the role of PP2A in the regulation of apoptosis is not clear, results from a gene knockout study of PP2A/C imply that PP2A may play a critical role in the regulation of apoptotic signaling.[203] In support of this notion, several groups reported that PP2A, through the dephosphorylation of the key oncogenic survival factor Akt, participated in the regulation of apoptosis induced by ceramide, mistletoe lectin, and 4-hydroxynonenal, an aldehyde product of membrane lipid peroxidation.[204-208] Furthermore, it has been known that PP2A is a Bcl-2 phosphatase and that the PP2A holoenzyme colocalizes with Bcl-2 at the mitochondrial membrane.[209,210] PP2A can be activated by lipid ceramide to dephosphorylate Bcl-2, converting Bcl-2 to a pro-apoptotic molecule. In addition, PP2A can also dephosphorylate Bad and refurbish its pro-apoptotic activity, which is repressed by Akt-mediated phosphorylation and inactivation.[205] Considering the fact that p38 is known to activate PP2A phosphatase activity[211] and that both p38 and Akt can regulate FasL expression and cytochrome *c* release from the mitochondria,[97,174-176,212-215] we reasoned that p38 might also affect Akt activation. To test that possibility, we blocked p38 activation in 231-E1A cells with SB203580, a specific p38 inhibitor. When the level of phosphorylated p38 was inhibited by SB20358, the level of phosphorylated Akt increased (Fig. 4A). Similar results were also observed by using an IPTG-inducible dominant-negative p38 in 231-E1A-expressing cells (Fig. 4B). The results suggested that PP2A/Akt/ASK1/p38 may form a feed-forward loop to regulate cellular response to apoptosis (Fig. 5). To test the notion of this feed-forward regulatory mechanism, we asked whether blockage

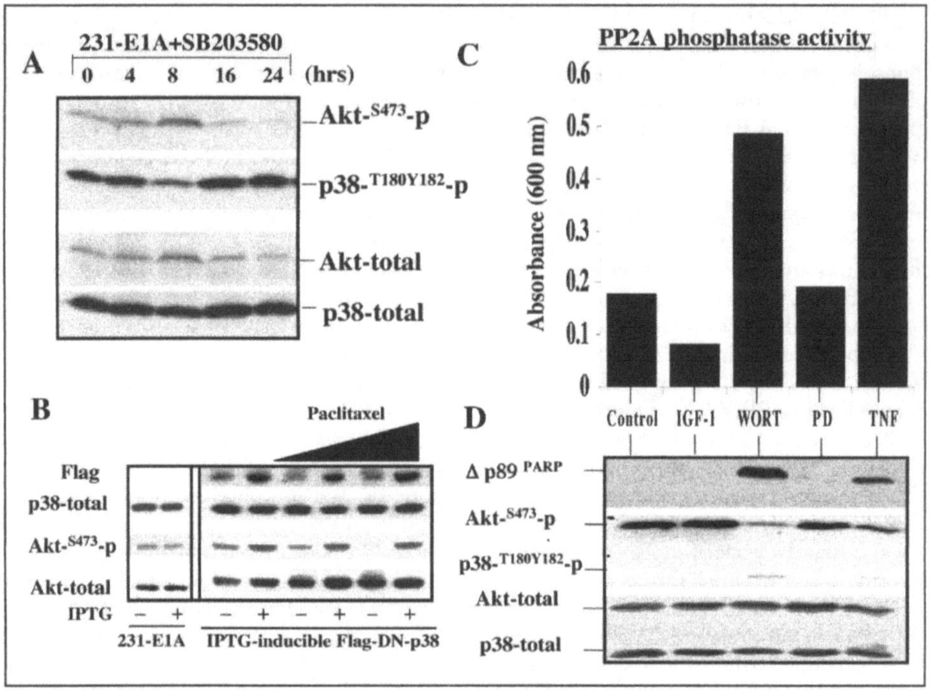

Figure 4. PP2A is involved in the cross-regulation between the Akt and p38 pathways. A) Expression of phospho-Akt and -p38 in stable E1A-expressing or parental MDA-MB-231 cells that were serum-starved for 24 hours before exposure to 20.0 μM SB203580, a specific p38 inhibitor. B) Repression of p38 activity by IPTG-inducible DN-p38 enhances Akt phosphorylation and abrogates E1A-mediated sensitization to paclitaxel in E1A-expressing cells in the presence of 5 μM IPTG for 24 hours. C) Regulation of PP2A phoaphatase activity by IGF-1, wortmannin, and TNF-α in MDA-MB-231 cells. D) PARP cleavage and Akt and p38 phosphorylation in MDA-MB-231 cells after exposure to IGF-1, wortmannin, and TNF-α. **WORT**: wortmannnin; **PD**: PD98058; **Flag**: anti-Flag tag. B and C-D) Adapted with permission from reference 37 (Liao Y, Hung MC. Mol Cell Bio 2003; 23:6836) and reference 152 (Liao Y, Hung MC. Cancer Res 2004; 64:5938), respectively.

of Akt activation might also activate PP2A activity. We found that blocking Akt activity by wortmannin indeed increased PP2A activity and induced PARP cleavage, while stimulating Akt activation by growth factor IGF-1 repressed PP2A activity (Fig. 4C). In addition, we also observed that treatment with TNF-α, which has been shown to inhibit Akt phosphorylation, also enhanced PP2A activity and PARP cleavage (Fig. 4D). However, the MEK1/2 kinase inhibitor PD98059 did not affect PP2A activity (Fig. 4D). Blocking Akt activation by using LY249002, another PI3K inhibitor, also increased PP2A activity. Thus, the results support the feed-forward mechanism and E1A, through upregulation of PP2A activity, may turn on this feed-forward mechanism and induce sensitization to apoptosis (Fig. 4D).

This feed-forward loop may represent a highly efficient and coordinated way to drive cells to death. This model is very consistent with all the results presented in this review and the data in the literatures.[37,91,152,154-158,187,199,205,206,211,216,217] For instance, PP2A directly dephosphorylated Akt and blockade of PP2A activity enhanced Akt-phosphorylation and inhibited p38-phosphorylation. Inhibition of Akt activity increased p38-phosphorylation and stimulated PP2A activity, whereas activation of Akt by growth factor IGF-1 repressed PP2A activity. Blocking p38 resulted in elevated Akt phosphorylation. Thus, the feed-forward loop is likely to

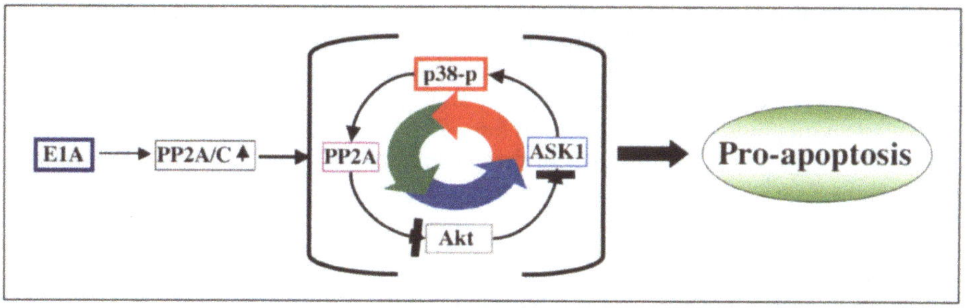

Figure 5. A feed-forward model for E1A-mediated sensitization to apoptosis.

be a general mechanism for cellular response to apoptosis, as it can be observed in sensitization to apoptosis induced by serum starvation, TNF-α, UV irradiation, and different categories of anticancer drugs. It also suggests that chemotherapeutic drug-induced apoptosis may utilize this feed-forward mechanism by upregulation of p38 and downregualtion of Akt activity and that E1A, by increasing PP2A activity, turns on this feed-forward loop and makes cells more sensitive to a variety of apoptotic stimuli.

Conclusion

The adenoviral E1A-mediated sensitization to anticancer drug-induced apoptosis is a well-established phenomenon in different types of cell systems. However, the molecular mechanisms underlying E1A-mediated chemosensitization are still not yet fully understood. Ever since the first report by Lowe et al[28,40] demonstrates that E1A-mediated activation of p53 is critical for its chemosensitization, the pRb-E2F-ARF-p53-dependent pathway is subsequently delineated. This was further amended by their recent report that an E2F-1-dependent, p53-independent mechanism that controls the transcriptional activation of pro-caspase enzymes also contributes to E1A-mediated chemosensitization in diploid fibroblasts.[130] Mechanisms underlying E1A-mediated chemosensitization in human cancer cells may be more complicated than those observed in normal diploid fibroblasts. For example, downregulation of Her-2/neu expression is critical for E1A-mediated chemosensitization in Her-2/neu-overexpressing cancer cells, but it may not so critical for E1A-mediated sensitization in diploid normal fibroblast cells, which do not overexpress Her-2/neu. Almost half of human cancers lose functional p53; however, expression of E1A in p53-mutated or functionally inactive cancer cells, such as MDA-MB-231, can also sensitize them to apoptosis induced by different types of anticancer drugs. A similar situation can also apply to ARF in human cancer; in addition, an ARF- and p53-independent mechanism for E1A-mediated chemosensitization is also observed in normal diploid fibroblasts. The apoptotic processes are tightly regulated by both pro- and anti-apoptotic signals, and either external or internal stresses can directly trigger apoptotic cell death program; however, growth factors and growth-stimulating hormones may counteract these death-stimulating stress signals by activation of intracellular cell survival pathways. For example, Akt can block apoptotic processes at different levels through direct phosphorylation and inactivation of Bad (which controls the release of cytochrome *c* from mitochondria) and caspase-9 (which controls downstream caspases activation after cytochrome *c* release from mitochondria). Therefore, it is not surprising for E1A to target such critical intracellular signaling molecules as Akt.

To date, most of domain analysis has pinpointed that the pRB-binding domain of E1A is critical for its sensitization to apoptosis, i.e., ARF and p53 stabilization, downregulation of Akt and upregulation of PP2A and p38, all depend on an intact pRb binding domain, suggesting that disrupting the pRb-E2F-1 pathway is critical in most cases of apoptosis, including those human cancer cells. Therefore, it would be helpful to further defining the roles of pRb and

E2F-1 played in E1A-mediated chemosensitization by studying E1A in pRb inactivated human cancer cells or E2F-1-null fibroblast cells.

Understanding the molecular mechanisms underlying E1A-mediated sensitization to anticancer agents can help us to better design future E1A clinical trials and help to select an appropriate patient subpopulation to be recruited into or excluded from the trials. Moreover, results obtained from Phase I/II clinical trials studying combination of E1A gene therapy with chemotherapy may help us to determine whether such regimens may benefit patients who do not respond to conventional chemotherapy.

References

1. American Cancer Society. Cancer statistics. Atlanta: 2005.
2. Gottesman MM, Fojo T, Bates SE. Multidrug resistance in cancer: Role of ATP-dependent transporters. Nat Rev Cancer 2002; 2:48-58.
3. Yan D, Shao RP, Hung MC. E1A cancer gene therapy. In: Lattime EC, Gerson SL, eds. Gene Therapy of Cancer. 2nd ed. San Diego: Academic Press, 2002:465-477.
4. Pozzatti R, McCormick M, Thompson MA et al. The E1a gene of adenovirus type 2 reduces the metastastic potential of ras-transformed rat embryo cells. Mol Cell Biol 1988; 8:2984-2988.
5. Steeg P, Bevilacqua G, Pozzatti R et al. Altered expression of NM23, a gene associated with low tumor metastatic potential, during adenovirus 2 E1a inhibition of experimental metastasis. Cancer Res 1988; 48:6550-6554.
6. Pozzatti R, McCormick M, Thompson MA et al. Regulation of the metastatic phenotype by the E1A gene of adenovirus-2. Adv Exp Med 1988; 233:293-301.
7. Yu D, Suen TC, Yan DH et al. Transcriptional repression of the neu protooncogene by the adenovirus 5 E1A gene products. Proc Natl Acad Sci USA 1990; 87:4499-4503.
8. Yu DH, Scorsone K, Hung MC. Adenovirus type 5 E1A gene products act as transformation suppressors of the neu oncogene. Mol Cell Biol 1991; 11:1745-1750.
9. Yu D, Hamada J, Zhang H et al. Mechanisms of c-erbB2/neu oncogene-induced metastasis and repression of metastatic properties by adenovirus 5 E1A gene products. Oncogene 1992; 7:2263-2270.
10. Frisch SM, Reich R, Collier IE et al. Adenovirus E1A represses protease gene expression and inhibits metastasis of human tumor cells. Oncogene 1990; 5:75-83.
11. Yan D, Rau KM, Hung MC. E1A and p202 as anti-metastasis genes. In: Curiel D, Douglas J, eds. Cancer Gene Therapy. 2nd ed. Totowa: Humana Press, 2004:87-98.
12. Frisch SM. Antioncogenic effect of adenovirus E1A in human tumor cells. Proc Natl Acad Sci USA 1991; 88:9077-9081.
13. Chinnadurai G. Adenovirus E1a as a tumor-suppressor gene. Oncogene 1992; 7:1255-1258.
14. Frisch SM. E1A—oncogene or tumor suppressor? Bioessays 1995; 17:1002.
15. Frisch SM. Reversal of malignancy by the adenovirus E1a gene. Mutat Res 1996; 350:261-266.
16. Frisch SM. The epithelial cell default-phenotype hypothesis and its implications for cancer. Bioessays 1997; 19:705-709.
17. Yu DH, Hung MC. The erbB2 gene as a cancer therapeutic target and the tumor- and metastasis-suppressing function of E1A. Cancer Metast Rev 1998; 17:195-202.
18. Frisch SM. Tumor suppression activity of adenovirus E1a protein: Anoikis and the epithelial phenotype. Adv Cancer Res 2001; 80:39-49.
19. Frisch SM, Mymryk JS. Adenovirus-5 E1A: Paradox and paradigm. Nat Rev Mol Cell Biol 2002; 3:441-452.
20. Frisch SM. E1A as a tumor suppressor gene: Commentary re S. Madhusudan et al. A multicenter Phase I gene therapy clinical trial involving intraperitoneal administration of E1A-lipid complex in patients with recurrent epithelial ovarian cancer overexpressing HER-2/neu oncogene. Clin Cancer Res 2004; 10:2905-2907.
21. Hortobagyi GN, Hung MC, Lopez-Berestein G. A Phase I multicenter study of E1A gene therapy for patients with metastatic breast cancer and epithelial ovarian cancer that overexpresses HER-2/neu or epithelial ovarian cancer. Hum Gene Ther 1998; 9:1775-1798.
22. Hung MC, Hortobagyi GN, Ueno NT. Development of clinical trial of E1A gene therapy targeting HER-2/neu-overexpressing breast and ovarian cancer. Adv Exp Med Biol 2000; 465.
23. Yoo GH, Hung MC, Lopez-Berestein G et al. Phase I trial of intratumoral liposome E1A gene therapy in patients with recurrent breast and head and neck cancer. Clin Cancer Res 2001; 7:1237-1245.
24. Hortobagyi GN, Ueno NT, Xia W et al. Cationic liposome-mediated E1A gene transfer to human breast and ovarian cancer cells and its biologic effects: A phase I clinical trial. J Clin Oncol 2001; 19:3422-3433.

25. Benjamin R, Helman L, Meyers P et al. A phase I/II dose escalation and activity study of intravenous injections of OCaP1 for subjects with refractory osteosarcoma metastatic to lung. Hum Gene Ther 2001; 12:1591-1593.
26. Villaret D, Glisson B, Kenady D et al. A multicenter phase II study of tgDCC-E1A for the intratumoral treatment of patients with recurrent head and neck squamous cell carcinoma. Head Neck 2002; 24:661-669.
27. Madhusudan S, Tamir A, Bates N et al. A multicenter Phase I gene therapy clinical trial involving intraperitoneal administration of E1A-lipid complex in patients with recurrent epithelial ovarian cancer overexpressing HER-2/neu oncogene. Clin Cancer Res 2004; 10:2986-2996.
28. Lowe SW, Ruley HE, Jacks T et al. p53-dependent apoptosis modulates the cytotoxicity of anticancer agents. Cell 1993; 74:957-967.
29. Sanchez-Prieto R, Carnero A, Marchetti E et al. Modulation of cellular chemoresistance in keratinocytes by activation of different oncogenes. Int J Cancer 1995; 60:235-243.
30. Frisch SM, Dolter KE. Adenovirus E1a-mediated tumor suppression by a c-erbB-2/neu-independent mechanism. Cancer Res 1995; 55:5551-5555.
31. Sanchez-Prieto R, Lleonart M, Ramon Y et al. Lack of correlation between p53 protein level and sensitivity of DNA- damaging agents in keratinocytes carrying adenovirus E1a mutants. Oncogene 1995; 11:675-682.
32. Ueno N, Yu D, Hung MC. Chemosensitization of Her-2/neu-overexpressing human breast cancer cells to paclitaxel (Taxol) by adenovirus type 5 E1A. Oncogene 1997; 15:953-960.
33. Brader KR, Wolf JK, Hung MC et al. Adenovirus E1A expression enhances the sensitivity of an ovarian cancer cell line to multiple cytotoxic agents through an apoptotic mechanism. Clin Cancer Res 1997; 3:2017-2024.
34. Ueno NT, Bartholomeusz C, Herrmann JL et al. E1A-mediated paclitaxel sensitization in Her-2/neu-overexpressing ovarian cancer SKOV3.ip1 through apoptosis involving the caspase-3 pathway. Clin Cancer Res 2000; 6:250-259.
35. Zhou Z, Jia SF, Hung MC et al. E1A sensitizes HER2/neu-overexpressing Ewing's sarcoma cells to topoisomerase II-targeting anticancer drugs. Cancer Res 2001; 61:3394-3398.
36. Viniegra JG, Losa JH, Sanchez-Arevalo VJ et al. Modulation of PI3K/Akt pathway by E1a mediates sensitivity to cisplatin. Oncogene 2002; 21:7131-7136.
37. Liao Y, Hung MC. Regulation of the activity of p38 mitogen-activated protein kinase by Akt in cancer and adenoviral protein E1A-mediated sensitization to apoptosis. Mol Cell Biol 2003; 23:6836-6848.
38. Lee W, Tai DI, Tsai SL et al. Adenovirus type 5 E1A sensitizes hepatocellular carcinoma cells to gemcitabine. Cancer Res 2003; 63:6229-6236.
39. Liao Y, Zou YY, Xia WY et al. Enhanced paclitaxel cytotoxicity and prolonged animal survival rate by a nonviral mediated systemic delivery of E1A gene in orthotopic xonograft human breast cancer. Cancer Gene Ther 2004; 11:594-602.
40. Lowe SW, Ruley HE. Stabilization of the p53 tumor suppressor is induced by adenovirus 5 E1A and accompanies apoptosis. Genes and Development 1993; 7:535-545.
41. White E. Regulation of p53-dependent apoptosis by E1A and E1B. Curr Top Microbiol Immunol 1995; 199:34-58.
42. Teodoro J, Shore GC, Branton PE. Adenovirus E1A proteins induce apoptosis by both p53-dependent and p53-independent mechanisms. Oncogene 1995; 11:467-474.
43. Attardi LD, Lowe SW, Brugarolas J et al. Transcriptional activation by p53, but not induction of the p21 gene, is essential for oncogene-mediated apoptosis. EMBO J 1996; 15:3693-3701.
44. de Stanchina E, McCurrach ME, Zindy F et al. E1A signaling to p53 involves the p19(ARF) tumor suppressor. Genes Dev 1998; 12:2434-2442.
45. Putzer BM, Stiewe T, Parssanedjad K et al. E1A is sufficient by itself to induce apoptosis independent of p53 and other adenoviral gene products. Cell Death Differ 2000; 7:177-188.
46. Schmitt CA, Lowe SW. Apoptosis and therapy. J Pathol 1999; 187:127-137.
47. Houghton JA. Apoptosis and drug response. Curr Opin Oncol 1999; 11:475-481.
48. Brown JM, Wouters BG. Apoptosis, p53, and tumor cell sensitivity to anticancer agents. Cancer Res 1999; 59:1391-1399.
49. Kaufmann SH, Earnshaw WC. Induction of apoptosis by cancer chemotherapy. Exp Cell Res 2000; 256:42-49.
50. Bamford M, Walkinshaw G, Brown R. Therapeutic applications of apoptosis research. Exp Cell Res 2000; 256:1-11.
51. Makin G, Dive C. Apoptosis and cancer chemotherapy. Trends Cell Biol 2001; 11:S22-S26.
52. Johnstone RW, Ruefli AA, Lowe SW. Apoptosis: A link between cancer genetics and chemotherapy. Cell 2002; 108:153-164.

53. Reed JC. Mechanisms of apoptosis avoidance in cancer. Curr Opin Oncol 1999; 11:68-75.
54. Igney FI, Krammer PH. Death and anti-death: Tumor resistance to apoptosis. Nature Rev Cancer 2002; 2:277-288.
55. Cryns V, Yuan J. Proteases to die for. Genes Dev 1998; 12:1551-1570.
56. Earnshaw WC, Martins LM, Kaufmann SH. Mammalian caspases: Structure, activation, substrates, and functions during apoptosis. Annu Rev Biochem 1999; 68:383-424.
57. Strasser A, O'Conner L, Dixit VM. Apoptosis signaling. Annu Rev Biochem 2000; 69:217-245.
58. Deveraux QL, Reed JC. IAP family proteins—supressors of apoptosis. Genes Dev 1999; 13:239-252.
59. Goyal L. Cell death inhibition: Keeping caspases in check. Cell 2001; 104:805-808.
60. Salvesen GS, Duckett CS. IAP proteins: Blocking the road to death's door. Nat Rev Mol Cell Biol 2002; 3:401-410.
61. Budihardjo I, Oliver H, Lutter M et al. Biochemical pathway of caspase activation during apoptosis. Annu Rev Cell Dev 1999; 15:269-290.
62. Adrain C, Martin SJ. The mitchondrial apoptosome: A killer unleashed by the cytochrome seas. Trends in Biochem Sci 2001; 26:390-397.
63. Danial NN, Korsmeyer SJ. Cell death: Critical control points. Cell 2004; 116:205-219.
64. Heiden MG, Thompson CB. Bcl-2 proteins: Regulators of apoptosis or of mitochondrial homeostasis? Nat Cell Biol 1999; 1:E209-E216.
65. Huang DCS, Strasser A. BH3-only proteins—essential initiators of apoptotic cell death. Cell 2000; 103:839-842.
66. Fesik SW. Insights into programmed cell death through structural biology. Cell 2000; 103:273-282.
67. Alaoui-Jamali MA, Paterson J, Al Moustafa AE et al. The role of ErbB2 tyrosine kinase receptor in cellular intrinsic chemoresistance: Mechanisms and implications. Biochem Cell Biol 1997; 75:315-325.
68. Yu D, Hung MC. Overexpression of ErbB2 in cancer and ErbB2-targeting strategies. Oncogene 2000; 19:6115-6121.
69. Yu D, Hung MC. Role of erbB2 in breast cancer chemosensitivity. Bioessays 2000; 22:673-680.
70. Weller M. Predicting response to cancer chemotherapy: The role of p53. Cell Tissue Res 1998; 292:435-445.
71. Mayo LD, Donner DB. The PTEN, Mdm2, p53 tumor suppressor-oncoprotein network. Trands Biology Sci 2002; 27:462-467.
72. Oren M, Damals A, Gottlieb T et al. Regulation of p53: Intricate loops and delicate balances. Biochem Pharm 2002; 64:865-871.
73. Fojo T, Bates S. Strategies for reversing drug resistance. Oncogene 2003; 22:7512-7523.
74. Sax JK, El-Deiry WS. p53 downstream targets and chemosensitivity. Cell Death Differ 2003; 10:413-417.
75. Haldar S, Basu A, Croce CM. Bcl-2 is the Gardian of Microtubule integrity. Cancer Res 1997; 57:229-233.
76. Krajewski S, Krajewska M, Turner BC et al. Prognostic significance of apoptosis regulators in breast cancer. Endocrine-Related Cancer 1999; 6:29-40.
77. Schorr K, Li M, Krajewski S et al. Bcl-2 gene family and related proteins in mammary gland involution and breast cancer. J Mammary Gland Biol Neoplasia 1999; 4:153-164.
78. Reed JC. Apoptosis-regulating proteins as targets for drug discovery. Trends in Molecular Medicine 2001; 7:314-319.
79. Simstein R, Burow M, Parker A et al. Apoptosis, chemoresistance, and breast cancer: Insights from the MCF-7 cell model system. Exp Biol Med 2003; 228:995-1003.
80. Debatin K. Apoptosis pathways in cancer and cancer therapy. Cancer Immmunol Immunother 2004; 53:153-159.
81. Giai M, Biglia N, Sismondi P. Chemoresistance in breast tumors. Eur J Gynaecol Oncol 1991; 12:359-373.
82. Dicato M, Duhem C, Pauly M et al. Multidrug resistance: Molecular and clinical aspects. Cytokines Cell Mol Ther 1997; 3:91-99.
83. Ruvolo PP. Ceramide regulates cellular homeostasis via diverse stress signaling pathways. Leukemia 2001; 15:1153-1160.
84. Hanada M, Feng J, Hemmings BA. Structure, regulation and function of PKB/AKT—a major therapeutic target. Biochim Biophys Acta 2004; 1697:3-16.
85. Luo J, Manning BD, Cantley LC. Targeting the PI3K-Akt pathway in human cancer: Rationale and promise. Cancer Cell 2003; 4:257-262.
86. Fresno Vara JA, Casado E, de Castro J et al. PI3K/Akt signalling pathway and cancer. Cancer Treat Rev 2004; 30:193-204.
87. Downward J. PI 3-kinase, Akt and cell survival. Semin Cell Dev Biol 2004; 15:177-182.

88. Zhou BP, Liao Y, Xia W et al. HER-2/neu induces p53 ubiquitination via Akt-mediated MDM2 phosphorylation. Nat Cell Biol 2001; 3:973-982.
89. Mayo L, Donner DB. A phosphatidylinositol 3-kinase/Akt pathway promotes translocation of Mdm2 from the cytoplasm to the nucleus. Proc Natl Acad Sci USA 2001; 98:11598-11603.
90. Zhou BP, Liao Y, Xia W et al. Cytoplasmic localization of p21Cip1/WAF1 by Akt-induced phosphorylation in HER-2/neu-overexpressing cells. Nat Cell Biol 2001; 3:245-252.
91. Kim AH, Khursigara G, Sun X et al. Akt phosphorylates and negatively regulates apoptosis signal-regulating kinase 1. Mol Cell Biol 2001; 21:893-901.
92. Potter CJ, Pedraza LG, Xu T. Akt regulates growth by directly phosphorylating Tsc2. Nat Cell Biol 2002; 4:658-665.
93. Inoki K, Li Y, Zhu T et al. TSC2 is phosphorylated and inhibited by Akt and suppresses mTOR signalling. Nat Cell Biol 2002; 4:648-657.
94. Clark AS, West K, Streicher S et al. Constitutive and inducible Akt activity promotes resistance to chemotherapy, Trastuzumab, or Tamoxifen in breast cancer cells. Mol Cancer Ther 2002; 1:707-717.
95. West KA, Castillo SS, Dennis PA. Activation of the PI3K/Akt pathway and chemotherapeutic resistance. Drug Resist Updat 2002; 5:234-248.
96. Hill MM, Hemmings BA. Inhibition of protein kinase B/Akt: Implications for cancer therapy. Pharmacol Ther 2002; 93:243-251.
97. Burgering BM, Medema RH. Decisions on life and death: FOXO Forkhead transcription factors are in command when PKB/Akt is off duty. J Leukoc Biol 2003; 73:689-701.
98. Mitsiades CS, Mitsiades N, Koutsilieris M. The Akt pathway: Molecular targets for anti-cancer drug development. Curr Cancer Drug Targets 2004; 4:235-256.
99. Piette J, Piret B, Bonizzi G et al. Multiple redox regulation in NF-kappaB transcription factor activation. Biol Chem 1997; 378:1237-1245.
100. Arlt A, Schafer H. NFkappaB-dependent chemoresistance in solid tumors. Int J Clin Pharmacol Ther 2002; 40:336-347.
101. Branton PE. Early gene expression. In: Seth P, ed. Adenoviruses: Basic biology to gene therapy. Georgetown, TX: R.G. Landes Company, 1999.
102. Fuchs M, Gerber J, Drapkin R et al. The p400 complex is an essential E1A transformation target. Cell 2001; 106:297-307.
103. Moran E, Mathews MB. Multiple functional domains in the adenovirus E1A gene. Cell 1987; 48:177-178.
104. Goodman RH, Smolik S. CBP/p300 in cell growth, transformation, and development. Genes and Devel 2000; 14:1553-1577.
105. Chakravarti D, Ogryzko V, Kao HY et al. A viral mechanism for inhibition of p300 and PCAF acetyltransferase activity. Cell 1999; 96:393-403.
106. Zhang Q, Yao H, Vo N et al. Acetylation of adenovirus E1A regulates binding of the transcriptional corepressor CtBP. Proc Natl Acad Sci USA 2000; 97:14323-14328.
107. Chinnadurai G. CtBP, an unconventional transcriptional corepressor in development and oncogenesis. Mol Cell 2002; 9:213-224.
108. Flint J, Shenk T. Viral transactivating proteins. Annu Rev Genet 1997; 31:177-212.
109. Ginsberg D. E2F1 pathways to apoptosis. FEBS Lett 2002; 529:122-125.
110. Kaelin Jr WG. E2F1 as a target: Promoter-driven suicide and small molecule modulators. Cancer Biol Ther 2003; 2:S48-54.
111. Classon M, Harlow E. The retinoblastoma tumor suppressor in development and cancer. Nat Rev Cancer 2002; 2:910-917.
112. Debbas M, White E. Wild-type p53 mediates apoptosis by E1A, which is inhibited by E1B. Genes Dev 1993; 7:546-554.
113. Chiou SK, Rao L, White E. Bcl-2 blocks p53-dependent apoptosis. Mol Cell Biol 1994; 14:2556-2563.
114. Pan H, Griep AE. Temporally distinct patterns of p53-dependent and p53-independent apoptosis during mouse lens development. Genes Dev 1995; 9:2157-2169.
115. Querido E, Teodoro JG, Branton PE. Accumulation of p53 induced by the adenovirus E1A protein requires regions involved in the stimulation of DNA synthesis. J Virol 1997; 71:3526-3533.
116. Ding HF, McGill G, Rowan S et al. Oncogene-dependent regulation of caspase activation by p53 protein in a cell-free system. J Biol Chem 1998; 273:28378-28383.
117. Lowe SW. Activation of p53 by oncogenes. Endocr Relat Cancer 1999; 6:45-48.
118. Breckenridge DG, Shore GC. Regulation of apoptosis by E1A and Myc oncoproteins. Crit Rev Eukary Gene Exp 2000; 10:273-280.
119. Pomerantz J, Schreiber-Agus N, Liegeois NJ et al. The Ink4a tumor suppressor gene product, p19ARF, interacts with MDM2 and neutralizes MDM2's inhibition of p53. Cell 1998; 92:713-723.

120. Zhang Y, Xiong Y, Yarbrough WG. ARF promotes MDM2 degradation and stabilizes p53: ARF-INK4a locus deletion impairs both the Rb and p53 tumor supression pathways. Cell 1998; 92:725-734.
121. Chiou SK, White E. p300 binding by E1A cosegregates with p53 induction but is dispensable for apoptosis. J Virol 1997; 71:3515-3525.
122. Li Z, Day CP, Yang JY et al. Adenoviral E1A targets Mdm4 to stabilize tumor suppressor p53. Cancer Res 2004; 64:9080-9085.
123. McCurrach ME, Connor TMF, Knudson CM et al. Bax-deficiency promotes drug resistance and oncogenic transformation by attenuating p53-dependent apoptosis. Proc Natl Acad Sci USA 1997; 94:2345-2349.
124. Fearnhead H, Rodriguez J, Govek EE et al. Oncogene-dependent apoptosis is mediated by caspase-9. Proc natl Acad Sci USA 1998; 95:13664-13669.
125. Harbour JW, Dean DC. Rb function in cell-cycle regulation and apoptosis. Nat Cell Biol 2000; 2:E65-E67.
126. Harbour JW, Dean DC. The Rb/E2F pathway: Expanding roles and emerging paradigms. Genes and Devel 2000; 14:2393-2409.
127. Classon M, Salama S, Gorka C et al. Combinatorial roles for pRB, p107, and p130 in E2F-mediated cell cycle control. Proc Natl Acad Sci USA 2000; 97:10820-10825.
128. Nicholson S, Okby NT, Khan MA et al. Alterations of p14ARF, p53, and p73 genes involved in the E2F-1-mediated apoptotic pathways in nonsmall cell lung carcinoma. Cancer Res 2001; 61:5636-5643.
129. Phillips A, Vousden KH. E2F-1 induced apoptosis. Apoptosis 2001; 6:173-182.
130. Nahle Z, Polakoff J, Davuluri RV et al. Direct coupling of the cell cycle and cell death machinery by E2F. Nature Cell Biology 2002; 4:859-864.
131. Irwin M, Marin MC, Phillips AC et al. Role for the p53 homologue p73 in E2F-1-induced apoptosis. Nature 2000; 407:645-648.
132. Furukawa Y, Nishimura N, Furukawa Y et al. Apaf-1 is a mediator of E2F-1-induced apoptosis. J Biol Chem 2002; 277:39760-39768.
133. Phillips A, Bates S, Ryan KM et al. Induction of DNA synthesis and apoptosis are separable functions of E2F-1. Genes Dev 1997; 11:1853-1863.
134. Phillips A, Ernst MK, Bates S et al. E2F-1 potentiates cell death by blocking antiapoptotic signaling pathways. Mol Cell 1999; 4:771-781.
135. Chattopadhyay D, Ghosh MK, Mal A et al. Inactivation of p21 by E1A leads to the induction of apoptosis in DNA-damaging cells. J Virol 2001; 75:9844-9856.
136. Yu D, Wolf JK, Scanlon M et al. Enhanced c-erbB-2/neu expression in human ovarian cancer cells correlates with more severe malignancy that can be suppressed by E1A. Cancer Res 1993; 53:891-898.
137. Zhang Y, Yu D, Xia W et al. HER-2/neu-targeting cancer therapy via adenovirus-mediated E1A delivery in an animal model. Oncogene 1995; 10:1947-1954.
138. Yu D, Matin A, Xia W et al. Liposome-mediated in vivo E1A gene transfer suppressed dissemination of ovarian cancer cells that overexpress HER-2/neu. Oncogene 1995; 11:1383-1388.
139. Hung MC, Matin A, Zhang Y et al. HER-2/neu-targeting gene therapy—a review. Gene Ther 1995; 159:65-71.
140. Chen H, Yu D, Chinnadurai G et al. Mapping of adenovirus 5 E1A domains responsible for suppression of neu-mediated transformation via transcriptional repression of neu. Oncogene 1997; 14:1965-1971.
141. Chen H, Hung MC. Involvement of coactivator p300 in the transcriptional regulation of the HER-2/neu gene. J Biol Chem 1997; 272:6101-6104.
142. Yu D, Liu B, Tan M et al. Overexpression of c-erbB-2/neu in breast cancer cells confers increased resistance to Taxol via mdr-1-independent mechanisms. Oncogene 1996; 13:1359-1365.
143. Slamon DJ, Leyland-Jones B, Shak S et al. Use of chemotherapy plus a monoclonal antibody against HER2 for metastatic breast cancer that overexpresses HER2. N Engl J Med 2001; 344:783-792.
144. Hayes DF, Thor AD. c-erbB-2 in breast cancer: Development of a clinically useful marker. Semin Oncol 2002; 29:231-245.
145. Ueno NT, Bartholomeusz C, Xia W et al. Systemic gene therapy in human xenograft tumor models by liposomal delivery of the E1A gene. Cancer Res 2002; 62:6712-6716.
146. Sabbatini AR, Basolo F, Valentini P et al. Induction of multidrug resistance (MDR) by transfection of MCF-10A cell line with c-Ha-ras and c-erbB-2 oncogenes. Int J Cancer 1994; 59:208-211.
147. Yu D, Liu B, Jing T et al. Overexpression of both p185c-erbB2 and p170mdr1 renders breast cancer cells highly resistant to taxol. Oncogene 1998; 16:2087-2094.

148. Lee WP, Liao Y, Robinson D et al. Axl-gas6 interaction counteracts E1A-mediated cell growth suppression and proapoptotic activity. Mol Cell Biol 1999; 19:8075-8082.
149. Lee WP, Wen Y, Varnum B et al. Akt is required for Axl-Gas6 signaling to protect cells from E1A-mediated apoptosis. Oncogene 2002; 21:329-336.
150. Deng J, Xia W, Hung MC. Adenovirus 5 E1A-mediated tumor suppression associated with E1A-mediated apoptosis in vivo. Oncogene 1998; 17.
151. Deng J, Kloosterbooer F, Xia W et al. The NH(2)-terminal and conserved region 2 domains of adenovirus E1A mediate two distinct mechanisms of tumor suppression. Cancer Res 2002; 62:346-350.
152. Liao Y, Hung MC. A new role of protein phosphatase 2A in adenoviral E1A protein-mediated sensitization to anticancer drug-induced apoptosis in human breast cancer cells. Cancer Res 2004; 64:5938-5942.
153. Meric F, Lee WP, Sahin A et al. Expression profile of tyrosine kinases in breast cancer. Clin Cancer Res 2002; 8:361-367.
154. Cross TG, Toellner DS, Henriquez NV et al. Serine/theonine protein kinases and apoptosis. Exp Cell Res 2000; 256:34-41.
155. Gratton JP, Kureishi Y, Fulton D et al. Akt down-regulation of p38 signaling provides a novel mechanism of vascular endothelial growth factor-mediated cytoprotection in endothelial cells. J Biol Chem 2001; 276:30359-30365.
156. Tobiume K, Matsuzawa A, Takahashi T et al. ASK1 is required for sustained activations of JNK/p38 MAP kinases and apoptosis. EMBO Rep 2001; 2:222-228.
157. Ichijo H, Nishida E, Irie K et al. Induction of apoptosis by ASK1, a mammalian MAPKKK that activates SAPK/JNK and p38 signaling pathways. Science 1997; 275:90-94.
158. Yuan ZQ, Feldman RI, Sussman GE et al. AKT2 inhibition of cisplatin-induced JNK/p38 and Bax activation by phosphorylation of ASK1: Implication of AKT2 in chemoresistance. J Biol Chem 2003; 278:23432-23440.
159. Brazil DP, Yang ZZ, Hemmings BA. Advances in protein kinase B signalling: AKTion on multiple fronts. Trends Biochem Sci 2004; 29:233-242.
160. Webster KA. Aktion in the nucleus. Circ Res 2004; 94:856-859.
161. Chang F, Lee JT, Navolanic PM et al. Involvement of PI3K/Akt pathway in cell cycle progression, apoptosis, and neoplastic transformation: A target for cancer chemotherapy. Leukemia 2003; 17:590-603.
162. Nicholson KM, Anderson NG. The protein kinase B/Akt signaling pathway in human malignancy. Cellular Signalling 2002; 14:381-395.
163. Vivanco I, Sawyers CL. The phosphatidylinositol 3-Kinase AKT pathway in human cancer. Nat Rev Cancer 2002; 2:489-501.
164. Arch RH, Gedrich RW, Thompson CB. Tumor necrosis factor receptor-associated factors (TRAFs)—a family of adapter proteins that regulates life and death. Genes Dev 1998; 12:2821-2830.
165. Madrid LV, Mayo MW, Reuther JY et al. Akt stimulates the transactivation potential of the RelA/p65 subunit of NF-{kappa}B through utilization of the I{kappa}B kinase and activation of the mitogen activated protein kinase p38. J Biol Chem 2001; 20:20.
166. Krueger A, Baumann S, Krammer PH et al. FLICE-inhibitory proteins: Regulators of death receptor-mediated apoptosis. Mol Cell Biol 2001; 21:8247-8254.
167. Shao R, Karunagaran D, Zhou BP et al. Inhibition of nuclear factor-kB activity is involved in E1A-mediated sensitization of radiation-induced apoptosis. J Biol Chem 1997; 272:32739-32742.
168. Shao R, Hu MCT, Zhou BP et al. E1A sensitizes cells to tumor necrosis factor-induced apoptosis through inhibition of IkB kinases and nuclear factor kB activities. J Biol Chem 1999; 274:21495-21498.
169. Perez D, White E. E1A sensitizes cells to tumor necrosis factor alpha by downregulating c-FLIP$_S$. J Virol 2003; 77:2651-2662.
170. Tanaka H, Matsumura I, Ezoe S et al. E2F1 and c-Myc potentiate apoptosis through inhibition of NF-kappaB activity that facilitates MnSOD-mediated ROS elimination. Mol Cell 2002; 9:1017-1029.
171. Chen M, Capps C, Willerson JT et al. E2F-1 regulates nuclear factor-kappaB activity and cell adhesion: Potential antiinflammatory activity of the transcription factor E2F-1. Circulation 2002; 106:2707-2713.
172. Zhou BP, Hung MC. Novel targets of Akt, p21(Cipl/WAF1), and MDM2. Semin Oncol 2002; 29:62-70.
173. Duelli DM, Lazebnik YA. Primary cells suppress oncogene-dependent apoptosis. Nat Cell Biol 2000; 2:859-862.

174. Page C, Lin HJ, Jin Y et al. Overexpression of Akt/AKT can modulate chemotherapy-induced apoptosis. Anticancer Res 2000; 20:407-416.
175. Hsu SC, Gavrilin MA, Tsai MH et al. p38 mitogen-activated protein kinase is involved in Fas ligand expression. J Biol Chem 1999; 274:25769-25776.
176. Van Laethem A, Van Kelst S, Lippens S et al. Activation of p38 MAPK is required for Bax translocation to mitochondria, cytochrome c release and apoptosis induced by UVB irradiation in human keratinocytes. FASEB J 2004; 18:1946-1948.
177. Schmidt M, hovelmann S, Beckers TL. A novel form of constitutively active farnesylated Akt prevents mammary epithelial cells from anoikis and suppresses chemotherapy-induced apoptosis. Brit J Cancer 2002; 87:924-932.
178. Cardone MH, Salvesen GS, Widmann C et al. The regulation of Anoikis: MEKK-1 activation requires cleavage by caspases. Cell 1997; 90:315-323.
179. Metcalfe A, Streuli C. Epithelial apoptosis. BioEssays 1997; 19:711-720.
180. Frisch SM, Francis H. Disruption of epithelial cell-matrix interactions induces apoptosis. J Cell Biol 1994; 124:619-626.
181. Frisch SM. E1a induces the expression of epithelial characteristics. J Cell Biol 1994; 127:1085-1096.
182. Janes SM, Watt FM. Switch from alphavbeta5 to alphavbeta6 integrin expression prodects squanmous cell carcinomas from anoikis. J Cell Biol 2004; 166:419-431.
183. Lei QY, Wang LY, Dai ZY et al. The relationship between PTEN expression and anoikis in human lung carcinoma cell lines. Sheng Wu hua Xue Yu Sheng Wu Wu Li Xue Bao (Shanghai) 2002; 34:463-468.
184. Lu Y, Lin YZ, LaPushin R et al. The PTEN/MMAC1/TEP tumor supressor gene decreases cell growth and induces apoptosis and anoikis in breast cancer cells. Oncogene 1999; 18:7034-7045.
185. Davies MA, Lu Y, Sano T et al. Adenoviral transgene expression of MMAC/PTEN in human glioma cells inhibits Akt activation and induces anoikis. Cancer Res 1998; 58:5285-5290.
186. Nagata Y, Lan KH, Zhou X et al. PTEN activation contributes to tumor inhibition by trastuzumab, and loss of PTEN predicts trastuzumab resistance in patients. Cance Cell 2004; 6:117-127.
187. Berra E, Diaz-Meco MT, Moscat J. The activation of p38 and apoptosis by the inhibition of Erk is antagonized by the phosphoinositide 3-kinase/Akt pathway. J Biol Chem 1998; 273:10792-10797.
188. Schonthal AH. Role of PP2A in intracellular signal transduction pathways. Front Biosci 1998; 3:D1262-1273.
189. Cantley LC, Neel BG. New insights into tumor suppression: PTEN suppresses tumor formation by restraining the phosphoinositide 3-kinase/AKT pathway. Proc Natl Acad Sci USA 1999; 96:4240-4245.
190. Millward TA, Zolnierowicz S, Hemmmings BA. Regulation of protien kinase cascades by protein phosphatase 2A. TIBS 1999; 24:186-191.
191. Sato S, Fujita N, Tsuruo T. Modulation of Akt kinase activity by binding to Hsp90. Proc Natl Acad Sci USA 2000; 97:10832-10837.
192. Virshup DM. Protein phosphatase 2A: A panoply of enzymes. Curr Opin Cell Biology 2000; 12:180-185.
193. Cristofano AD, Pandolfi PP. The multiple roles of PTEN in tumor suppression. Cell 2000.
194. Zolnierowicz S. Type 2A protein phosphatases, the complex regulator of numerous signaling pathways. Biochem Pharm 2000; 60:1225-1235.
195. Simpson L, Parsons R. PTEN: Life as a tumor suppressor. Exp Cell Res 2001; 264:29-41.
196. Janssens V, Goris J. Protein phosphatase 2A: A highly regulated family of serine/threonine phosphatases implicated in cell growth and signaling. Biochem J 2001; 353:417-439.
197. Schonthal AH. Role of serine/threonine protein phosphatase 2A in cancer. Cancer Lett 2001; 170:1-13.
198. Sontag E. Protein phosphatase 2A: The Trojan Horse of cellular signaling. Cell Signal 2001; 13:7-16.
199. Garcia A, Cayla X, Guergnon J et al. Serine/threonine protein phosphatases PP1 and PP2A are key players in apoptosis. Biochemie 2003; 85:721-726.
200. Kowluru A, Metz SA. Ceramide-activated protein phosphatase-2A activity in insulin-secreting cells. FEBS Lett 1997; 418:179-182.
201. Paez J, Sellers WR. PI3K/PTEN/AKT pathway. A critical mediator of oncogenic signaling. Cancer Treat Res 2003; 115:145-167.
202. Barford D, Das AK, Egloff MP. The structure and mechanism of protein phosphatases: Insights into catalysis and regulation. Annu Rev Biophys Biomol Struct 1998; 27:133-164.
203. Gotz J, Probst A, Ehler E et al. Delayed embryonic lethality in mice lacking protein phosphatase 2A catalytic subunit Calpha. Proc Natl Acad Sci USA 1998; 95:12370-12375.
204. Mills JC, Lee VM, Pittman RN. Activation of a PP2A-like phosphatase and dephosphorylation of tau protein characterize onset of the execution phase of apoptosis. J Cell Sci 1998; 111:625-636.

205. Chiang CW, Harris G, Ellig C et al. Protein phosphatase 2A activates the proapoptotic function of BAD in interleukin- 3-dependent lymphoid cells by a mechanism requiring 14-3-3 dissociation. Blood 2001; 97:1289-1297.
206. Liu W, Akhand AA, Takeda K et al. Protein phosphatase 2A-linked and -unlinked caspase-dependent pathways for downregulation of Akt kinase triggered by 4-hydroxynonenal. Cell Death Differ 2003; 10:772-781.
207. Nakashima I, Liu W, Akhand AA et al. 4-hydroxynonenal triggers multistep signal transduction cascades for suppression of cellular functions. Mol Aspects Med 2003; 24:231-238.
208. Choi SH, Lyu SY, Park WB. Mistletoe lectin induces apoptosis and telomerase inhibition in human A253 cancer cells through dephosphorylation of Akt. Arch Pharm Res 2004; 27:68-76.
209. Ruvolo PP, Deng X, Ito T et al. Ceramide induces Bcl-2 dephosphorylation via a mechanism involving mitochondria PP2A. J Biol Chem 1999; 274:20296-20300.
210. Deng X, Ito T, Carr B et al. Reversible phosphorylation of Bcl2 following interleukin 3 or bryostatin 1 is mediated by direct interaction with protein phosphatase 2A. J Biol Chem 1998; 273:34157-34163.
211. Westermarck J, Li SP, Kallunki T et al. p38 mitogen-activated protein kinase-dependent activation of protein phosphatases 1 and 2 A inhibits MEK1 and MEK2 activity and collagenase 1 (MMP-1) gene expression. Mol Cell Biol 2001; 21:2373-2383.
212. De Zutter GS, Davis RJ. Pro-apoptotic gene expression mediated by the p38 mitogen-activated protien kinase signal transduction pathway. Proc Natl Acad Sci USA 2001; 98:6168-6173.
213. Assefa Z, Vantieghem A, Garmyn M et al. p38 mitogen-activated protein kinase regulates a novel, caspase-independent pathway for the mitochondrial cytochrome c release in ultraviolet B radiation-induced apoptosis. J Biol Chem 2000; 275:21416-21421.
214. Kennedy SG, Kandel ES, Cross TK et al. Akt/protein kinase B inhibits cell death by preventing the release of cytochrome c from mitochondria. Mol Cell Biol 1999; 19:5800-5810.
215. Kops GJ, Burgering BM. Forkhead transcription factors: New insights into protein kinase B (c-akt) signaling. J Mol Med 1999; 77:656-665.
216. Brazil DP, Park J, Hemmings BA. PKB binding proteins. Getting in on the Akt. Cell 2002; 111:293-303.
217. Chen D, Fucini RV, Olson AL et al. Osmotic shock inhibits insulin signaling by maintaining Akt/protein kinase B in an inactive dephosphorylated state. Mol Cell Biol 1999; 19:4684-4694.

Index

G

H

I

K

L

M

S

T

U

V

X

MIX
Papier aus verantwortungsvollen Quellen
Paper from responsible sources
FSC® C105338

If you have any concerns about our products,
you can contact us on
ProductSafety@springernature.com

In case Publisher is established outside the EU,
the EU authorized representative is:
Springer Nature Customer Service Center GmbH
Europaplatz 3, 69115 Heidelberg, Germany

Printed by Libri Plureos GmbH
in Hamburg, Germany